AF411221

Nonlinear Systems: Dynamics, Control, Optimization and Applications to the Science and Engineering

Nonlinear Systems: Dynamics, Control, Optimization and Applications to the Science and Engineering

Editor

Quanxin Zhu

MDPI • Basel • Beijing • Wuhan • Barcelona • Belgrade • Manchester • Tokyo • Cluj • Tianjin

Editor
Quanxin Zhu
Hunan Normal University
China

Editorial Office
MDPI
St. Alban-Anlage 66
4052 Basel, Switzerland

This is a reprint of articles from the Special Issue published online in the open access journal *Mathematics* (ISSN 2227-7390) (available at: https://www.mdpi.com/journal/mathematics/special_issues/Nonlinear_Systems_Dynamics_Control_Optimization_Applications_Science_Engineering).

For citation purposes, cite each article independently as indicated on the article page online and as indicated below:

LastName, A.A.; LastName, B.B.; LastName, C.C. Article Title. *Journal Name* **Year**, *Volume Number*, Page Range.

ISBN 978-3-0365-6322-0 (Hbk)
ISBN 978-3-0365-6323-7 (PDF)

Contents

About the Editor

Quanxin Zhu

Professor Quanxin Zhu received his Ph.D. degree from Sun Yatsen (Zhongshan) University, Guangzhou, China, in 2005. He is currently a Professor at Hunan Normal University, a distinguished professor of Hunan Province, Leading talent of scientific and technological innovation in Hunan Province, and deputy director of the Key Laboratory of computing and stochastic mathematics of the Ministry of education. He won the Alexander von Humboldt Foundation of Germany award and the Highly Cited Researcher Award from Clarivate Analytics in 2018 and 2022. He won the first prize of the Hunan Natural Science Award, and was ranked in the top two percent of the world's top scientists in 2020 to 2022. In addition, Professor Zhu is a senior member of the IEEE and the Lead Guest Editor of several international journals. Professor Zhu obtained the 2011 Annual Chinese "The One Hundred Most Influential International Academic Papers" Award and wasone of the most cited Chinese researchers from 2014–2021 (Elsevier). He is an associate editor of various international journals, including *IEEE Transactions on Automation Science and Engineering, Application Analysis*, etc., and a reviewer for more than 50 other journals. He is the author or co-author of more than 300 journal articles. His research interests include stochastic control, stochastic systems, stochastic stability, stochastic delayed systems, Markovian jump systems, and stochastic complex networks.

mathematics

Editorial

Nonlinear Systems: Dynamics, Control, Optimization and Applications to the Science and Engineering

Quanxin Zhu

School of Mathematics and Statistics, Hunan Normal University, Changsha 410081, China; zqx22@hunnu.edu.cn

MSC: 34D23; 34H05; 37M05; 37M15; 49M05; 65P40; 92B05; 93D05; 93D21; 93E03; 93E15

Citation: Zhu, Q. Nonlinear Systems: Dynamics, Control, Optimization and Applications to the Science and Engineering. *Mathematics* **2022**, *10*, 4837. https://doi.org/10.3390/math10244837

Received: 28 November 2022
Accepted: 14 December 2022
Published: 19 December 2022

Publisher's Note: MDPI stays neutral with regard to jurisdictional claims in published maps and institutional affiliations.

Nonlinear phenomena frequently occur in many fields, such as physics, biology, and engineering. The mathematical models of nonlinear factors are complicated for theoretical and numerical analysis due to the underlying evolution over a large range of time scales and length scales. With the rapid development of advanced nonlinear dynamics methods, numerous physical, biological, or technologically complex systems and stochastic systems, such as mechanical or electronic devices, can be managed through nonlinear dynamics methods, both analytically and through computer simulation.

This Special Issue aimed to highlight the newest results on the dynamics, control, optimization, and applications of nonlinear systems. Several advanced nonlinear dynamics and numerical methods were covered in this Special Issue.

The Special Issue contains more than ten successful invited submissions [1–14], which are highly related to the potential topics. In [1], new stability criteria based on the Razumikhin technique were developed for impulsive switched delay systems subject to stochastic disturbances. In particular, the delay in the studied system was assumed to be a Markov chain rather than a deterministic delay. An optimal timing fault tolerant control algorithm was presented in [2] for switched stochastic systems with switched drift faults. In [3], the existence and stability of equilibrium points for quantized Hill systems were studied. By dividing three different cases that provide all possible locations, the locations of the equilibrium points were analyzed. For a class of discrete weakly nonlinear state-dependent coefficient control systems, [4] investigated the asymptotic solution of the initial singularly perturbed control problem for the matrix discrete Riccati equation with coefficients weakly dependent on the state and proposed a one-point PA regulator. In [5], the study of finite-time passivity analysis for neural-type neural networks was investigated. Applications of the fixed-time control approach to flexible spacecraft and third-order sliding mode control to single-rotor wind turbines were addressed in [6,7].

The nonlinear dynamics in ecological and biological complex systems are also of high interest in this Special Issue. In [8], the global stability of a delayed ecosystem, namely, a delayed feedback Gilpin–Ayala competition model with impulsive disturbance, was reported. In [9,10], several kinds of nonlinear dynamics problems in biological systems (such as epidemic systems with delayed impulse and delayed virus dynamic models) were studied. The topic of the optimization methods of complex systems is also included in this Special Issue. An adaptive evolutionary computation algorithm was proposed in [11] to overcome the overparameterization issue in traditional evolutionary and swarm computing paradigms. A new iterative method for finding extreme equations based on the maximum principle was developed for quantum systems in [12]. Two published works [13,14] focused on the applications of optimization methods in complex engineering systems.

The range of topics addressed in the current issue is not exhaustive. Further research on the dynamics, control, optimization, and applications of nonlinear systems is needed. We hope that some new insights into nonlinear systems are provided in this Special Issue.

We would like to end this preface by thanking all of the reviewers and editors that helped us in the completion of this Special Issue.

The response to our call had the following statistics:

Submissions (36);

Publications (14);

Rejections (22);

Article types: Research Article (14);

We found the edition and selections of papers for this Special Issue very inspiring and rewarding. We also thank the editorial staff and reviewers for their efforts and help during the process.

Data Availability Statement: Not applicable.

Conflicts of Interest: The author declares no conflict of interest.

References

1. Hu, W. Stability of Impulsive Stochastic Delay Systems with Markovian Switched Delay Effects. *Mathematics* **2022**, *10*, 1110. [CrossRef]
2. Zhu, C.; He, L.; Zhang, K.; Sun, W.; He, Z. Optimal Timing Fault Tolerant Control for Switched Stochastic Systems with Switched Drift Fault. *Mathematics* **2022**, *10*, 1880. [CrossRef]
3. Ansari, A.; Alhowaity, S.; Abouelmagd, E.; Sahdev, S. Analysis of Equilibrium Points in Quantized Hill System. *Mathematics* **2022**, *10*, 2186. [CrossRef]
4. Danik, Y.; Dmitriev, M. Symbolic Regulator Sets for a Weakly Nonlinear Discrete Control System with a Small Step. *Mathematics* **2022**, *10*, 487. [CrossRef]
5. Khonchaiyaphum, I.; Samorn, N.; Botmart, T.; Mukdasai, K. Finite-Time Passivity Analysis of Neutral-Type Neural Networks with Mixed Time-Varying Delays. *Mathematics* **2021**, *9*, 3321. [CrossRef]
6. Yao, Q.; Jahanshahi, H.; Moroz, I.; Alotaibi, N.; Bekiros, S. Neural Adaptive Fixed-Time Attitude Stabilization and Vibration Suppression of Flexible Spacecraft. *Mathematics* **2022**, *10*, 1667. [CrossRef]
7. Benbouhenni, H.; Bizon, N. Improved Rotor Flux and Torque Control Based on the Third-Order Sliding Mode Scheme Applied to the Asynchronous Generator for the Single-Rotor Wind Turbine. *Mathematics* **2021**, *9*, 2297. [CrossRef]
8. Rao, R. Global Stability of Delayed Ecosystem via Impulsive Differential Inequality and Minimax Principle. *Mathematics* **2021**, *9*, 1943. [CrossRef]
9. Rao, R.; Lin, Z.; Ai, X.; Wu, J. Synchronization of Epidemic Systems with Neumann Boundary Value under Delayed Impulse. *Mathematics* **2022**, *10*, 2064. [CrossRef]
10. Li, M.; Guo, K.; Ma, W. Uniform Persistence and Global Attractivity in a Delayed Virus Dynamic Model with Apoptosis and Both Virus-to-Cell and Cell-to-Cell Infections. *Mathematics* **2022**, *10*, 975. [CrossRef]
11. Altaf, F.; Chang, C.; Chaudhary, N.; Raja, M.; Cheema, K.; Shu, C.; Milyani, A. Adaptive Evolutionary Computation for Nonlinear Hammerstein Control Autoregressive Systems with Key Term Separation Principle. *Mathematics* **2022**, *10*, 1001. [CrossRef]
12. Buldaev, A.; Kazmin, I. Operator Methods of the Maximum Principle in Problems of Optimization of Quantum Systems. *Mathematics* **2022**, *10*, 507. [CrossRef]
13. Pérez-Aracil, J.; Camacho-Gómez, C.; Pereira, E.; Vaziri, V.; Aphale, S.; Salcedo-Sanz, S. Eliminating Stick-Slip Vibrations in Drill-Strings with a Dual-Loop Control Strategy Optimised by the CRO-SL Algorithm. *Mathematics* **2021**, *9*, 1526. [CrossRef]
14. Yang, S.; Wang, J.; Li, M.; Yue, H. Research on Intellectualized Location of Coal Gangue Logistics Nodes Based on Particle Swarm Optimization and Quasi-Newton Algorithm. *Mathematics* **2022**, *10*, 162. [CrossRef]

Eliminating Stick-Slip Vibrations in Drill-Strings with a Dual-Loop Control Strategy Optimised by the CRO-SL Algorithm

Jorge Pérez-Aracil [1,*], Carlos Camacho-Gómez [2], Emiliano Pereira [1], Vahid Vaziri [3], Sumeet S. Aphale [3] and Sancho Salcedo-Sanz [1]

[1] Department of Signal Processing and Communications, Universidad de Alcalá, Alcalá de Henares, 28805 Madrid, Spain; emiliano.pereira@uah.es (E.P.); sancho.salcedo@uah.es (S.S.-S.)
[2] Department of Information Systems, Universidad Politécnica de Madrid, Campus Sur, 28031 Madrid, Spain; carlos.camacho@upm.es
[3] School of Engineering, University of Aberdeen, Aberdeen AB24 3UE, UK; vahid.vaziri@abdn.ac.uk (V.V.); s.aphale@abdn.ac.uk (S.S.A.)
* Correspondence: jorge.perezraracil@uah.es

Abstract: Friction-induced stick-slip vibrations are one of the major causes for down-hole drill-string failures. Consequently, several nonlinear models and control approaches have been proposed to solve this problem. This work proposes a dual-loop control strategy. The inner loop damps the vibration of the system, eliminating the limit cycle due to nonlinear friction. The outer loop achieves the desired velocity with a fast time response. The optimal tuning of the control parameters is carried out with a multi-method ensemble meta-heuristic, the Coral Reefs Optimisation algorithm with Substrate Layer (CRO-SL). It is an evolutionary-type algorithm that combines different search strategies within a single population, obtaining a robust, high-performance algorithm to tackle hard optimisation problems. An application example based on a real nonlinear dynamics model of a drill-string illustrates that the controller optimised by the CRO-SL achieves excellent performance in terms of stick-slip vibrations cancellation, fast time response, robustness to system parameter uncertainties and chattering phenomenon prevention.

Keywords: stick-slip; drill-strings; vibration control; coral reefs optimisation; meta-heuristics

Citation: Pérez-Aracil, J.; Camacho-Gómez, C.; Pereira, E.; Vaziri, V.; Aphale, S.S.; Salcedo-Sanz, S. Eliminating Stick-Slip Vibrations in Drill-Strings with a Dual-Loop Control Strategy Optimised by the CRO-SL Algorithm. *Mathematics* **2021**, *9*, 1526. https://doi.org/10.3390/math9131526

Academic Editor: Stevan Dubljevic

Received: 1 June 2021
Accepted: 25 June 2021
Published: 29 June 2021

Publisher's Note: MDPI stays neutral with regard to jurisdictional claims in published maps and institutional affiliations.

1. Introduction

Stick-slip phenomena occur when, due to a particular confluence of parameters, the bit–rock interaction is such that rather than a constant (ideally synchronous) rotating speed of the top drive and the drill-bit, the drill-bit speed varies between zero and six times the speed of the top drive at the top of the drill-string. This unwanted phenomenon severely impacts the achieved rate of penetration during most drilling operations. In severe cases, this can lead to faster degradation of drill-bits and in rare cases warping or breaking of the drill-string [1]. To thoroughly understand this highly nonlinear phenomena, several drill-string models have been proposed in the literature [2–5]. Consequently, several bit–rock interaction models have also been proposed and validated [4,6,7]. The importance of the stick-slip vibrations problem has been widely analysed in [8–11]. In these previous works, researchers focus on understanding of the drill-string dynamics, the bit–rock interaction and the quantification of stick-slip vibrations. The current research aims to reduce the effects of the stick-slip phenomena during the drilling operation.

Vibration control in nonlinear systems is a problem that can be solved by linear and nonlinear controllers [3,12]. These controllers may be designed by using optimisation algorithms [13]. The problem of mitigating stick-slip vibrations in drill-strings is a well-known nonlinear control problem [3]. Some of the more recent and noteworthy approaches include the bit speed and torque independent stick-slip compensator [14], sliding-mode controllers [15], feedback controllers by including the axial and torsional motions in the numerical model [16], the use of robust controllers based on proportional-derivative feedback control design to keep a constant drill-bit rotation speed [17]. The use of Kalman filter based full-state feedback controllers was proposed in [18]. Sliding-mode controllers have shown very good results in the problem of reducing stick-slip vibration issues [19–21]. This fact arises from the good qualities of this controller, such as: robustness against inaccuracies related to the model and/or unpredicted dynamics. It also has a wide disturbance rejection and broad control bandwidth. It has been recently presented in [15,22] for eliminating stick-slip vibrations. The use of proportional-integral-derivative (PID) motivated linear controllers have also been proposed to address the stick-slip [23].

The complexity of the design of vibration controllers for drill-strings justifies the use of optimisation algorithms; in many cases, these are meta-heuristics approaches. Thus, the problem of stick-slip vibration has been solved by using optimisation algorithms [24,25]. In Reference [24], a torsional vibration control approach based on the genetic algorithm is studied. In [25], the authors show the use of a dynamic programming technique to design the optimal control of nonlinear systems. The Nelder–Mead method has been applied to design the controllers of multiple isolators situated on the same supporting structure [26]. In addition, some variants of the well-known genetic meta-heuristics algorithm have been applied to structural optimisation problems. In [27], the authors proposed a fuzzy control and a genetic algorithm for structural shape optimisation. In [28], the optimisation of 3D trusses was tackled with a Grouping Genetic Algorithm (GGA). In [29], a micro-genetic algorithm (μ-GA) was used for impact load identification and characterisation of concrete structures. For the plane stress problem, a variant of the Evolutionary Structural Optimisation (ESO) algorithm was proposed in [30]. Finally, in [31], the GA and among other meta-heuristic methods were implemented to obtain the optimal parameters in the welded beam structure problem. The particle swarm optimisation (PSO) algorithm introduced in [32] is also an important meta-heuristic that has been successfully applied to structural optimisation problems such as in [33], where the so-called democratic-PSO was proposed for truss layout with frequency constraints, or in [34], where optimal sizing design of truss structures was studied. The teaching-learning-based algorithm [35] has also been applied to optimise mechanical design problems [36,37]. The Harmony Search (HS) approach [38] has also been applied to structural optimisation; however, in [39], the authors aimed at two main ideas: firstly, they gathered a large number of meta-heuristics that have been used in structural optimisation; secondly, they reflected about the lack of mathematical background in meta-heuristics basics and the contributions of the HS. In the last decade, a new kind of meta-heuristics based on physics phenomena have been applied to structural optimisation problems. The Big-Bang Big-Crunch algorithm was updated in [40] to a memetic algorithm by mixing it with a local search process called the quasi-Newton method and was proposed to find the optimal weight of the structure. The Colliding Bodies Optimisation [41], the Ray Optimisation [42] or the Charged System Search [43] algorithms were created by A. Kaveh and tested in the welded beam design problem.

The concept of modified output, which was firstly introduced in [44], is used in this paper to propose a new control approach for eliminating stick-slip vibrations in drill-strings. The controller is based on a dual-loop linear control scheme. The inner loop, whose feedback signal is a combination of output signals, damps the vibration of the system, eliminating the limit cycle due to nonlinear friction. The outer loop achieves the desired velocity with a fast time response. The performance of the controller is improved by the use of a multi-method ensemble meta-heuristic, the Coral Reefs Optimisation algorithm with Substrate Layer (CRO-SL) [45,46], is adopted. This ensemble method was successfully applied in other engineering problems, such as [47–49]. In [47], the problem of optimal design and location of tuned mass dampers for structures subjected to earthquake ground motions is solved. In [48], the complex problem of design and location of inertial mass dampers for floor based structures subjected to human induced vibration is tackled. In [49], the design of submerged arch structures is approached. All the problems show that CRO-SL can be suitable for obtaining global optimum designs, with affordable computation time in complex vibration control problems. Consequently, the CRO-SL algorithm is applied to find the best configuration of the control parameters, i.e., obtaining the four controller parameters needed to eliminate stick-slip vibrations in a drill-string model with system uncertainties. The objective function to solve the problem is a functional that considers vibration, steady-state error and control effort.

The structure of the remainder of the paper is as follows: Section 2 presents the mathematical model of the drill-string, the bit–rock interaction and the model verification. Section 3 describes the proposed control scheme along with the CRO-SL optimisation algorithm, including the definition of a substrate layerand how this approach can be considered as a multi-method ensemble optimisation algorithm. Section 4 presents the optimisation results as well as the closed-loop results for the drill-string system, where the proposed algorithm's performance is evaluated. Section 5 closes the paper by giving some final conclusions and remarks on this research.

2. System Model

To realistically reproduce the stick-slip phenomena occurring in drill-strings and enable the implementation and testing of various control schemes, a scaled drill-string piece of equipment was set up at the Centre for Applied Dynamics Research, University of Aberdeen [22]. The following subsections describe the two-degrees-of-freedom (DOF) mathematical model that accurately captures the drill-string dynamics. The bit–rock interaction model is also described along with the system parameters on which the presented control design and analysis is based on.

2.1. Mathematical Model

As mentioned above, a 2-DOF model that accurately captures the stick-slip vibrations experienced by the drill-string is adopted. This is to ensure that the model is accurate, adequate and not overcomplicated. There are several works that have adopted a similar 2-DOF model [2,6,7,50].

The model lumps the drill-string components into two discs: (i) The upper disc that represents the top-drive system and (ii) the lower disc that represents the drill bit. Between the upper (top drive) and lower disc (drill bit), the drill-string is placed, mainly composed from a drill pipe and bottom hole assemble (BHA). A simple schematic of this model is shown in Figure 1. As labelled in Figure 1, the input torque that drives the system is denoted by U. The angular position of top drive is ϕ_r, its viscous damping is c_r, and the inertia J_r. The pipe, which connects the top drive with the BHA, is modelled by its torsional damping C and the torsional stiffness K. The torque of friction on the lower disc, which models the interaction between the bit and the rock, is denoted by T_b. The angular position of the BHA is ϕ_b, and its inertia is J_b.

The equation of motion that encapsulates the entire system behaviour is then given by:

$$\ddot{\phi}_r = \frac{U}{J_r} - \frac{C_r + C}{J_r}\dot{\phi}_r + \frac{C}{J_r}\dot{\phi}_b - \frac{K}{J_r}(\phi_r - \phi_b) \tag{1}$$

$$\ddot{\phi}_b = \frac{C}{J_b}\dot{\phi}_r - \frac{C}{J_b}\dot{\phi}_b + \frac{K}{J_b}(\phi_r - \phi_b) - \frac{T_b}{J_b} \tag{2}$$

However, to facilitate control design, the new state of the system can be initialised as follows:

$$x = \left(\dot{\phi}_r, \phi_r - \phi_b, \dot{\phi}_b\right)^T = (x_1, x_2, x_3)^T \tag{3}$$

which further simplifies the system equation to:

$$\dot{x}_1 = \frac{U}{J_r} - \frac{C_r + C}{J_r}x_1 + \frac{C}{J_r}x_3 - \frac{K}{J_r}x_2 \tag{4}$$

$$\dot{x}_2 = x_1 - x_3 \tag{5}$$

$$\dot{x}_3 = \frac{C}{J_b}x_1 - \frac{C}{J_b}x_3 + \frac{K}{J_b}x_2 - \frac{T_b}{J_b} \tag{6}$$

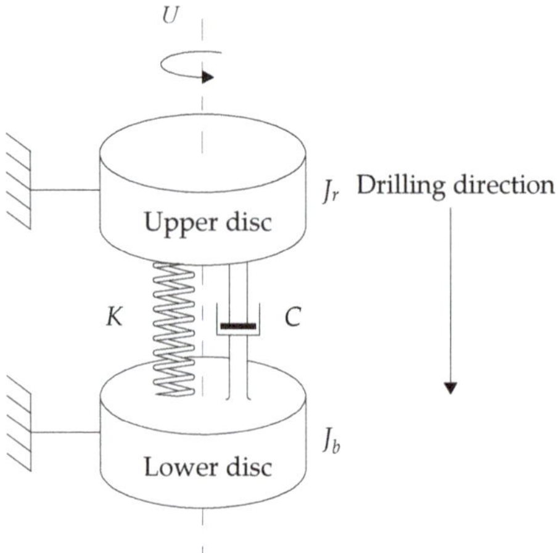

Figure 1. Schema of the 2-DOF lumped-parameter model of the drill-string.

Equations (3)–(6) can be re-described as the following state-space system:

$$\dot{x} = \begin{bmatrix} -\frac{C_r+C}{J_r} & -\frac{K}{J_r} & \frac{C}{J_r} \\ 1 & 0 & -1 \\ \frac{C}{J_b} & \frac{K}{J_b} & \frac{C}{J_b} \end{bmatrix} x + \begin{bmatrix} 1 \\ 0 \\ 0 \end{bmatrix} \frac{U}{J_r} + \begin{bmatrix} 0 \\ 0 \\ -1 \end{bmatrix} \frac{T_b}{J_b}. \tag{7}$$

The problem of drill-string can be modelled as a linear part (Equation (7)) connected with the hard nonlinear model of the bit–rock interaction. This interaction is modelled by the torque T_b in Equation (7). This is explained below.

2.2. Bit–Rock Interaction

It is possible to distinguish three different phases in the bit–rock interaction: (i) stick phase: the drill-bit is not rotation since it is stuck with the rock, (ii) stick-to-slip phase: to begin slipping, the drill-bit has to achieve enough torque, (iii) slip phase: the drill-bit rotates and it is actually drilling [4,15,51,52].

It is important to note that the system is locked between phases (ii) and (iii). The sticking phase ends when the reaction torque reaches the peak value. The slip phase starts when the drill-bit starts to rotate. For $x_3 = 0$, the dry friction is estimated by combining a zero band velocity introduced in [53] and the switch model in [54].

When $x_3 = 0$, the dry friction is approximated by combining a zero band velocity introduced in [53] and the switch model in [54]. This is as follows:

$$T_b = \left\{ \begin{array}{ll} \tau_r, & \text{if}\,|x_3| < \zeta \quad \text{and} \quad |\tau_r| \leq \tau_s \\ \tau_s \cdot sgn(\tau_r), & \text{if}\,|x_3| < \zeta \quad \text{and} \quad |\tau_r| > \tau_s \\ \mu_b \cdot R_b \cdot W_{ob} \cdot sgn(x_3), & \text{if}\,|x_3| \geq \zeta \end{array} \right\}, \tag{8}$$

where sgn stands for the *sign* function,

$$sgn(x) = \left\{ \begin{array}{ll} 1 & x > 0 \\ 0 & x = 0 \\ -1 & x < 0 \end{array} \right.,$$

The frictional torque when the drill–bit interaction occurs can be formulated as $\tau_r = C(x_1 - x_3) + K(x_2)$, which is the reaction torque, whilst $\tau_s = \mu_{sb} R_b W_{ob}$ is the friction torque, μ_{sb} is the static friction coefficient, W_{ob} is the Weight on Bit (WOB), R_b is the bit radius. A set of drill-string realistic simulation parameters was obtained in [2,22], and these parameters are tabulated in Table 1.

Table 1. The basic realistic drill-string physical parameters for the model.

Parameters	Values
J_r	$13.93 \text{ kg} \cdot \text{m}^2$
J_b	$1.1378 \text{ kg} \cdot \text{m}^2$
C	$0.005 \text{ N} \cdot \text{m} \cdot \text{s/rad}$
K	$10 \text{ N} \cdot \text{m/rad}$
C_r	$11.38 \text{ N} \cdot \text{m} \cdot \text{s/rad}$
μ_{sb}	0.0843
μ_{cb}	0.0685
R_b	0.0492 m
γ_b	0.3
ν_f	0.1935
ζ	10^{-4}

3. Control Design and Parameter Optimisation

The problem of control of stick-slip oscillations in drill-strings bears some similarity to the problem of control of one-link flexible manipulators. It is clear that as the sensors and actuators for these systems (flexible robot and drill-string) are not physically located at the same spot (not co-located). Therefore, these systems are non-minimum phase in nature. Moreover, the Coulomb friction present in the joint of the robotic manipulator can be treated as a complication similar to the drill-string's nonlinear friction in the bit–rock interaction model (presented in Equation (8)). In the case of the flexible manipulator, modified outputs, like the reflected tip position in one-link flexible manipulators [44,55], have been proposed in the past in order to solve this seemingly complicated problem. These modified outputs have been used to propose simple control schemes, which achieves the objectives of tracking and vibration control [44,55,56]. In addition, these controllers guarantee the stability of the highly unconsidered vibration mode (spillover).

The proposed linear control is inspired by the aforementioned controllers applied to flexible robotic manipulators, where the objective is to cancel the vibrations at the tip with an actuator placed at the joint. The proposed controller is shown in Figure 2. Note that there are two nested loops. The first one is used to stabilise the system, cancelling the vibration of drill-string, which is due to the vibration modes of the linear part and the nonlinear dynamics of the bit–rock interaction.

The controlled input of the inner loop (U_{IL}) is as follows:

$$U_{IL} = \frac{\beta}{\alpha}(\phi_r - \phi_b) - \frac{1}{\alpha}\ddot{U}_{IL}, \tag{9}$$

where α and β are the parameters used to obtain the modified output of the drill-string. The objective of the outer loop is to achieve a desired velocity of the top drive. Thus, if the inner loop (9) cancels the vibration, the outer loop can achieve the desired velocity of BHA. This outer loop is as follows:

$$U_{OL} = P(\dot{\phi}_{r,ref} - \dot{\phi}_r) + I(\phi_{r,ref} - \phi_r), \tag{10}$$

where P and I are the proportional and integral constants of the tracking controller. The proposed control law can then be written as follows:

$$U = U_{IL} + U_{OL}. \tag{11}$$

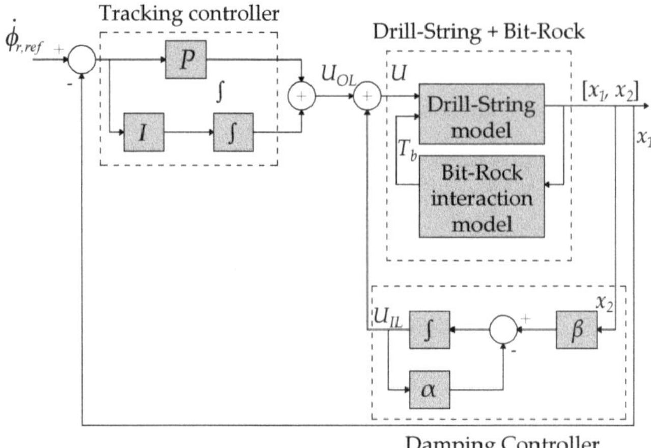

Figure 2. The adopted dual-loop control scheme. The inner loop consists of a damping controller, which uses the modified output $x_2\,C_d$, and the outer loop consists of a tracking controller (proportional and integral).

The optimal control parameters are tuned according to a functional, which considered the following issues:

- J_1: The steady-state error between the desired (top-drive) and achieved (drill-bit) angular velocity $(\dot{\phi}_r - \dot{\phi}_{r,ref})$.
- J_2: The residual vibration $(\dot{\phi}_b - \dot{\phi}_r)$.
- J_3: The settling time of the variables $\dot{\phi}_b$ and $\dot{\phi}_r$
- J_4: The control effort from a determined value of time, which rejects solutions with chattering.

The functional is defined as follows:

$$J(\mathbf{z}) = \gamma_1 J_1 + \gamma_2 J_2 + \gamma_3 J_3 + \gamma_4 J_4, \tag{12}$$

where γ_1, γ_2, γ_3, and γ_4 are the terms that ponderate each sub-functional. These sub-functionals are as follows:

$$
\begin{aligned}
J_1 &= \left| \dot{\phi}_{r,ref} - \dot{\phi}_r(t_f) \right|, \\
J_2 &= \max_{t \in (t_i, t_f)} |\dot{\phi}_b(t) - \dot{\phi}_r(t)|, \\
J_3 &= t_{set,r} + t_{set,b}, \\
J_4 &= \frac{1}{N} \sum_{t=t_{set,r}}^{t_f} (U(t) - U_{OL}),
\end{aligned}
\tag{13}
$$

where t_i is the time at which the controller is engaged, t_f is the final value of time considered in simulations and N is the number of samples considered between t_f and t_i. The variables $t_{set,r}$ and $t_{set,b}$ are the settling time of $\dot{\phi}_r$ and $\dot{\phi}_b$ after the value of t_i and considering a final value equal to $\dot{\phi}_{r,ref}$ with an error of ± 4 %. The variable $\mathbf{z}$ is defined by the controller parameters: P, I, β and α. Note that the initial conditions of Equation (7) are considered in J_2, where the $\max_{t \in (t_i, t_f)} |\dot{\phi}_b(t) - \dot{\phi}_r(t)|$ depends on the system response before t_i. The functional J_1 and J_4 consider the values of $\dot{\phi}_r$ and U after $t_{set,r}$.

Due to the nonlinear nature of the system, control parameter tuning is not straightforward. Thus, a suitable optimisation algorithm can be used in order to automate the process, ensure stable results and significantly improve the control performance. The Coral Reef Optimisation (CRO) is a meta-heuristic algorithm for optimisation that has been recently proposed as a highly-effective algorithm for such problems [48,57–59]. This work uses an improved version of the CRO algorithm to generate the controller parameters namely β, α, P and I.

Coral Reefs Optimisation Algorithm with Substrate Layer: A Multi-Method Ensemble Approach

The CRO-SL is a further version of the original CRO [57], which was founded on the processes that occur in a coral reef, including reproduction, fight for space or depredation [46]. The basics can be shown in the pseudo-code below, Algorithm 1, with the different CRO phases where the individuals (also called corals) are initialised in the population (or reef), along with all the operators applied to guide the search.

Algorithm 1 Pseudo-code for the original CRO

Require: Valid configuration of parameters controlling the CRO algorithm
Ensure: A single individual with near-optimal value of its *fitness*
  1: Initialize the corals and the reef
  2: **for** each iteration of the experiment **do**
  3:     Update values of the algorithm parameters
  4:     Sexual reproduction processes for new individuals (also larvae)
  5:     Settlement of new larvae
  6:     Predation of some of the weakest
  7:     The new population is ready for the next generation
  8: **end for**
  9: Return the best individual (final solution) from the reef

---

As has been mentioned before, the CRO-SL is an improved version of the CRO, firstly proposed in [60] and further developed in [45]. It is considered as a low-level ensemble method for optimisation, which combines several search processes in parallel over different subpopulations and with information exchange between them, leading to extremely good search capabilities. The CRO-SL has the same structure as shown in the pseudo code of the basic CRO, but including substrate layers or subpopulations, wherein each one is implementing a different search procedure or strategy. A substrate may represent different models, operators, parameters, constraints, repairing functions, etc., though this version, which represents different search operators, has been the most successful version so far. Thus, the CRO-SL is an ensemble approach that promotes cooperative co-evolution. The use of CRO-SL as a cooperative multi-method ensemble has been successfully tested in different applications and problems, such as battery scheduling and topology design in micro-grids [58,61], medical image registration [59], antenna design [62] and vibration cancellation problems in buildings [47] and open floors [48]. This wide application demonstrates the potency of the CRO-SL and deems it an ideal candidate for optimising controller parameters for eliminating stick-slip vibrations in drill-strings. Details of the overall CRO-SL algorithm and the mechanism to include substrate layers is well-reported in [45].

Although different search strategies have been defined at the practitioner's discretion, this work has taken into account a five-substrate construct of the CRO-SL. They are briefly described below:

1. Harmony Search (HS): It is a metaheuristic method based on stochastic optimisation [38]. It imitates the process found in music improvisation, which searches for better harmony. There are two parameters that determine the way in which new larvae are generated: (i) Harmony Memory Considering Rate (HMCR), which ranges from zero to one. If a uniformly spawned value is above the value of HMCR, then the encoded parameter value is uniformly drawn from the values in the coral, (ii) Pitch

Adjusting Rate (PAR), which ranges from zero to one, which sets the probability of choosing a neighbour value of the current larva.

2. Differential Evolution (DE): It is an Evolutionary Algorithm (EA), which has good abilities for global search [63]. The new larvae can be generated either by the mutation or crossover process. For a randomly selected encoded parameter, if a uniformly generated value is above the Crossover Probability (CR) value, which ranges from zero to one, the new value is obtained by $x_i' = x_i^1 + F(x_i^2 - x_i^3)$, in which F is the evolution factor. Otherwise, the value is crossed with the randomly selected encoded parameter.

3. Classical 2-points crossover (2Px): The crossover operator is the most used exploration mechanism in genetic and evolutionary algorithms [64] since its combination with an efficient mutation process allows achieving a suitable balance between exploration and exploitation. 2Px selects two random parents and exchanges the genetic material in-between two random points on them. Despite that each substrate is linked to a searching process, when another parent must be picked, the selection is not limited to their substrate, but it can be chosen from any part of the population instead. The reason is to contribute to genetic information exchange among substrates so they can easily cooperate.

4. Multi-points crossover (MPx): This search method is a generalisation of the 2Px. In this case, k points are selected in the parents. In this work, due to the dimensionality of the problem, the value of k has been chosen to be three. Thus, a binary vector decides whether the parts of each parent are exchanged or not for the new offspring generation.

5. Gaussian Mutation (GM): The Gaussian Mutation is a noisy search method based on adding random values from the Gaussian distribution to the encoded parameter values, thus generating an offspring. The standard deviation σ value in this work is linearly decreasing during the run, from $0.2 \cdot (A - B)$ to $0.02 \cdot (A - B)$, where $[B, A]$ is the domain search. The Gaussian probability density function is:

$$f_{G(0,\sigma^2)}(x) = \frac{1}{\sigma\sqrt{2\pi}}e^{-\frac{x^2}{2\sigma^2}}.$$

With the aim of exploring the search space at the beginning of the optimisation process and exploiting it at the end, the parameter σ is adapted during the simulation.

4. Computational Evaluation

This section presents the computational evaluation of the proposed control algorithm for stick-slip vibrations cancellation, optimised with the CRO-SL algorithm. The open-loop parameters of the drill-string system are given in Table 1. The torsional stiffness (K) is considered as a variable parameter and the velocity reference ($\dot{\phi}_{r,ref}$) is selected to be within 2 and 5 rad/s (acceptable practical range). The system is simulated for 40 s in an open-loop after which the loop is closed and the proposed dual-loop control scheme is engaged. The value of the control input (U) in Equation (11) is limited to $U \in [20, 80]$ N·m. This is to ensure that the control input lies within the practical limits [7,22].

The CRO-SL algorithm searches for the best solution in terms of Equation (12) for $K \in \{5, 10, 20\}$ N·m/rad and $\dot{\phi}_{r,ref} \in \{2, 3, 4, 5\}$. The values of torque needed to obtain this velocity are $U \in \{30.2, 41.42, 52.65, 63.9\}$ N·m. The parameters of Equations (12) and (13) are:

- $\gamma_1 = 1$, $\gamma_2 = 0.5$, $\gamma_3 = 0.25$ and $\gamma_3 = 2$;
- $t_i = 40$ s and $t_f = 120$ s;
- N is the result of using a 0.001 s of sampling time. The simulation is carried out by a fourth-order ordinal differential equation solver (Runge–Kutta) with fixed time of 0.001 s (i.e., $N = (120 - 40)/0.001 = 80 \times 10^3$);
- The input torque is, as mentioned earlier, limited to $U \in [20, 80]$ N·m.

The CRO-SL parameters used in the experiments are shown in Table 2.

Table 2. Parameters values used in the CRO-SL.

Parameter	Description	Value
Reef	Reef size	120
F_b	Frequency of broadcast spawning	97%
Substrates	HS, DE, 2Px, MPx, GM	5
$\mathcal{N}_{att}$	Number of tries for larvae settlement	3
F_a	Percentage of asexual reproduction	5%
F_d	Fraction of corals for depredation	5%
P_d	Probability of depredation	5%
α	Maximum number of iterations	50

The optimisation results obtained with the CRO-SL are presented in Figure 3. Figure 3a shows the ratio for which each substrate generates the best larva in the CRO-SL approach.

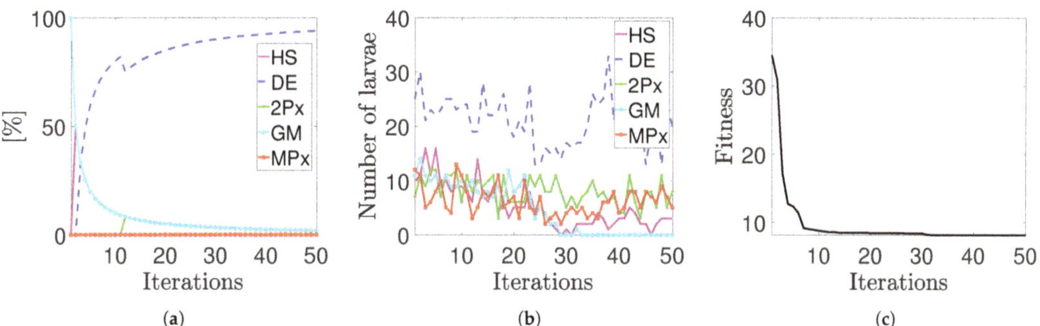

Figure 3. Evolution of different quality parameters in the CRO-SL; (**a**) evolution of the number of larvae that are finally allocated into the reef as new solutions; (**b**) ratio of times that each substrate produces the best larva in each iteration of the CRO-SL algorithm; (**c**) evolution of the best solution within the CRO-SL.

It is possible to see how DE is the one exploration operator that contributes the most to the CRO-SL search. The contribution of the GM and 2Px are also significant, but they decrease during the search. Figure 3b shows the number of larvae that are produced in the reef for each iteration, as new solutions. Figure 3c shows the evolution of the best solution within the CRO-SL. It is important to note that around 20 iterations, the proposed approach achieves the optimal solution. The final value of the functional (Figure 3c) is $J = 8.1560$, which corresponds with $K = 10$ N·m and $\dot{\phi}_{r,ref} = 5$ rad/s. The controller parameters are shown in Table 3.

Table 3. Optimal control parameters obtained by means of CRO-SL.

Optimal Control Parameters			
P	I	β	α
3.33×10^4	3.51×10^4	9.15×10^4	25×10^{-4}

The next subsection shows the performance in terms of stick-slip vibrations cancellation, fast time response, robustness to system parameter uncertainties and chattering phenomenon prevention.

Results

First of all, the closed-loop system configured with the above shown parameters is simulated to generate time responses for the nominal conditions ($K = 10$ N·m/rad and a $\dot{\phi}_{r,ref} = 2.5$ rad/s). The controller starts at 40 s, the value of t_i used in optimal design.

Two values of restrictions for control input torque are used. The first one corresponds to $U \in [20, 80]$ N·m, which is the restriction used in the optimisation process. The second one is a naive selection corresponding to $U \in [-100, 100]$ N·m. Figure 4 shows the angular velocity of the top-drive, drill-bit and input torque signal against the time variable. It is clear after some careful understanding of the system that negative values of control input torque require the top-drive to rotate in the opposite direction to the standard drilling. This is, in practice, an unacceptable scenario as physically reversing the motor's direction of rotation is an almost impossible task on the field. However, it is likely that this might be possible or even required in some scenarios (drill-bit retrieval). The objective of these experiments is to show that the controlled system optimised with the CRO-SL scheme is stable in both scenarios.

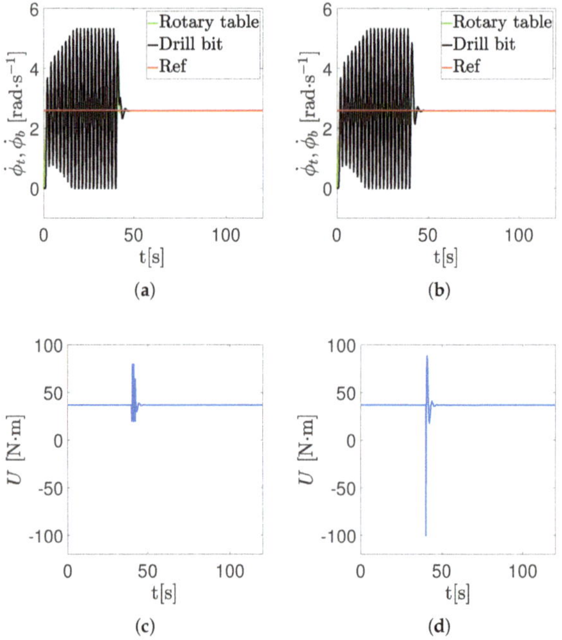

Figure 4. Time histories of the angular velocity of the top-drive, drill-bit and input torque signal for a desired velocity (2.5 rad/s) and $K = 10$ N·m/rad: (**a**) angular velocity of the top-drive and drill-bit when $U \in [20, 80]$ N·m; (**b**) angular velocity of the top-drive and drill-bit when $U \in [-100, 100]$ N·m; (**c**) input torque when $U \in [20, 80]$ N·m; (**d**) input torque when $U \in [-100, 100]$ N·m.

Figure 5 shows how the controller of Figure 2 can eliminate the limit cycle of the stick-slip. The controller starts at $t = 75$ s in order to guarantee that the limit cycle is in steady-state for all the reference values ($\dot{\phi}_{r,ref}(t)$) when $U \in [20, 80]$ N·m (see Figure 5e). Figure 5a,b shows the stick-slip limit cycle in an open-loop for $U = 36$ N·m and its corresponding 2D phase-plane portrait, respectively. Figure 5c shows the corresponding 3D perspective with respect to time in order to better clarify the transient response. Figure 5d shows the motor torque U. In this example, the open loop torque is $U = 36$ N·m. The controller is turned on at $t = 75$ s, and the control input undergoes a brief period of transients, with its permanent value equal to $U = 36$ N·m. Figure 5e shows how the controller can stabilise the stick-slip for $U \in [30, 69.5]$ N·m. It is important to highlight that desired velocity can be achieved from any initial value of input torque in an open loop.

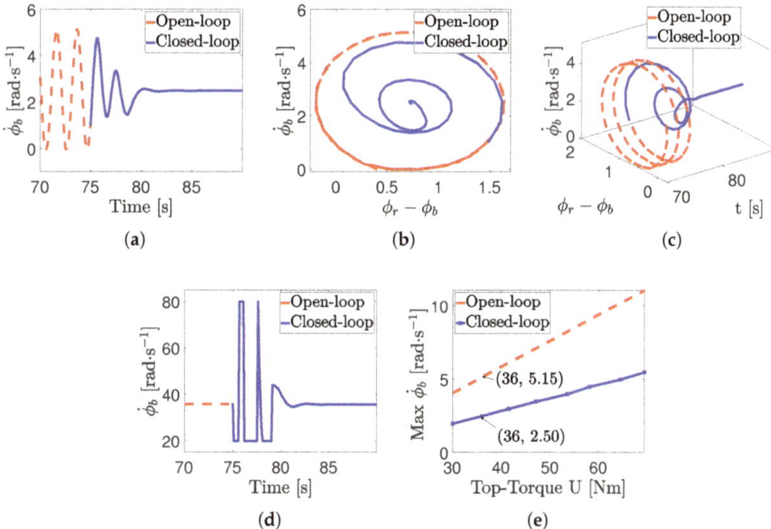

Figure 5. The red line and the blue line stand for uncontrolled and scheme controlled, respectively. (**a–c**) The time-history, 2D and 3D phase-portraits, respectively. (**d**) The control input versus time. (**e**) The complete bifurcation diagram showing all types of behaviour.

An additional robustness analysis is carried out. Figure 6a shows the value of the functional (J) as a function of the desired velocity and stiffness value. Figure 6a shows that the maximum value of J is achieved for $K = 10$ N·m/rad and a $\dot{\phi}_{r,ref} = 5$ rad/s. In addition, the most unfavourable case for each value of K is the maximum value of the desired velocity. Then, three examples are plotted in order to show how the stick-slip limit cycle is cancelled. In these examples, like in Figure 5, the controller starts at $t = 75$ s in order to guarantee that this vibration is in a steady-state for all the cases. Thus, Figure 6b–d shows how the controlled system eliminates the stick-slip vibrations for K equal to 5, 10 and 20 N·m/rad, respectively. Note that, although a large torque control input is needed, after the settling time, all time responses are the same, showing that the controller effectively eliminates the stick-slip vibrations for all these cases. In addition, Figure 6e–g shows the input torque when $U \in [20, 80]$ N·m for K equal to 5, 10 and 20 N·m/rad, respectively.

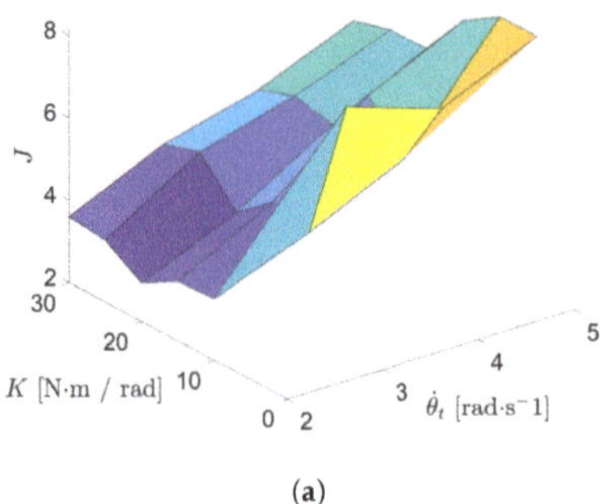

(a)

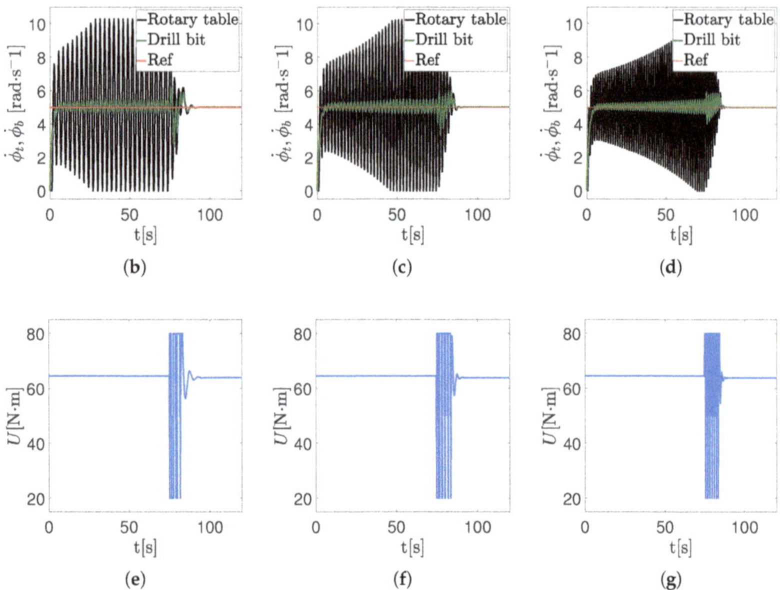

(b) (c) (d)

(e) (f) (g)

Figure 6. Robustness analysis: (**a**) value of the functional (J) for different values of the desired velocity and stiffness values; (**b**) angular velocity of the top-drive and drill-bit for $K = 5$ N·m/rad; (**c**) angular velocity of the top-drive and drill-bit for $K = 10$ N·m/rad; (**d**) angular velocity of the top-drive and drill-bit for $K = 20$ N·m/rad; (**e**) input torque when $U \in [20, 80]$ N·m for $K = 5$ N·m/rad; (**f**) input torque when $U \in [20, 80]$ N·m for $K = 10$ N·m/rad; (**g**) input torque when $U \in [20, 80]$ N·m for $K = 20$ N·m/rad.

Finally, a final case is simulated in order to show the problem of the control effort (chattering phenomenon), which is avoided by including J_4 (see Equations (12) and (13)), justifying the use of the optimisation algorithm. Thus, the tuning of P, I, β and γ is carried out according to the following practical guide: (i) increases the values of P and I in order to achieve a good time response for the top-drive, (ii) modified β in order to reduce the oscillation between drill-bit and top-drive angular velocities and (iii) tune a small, but not equal to zero, value for α, in order to avoid possible practical problems with offset in x_2. The control parameters are shown in Table 4.

Table 4. Control parameters for the non-optimised case.

Non-Optimal Control Parameters			
P	I	β	α
5.00×10^4	5.00×10^4	9.00×10^4	30.00×10^{-4}

Figure 7 shows the time histories for the angular velocity of the top-drive, drill-bit and input torque signal for a desired velocity of 2.5 rad/s and $K = 10$ N·m/rad (i.e., the nominal conditions). Note that the time responses of the top-drive and the drill-bit are good and comparable with Figure 4. However, the control effort (chattering phenomenon) for both cases is not admissible because it can damage the actuator. In addition, the design of this controller for a wider range of K (parameter uncertainty) and desired velocities is not obvious. This justifies the use of CRO-SL, which simplifies and improves the control parameter selection, consequently resulting in improved closed-loop performance.

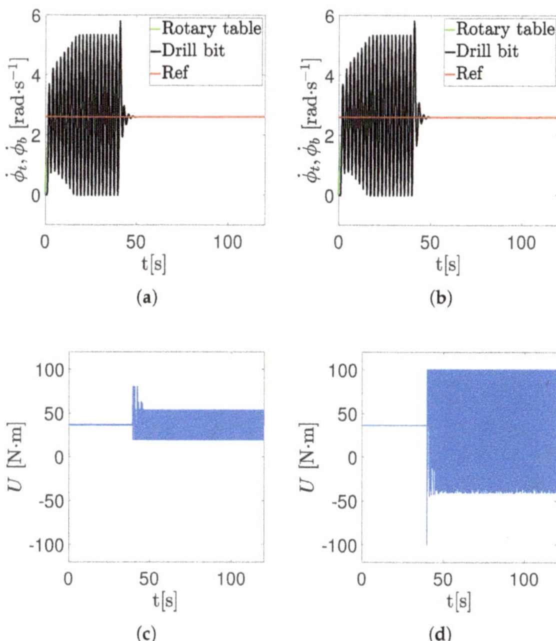

Figure 7. Non-optimised design: time histories of the angular velocity of the top-drive, drill-bit and input torque signal for a desired velocity (2.6 rad/s) and $K = 10$ N·m/rad. (**a**) angular velocity of the top-drive and drill-bit when $U \in [20, 80]$ N·m; (**b**) Angular velocity of the top-drive and drill-bit when $U \in [-100, 100]$ N·m; (**c**) input torque when $U \in [20, 80]$ N·m; (**d**) input torque when $U \in [-100, 100]$ N·m.

5. Conclusions

This work proposes a dual-loop control strategy to alleviate stick-slip vibration problems in drill-string applications. The inner loop controller is based on the concept of modified output, which has been used in the past to control non-collated systems. The results show that the combination of both velocities, the top-drive and the drill-bit, can eliminate the limit circle of the nonlinear system. The outer loop is used to achieve the desired drill-bit velocity within a low settling time and without steady-state error.

The advantages of optimising the controller parameters have also been shown in this work. The controller has been optimally tuned by using a multi-method ensemble meta-heuristic approach, the Coral Reefs Optimisation algorithm with Substrate Layers (CRO-SL). A functional function is defined in this work, which involves: (i) steady-state error, (ii) the residual vibration, (iii) the settling time and (iv) the control effort. Thus, the optimised controller is robust to parameter systems uncertainties, control input torque limitations and different tip velocities references, eliminating the chattering phenomenon in the control effort. Therefore, this dual-loop strategy, together with the optimisation procedure defined by the functional, has big potential in drill-string applications.

Author Contributions: Conceptualisation, E.P., S.S.A., J.P.-A., V.V., S.S.-S. and C.C.-G.; methodology, S.S.A., E.P., S.S.-S., J.P.-A., V.V.; software, J.P.-A., C.C.-G.; validation, S.S.A., E.P., S.S.-S., V.V. and J.P.-A.; formal analysis, J.P.-A., E.P., S.S.A, V.V.; investigation, J.P.-A., S.S.A., E.P., S.S.-S., C.C.-G. and V.V.; resources, S.S.A., V.V. and E.P.; data curation, J.P.-A., C.C.-G., E.P. and S.S.-S.; writing—original draft preparation, J.P.-A., E.P., S.S.-S. and S.S.A.; writing—J.P.-A., E.P., S.S.-S. and S.S.A.; visualisation, J.P.-A., E.P. and S.S.A.; supervision, S.S.-S., S.S.A. and E.P.; project administration, S.S.-S., S.S.A. and E.P.; funding acquisition, S.S.A. All authors have read and agreed to the published version of the manuscript.

Funding: This work was partially supported by the Spanish Ministerial Commission of Science and Technology (MICYT) through project number TIN2017-85887-C2-2-P.

Data Availability Statement: Not applicable.

Acknowledgments: The authors would like to thank Marian Wiercigroch and Vahid Vaziri from the Centre for Applied Dynamics Research, University of Aberdeen, for providing the realistic drill-string parameters used in this work.

Conflicts of Interest: The authors declare no conflict of interest. The funders had no role in the design of the study; in the collection, analyses, or interpretation of data; in the writing of the manuscript, or in the decision to publish the results.

References

1. Brett, J.F. The genesis of bit-induced torsional drillstring vibrations. *SPE Drill. Eng.* **1992**, *7*, 168–174. [CrossRef]
2. Hamaneh, S.V.V. Dynamics and Control of Nonlinear Engineering Systems. Ph.D. Thesis, University of Aberdeen, Aberdeen, UK, 2015.
3. Vromen, T.; Dai, C.-H.; van de Wouw, N.; Oomen, T.; Astrid, P.; Nijmeijer, H. Robust output-feedback control to eliminate stick-slip oscillations in drill-string systems. *IFAC Pap.* **2015**, *48*, 266–271. [CrossRef]
4. Navarro-López, E.M. An alternative characterization of bit-sticking phenomena in a multi-degree-of-freedom controlled drillstring. *Nonlinear Anal. Real World Appl.* **2009**, *10*, 3162–3174. [CrossRef]
5. Oladunjoye, I.O.; Vaziri, V.; Ing, J.; Aphale, S.S. Severity analysis of stick-slip bifurcation in drill-string dynamics under parameter variation. *Afr. Model. Simul.* **2016**. [CrossRef]
6. Navarro-López, E.M.; Cortés, D. Avoiding harmful oscillations in a drillstring through dynamical analysis. *J. Sound Vib.* **2007**, *307*, 152–171. [CrossRef]
7. Kapitaniak, M.; Hamaneh, V.V.; Chávez, J.P.; Nandakumar, K.; Wiercigroch, M. Unveiling complexity of drill-string vibrations: Experiments and modelling. *Int. J. Mech. Sci.* **2015**, *101*, 324–337. [CrossRef]
8. Sugiura, J.; Jones, S. Real-time stick-slip and vibration detection for 8 1/2″-hole-size rotary steerable tools in deeper wells and more aggressive drilling. In Proceedings of the AADE National Technical Conference, Houston, TX, USA, 10–12 April 2007.
9. Tucker, W.; Wang, C. On the effective control of torsional vibrations in drilling systems. *J. Sound Vib.* **1999**, *224*, 101–122. [CrossRef]
10. Sugiura, J. The use of the near-bit vibration sensor while drilling lead to optimized rotary-steerable drilling in push-and point-the-bit configurations. In Proceedings of the SPE Asia Pacific Oil and Gas Conference and Exhibition, Perth, Australia, 20–22 October 2008.
11. Robnett, E.; Hood, J.; Heisig, G.; Macpherson, J. Analysis of the stick-slip phenomenon using downhole drillstring rotation data. In Proceedings of the SPE/IADC Drilling Conference, Amsterdam, The Netherlands, 9–11 March 1999.
12. Vaziri, V.; O, I.; Kapitaniak, M.; Aphale, S.S.; Wiercigroch, M. Parametric Analysis of a Sliding-Mode Controller to Suppress Drill-String Stick-Slip Vibration. *Meccanica* **2020**, *55*, 2475–2492. [CrossRef]
13. Baradaran-Nia, M.; Afizadeh, G.; Khanmohammadi, S.; Azar, B.F. Optimal sliding mode control of single degree-of-freedom hysteretic structural system. *Commun. Nonlinear Sci. Numer. Simul.* **2012**, *17*, 4455–4466. [CrossRef]
14. Sassan, A.; Halimberdi, B. Design of a controller for suppressing the stick-slip oscillations in oil well drillstring. *Res. J. Recent Sci.* **2013**, *2*, 78–82.

15. Liu, Y. Suppressing stick-slip oscillations in underactuated multibody drill-strings with parametric uncertainties using sliding-mode control. *IET Control. Theory Appl.* **2014**, *9*, 91–102. [CrossRef]
16. Sairafi, F.A.; Ajmi, K.A.; Yigit, A.; Christoforou, A. Modeling and control of stick slip and bit bounce in oil well drill strings. In Proceedings of the SPE/IADC Middle East Drilling Technology Conference and Exhibition, Abu Dhabi, United Arab Emirates, 26–28 January 2016.
17. Lin, W.; Liu, Y. Proportional-derivative control of stick-slip oscillations in drill-strings. In Proceedings of the International Conference on Engineering Vibration, Sofia, Bulgaria, 4–7 September 2017.
18. Hong, L.; Girsang, I.P.; Dhupia, J.S. Identification and control of stick-slip vibrations using kalman estimator in oil-well drill strings. *J. Pet. Sci. Eng.* **2016**, *140*, 119–127. [CrossRef]
19. Navarro-López, E.M.; Cortés, D. Sliding-mode control of a multi-dof oilwell drillstring with stick-slip oscillations. In Proceedings of the American Control Conference, ACC'07, New York, NY, USA, 11–13 July 2007; IEEE: Piscataway, NJ, USA, 2007; pp. 3837–3842.
20. Biel, D.; Fossas, E.; Guinjoan, F.; Alarcón, E.; Poveda, A. Application of sliding-mode control to the design of a buck-based sinusoidal generator. *IEEE Trans. Ind. Electron.* **2001**, *48*, 563–571. [CrossRef]
21. Hernandez-Suarez, R.; Puebla, H.; Aguilar-Lopez, R.; Hernandez-Martinez, E. An integral high-order sliding mode control approach for stickslip suppression in oil drillstrings. *Pet. Sci. Technol.* **2009**, *27*, 788–800. [CrossRef]
22. Vaziri, V.; Kapitaniak, M.; Wiercigroch, M. Suppression of drill-string stick-slip vibration by sliding mode control: Numerical and experimental studies. *Eur. J. Appl. Math.* **2018**, *29*, 1–21. [CrossRef]
23. Ritto, T.; Ghandchi-Tehrani, M. Active control of stick-slip torsional vibrations in drill-strings. *J. Vib. Control.* **2019**, *25*, 194–202. [CrossRef]
24. Feng, T.; Zhang, H.; Chen, D. Dynamic Programming Based Controllers to Suppress Stick-Slip in A Drilling System. In Proceedings of the 2017 American Control Conference (ACC), Seattle, WA, USA, 24–26 May 2017.
25. Ke, C.; Song, X.Y. Control of Down-Hole Drilling Process Using a Computationally Efficient Dynamic Programming Method. *J. Dyn. Din. Systms Meas. Control Trans. ASME* **2018**, *140*. [CrossRef]
26. Pérez-Aracil, J.; Pereira, E.; Aphale, S.S.; Reynolds, P. Vibration Isolation and Alignment of Multiple Platforms on a Non-Rigid Supporting Structure. *Actuators* **2020**, *9*, 108. [CrossRef]
27. Soh, C.K.; Yang, J. Fuzzy controlled genetic algorithm search for shape optimization. *ASCE J. Comput. Civ. Eng.* **1996**, *10*, 143–150. [CrossRef]
28. Togan, V.; Daloglu, A.T. Optimization of 3D trusses with adaptive approach in genetic algorithms. *Eng. Struct.* **2006**, *28*, 1019–1027. [CrossRef]
29. Yan, G.; Zhou, L. Impact load identification of composite structure using genetic algorithms. *J. Sound Vib.* **2009**, *319*, 869–884. [CrossRef]
30. Simonetti, H.L.; Almeida, V.S.; de Neto, L. A smooth evolutionary structural optimization procedure applied to plane stress problem. *Eng. Struct.* **2014**, *75*, 248–258. [CrossRef]
31. Deb, K. Optimal design of a welded beam via genetic algorithms. *AIAA J.* **1991**, *29*, 2013–2015. [CrossRef]
32. Kennedy, J.; Eberhart, R.C. Particle swarm optimization. In Proceedings of the IEEE International Conference on Neural Networks, Perth, Australia, 27 November–1 December 1995; Volume IV, pp. 1942–1948.
33. Kaveh, A.; Zolghadr, A. Democratic PSO for truss layout and size optimization with frequency constraints. *Comput. Struct.* **2014**, *130*, 10–21. [CrossRef]
34. Schutte, J.J.; Groenwold, A.A. Sizing design of truss structures using particle swarms. *Struct. Multidiscip. Optim.* **2003**, *23*, 261–269. [CrossRef]
35. Rao, R.V.; Savsani, V.J.; Vakharia, D.P. Teaching?earning-based optimization: A novel optimization method for continuous non-linear large scale problems. *Inf. Sci.* **2011**, *183*, 1–15. [CrossRef]
36. Rao, R.V.; Savsani, V.J.; Vakharia, D.P. Teaching-learning-based optimization: A novel method for constrained mechanical design optimization problems. *Comput. Aided Des.* **2011**, *43*, 303-315. [CrossRef]
37. Degertekin, S.O.; Hayalioglu, M.S. Sizing truss structures using teaching-learning based optimization. *Comput. Struct.* **2013**, *119*, 177–188. [CrossRef]
38. Geem, Z.W.; Kim, J.H.; Loganathan, G.V. A new heuristic optimization algorithm: Harmony Search. *Simulation* **2001**, *76*, 60–68. [CrossRef]
39. Saka, M.P.; Hasancebi, O.; Geem, Z.W. Metaheuristics in structural optimization and discussions on harmony search algorithm. *Swarm Evol. Comput.* **2016**, *28*, 88–97. [CrossRef]
40. Kaveh, A.; Mahdavi, V.R. Optimal design of structures with multiple natural frequency constraints using a hybridized BB-BC/Quasi-Newton algorithm. *Period. Polytech. Civ. Eng.* **2013**, *57*, 27–38. [CrossRef]
41. Kaveh, A.; Mahdavi, V.R. Colliding bodies optimization: A novel meta-heuristic method. *Comput. Struct.* **2014**, *139*, 18–27. [CrossRef]
42. Kaveh, A.; Khayatazad, M. A new meta-heuristic method: Ray Optimization. *Comput. Struct.* **2012**, *112*, 283–294. [CrossRef]
43. Kaveh, A.; Talatahari, S.A. Novel heuristic optimization method: Charged system search. *Acta Mech.* **2010**, *213*, 267–289. [CrossRef]
44. Wang, D.; Vidyasagar, M. Passive control of a stiff flexible link International. *J. Robot. Res.* **1992**, *11*, 572–578. [CrossRef]

45. Salcedo-Sanz, S.; Camacho-Gómez, C.; Molina, D.; Herrera, F. A Coral Reefs Optimization algorithm with substrate layers and local search for large scale global optimization. In Proceedings of the IEEE Congress on Evolutionary Computation (CEC), Vancouver, BC, Canada, 24–29 July 2016; pp. 1–8.
46. Salcedo-Sanz, S. A review on the coral reefs optimization algorithm: New development lines and current applications. *Prog. Artif. Intell.* **2017**, *6*, 1–15. [CrossRef]
47. Camacho-Gómez, S.S.C.; Magdaleno, A.; Pereira, E.; Lorenzana, A. Structures vibration control via Tuned Mass Dampers using a co-evolution Coral Reefs Optimization algorithm. *J. Sound Vib.* **2017**, *393*, 62–75.
48. Camacho-Gómez, C.; Wang, X.; Pereira, E.; Díaz, I.M.; Salcedo-Sanz, S. Active vibration control design using the Coral Reefs Optimization with Substrate Layer algorithm. *Eng. Struct.* **2018**, *157*, 14–26. [CrossRef]
49. Pérez-Aracil, J.; Camacho-Gómez, C.; Hernández-Díaz, A.M.; Pereira, E.; Salcedo-Sanz, S. Submerged Arches Optimal Design with a Multi-Method Ensemble Meta-Heuristic Approach. *IEEE Access* **2020**, *8*, 215057–215072. [CrossRef]
50. Mihajlovic, N.; Veggel, A.V.; de Wouw, N.V.; Nijmeijer, H. Analysis of friction-induced limit cycling in an experimental drill-string system. *J. Dyn. Syst. Meas. Control* **2004**, *126*, 709–720. [CrossRef]
51. Navarro-López, E.M.; Suárez, R. Practical approach to modelling and controlling stick-slip oscillations in oilwell drillstrings. In Proceedings of the 2004 IEEE International Conference, Paris, France, 20–24 June 2004; IEEE: Piscataway, NJ, USA, 2004; Volume 2, pp. 1454–1460.
52. Navarro-López, E.M.; Licéaga-Castro, E. Non-desired transitions and sliding-mode control of a multi-dof mechanical system with stick-slip oscillations. *Chaos Solitons Fractals* **2009**, *41*, 2035–2044. [CrossRef]
53. Armstrong-Hélouvry, B.; Dupont, P.; Wit, C.C.D. A survey of models, analysis tools and compensation methods for the control of machines with friction. *Automatica* **1994**, *30*, 1083–1138. [CrossRef]
54. Leine, R.; Campen, D.V.; Kraker, A.D.; Steen, L.V.D. Stickslip vibrations induced by alternate friction models. *Nonlinear Dyn.* **1998**, *16*, 41–54. [CrossRef]
55. Liu, L.Y.; Yuan, K. Noncollocated passivity-based PD control of a single-link flexible manipulator. *Robotica* **2003**, *21*, 117–135. [CrossRef]
56. Ryu, J.H.; Kwon, D.S.; Hannaford, B. Control of a flexible manipulator with noncollocated feedback: Time-domain passivity approach. *IEEE Trans. Robot.* **2004**, *20*, 776–778. [CrossRef]
57. Salcedo-Sanz, S.; Del Ser, J.; Landa-Torres, I.; Gil-López, S.; Portilla-Figueras, J.A. The Coral Reefs Optimization algorithm: A novel metaheuristic for efficiently solving optimization problems. *Sci. World J.* **2014**, *2014*, 739768. [CrossRef] [PubMed]
58. Jiménez-Fernández, S.; Camacho-Gómez, C.; Mallol-Poyato, R.; Fernández, J.C.; Ser, J.D.; Portilla-Figueras, A.; Salcedo-Sanz, S. Optimal microgrid topology design and siting of distributed generation sources using a multi-objective substrate layer Coral Reefs Optimization algorithm. *Sustainability* **2019**, *11*, 169. [CrossRef]
59. Bermejo, E.; Chica, M.; Damas, S.; Salcedo-Sanz, S.; Cordón, O. Coral Reef Optimization with ubstrate layers for medical image registration. *Swarm Evol. Comput.* **2018**, *42*, 138–159. [CrossRef]
60. Salcedo-Sanz, S.; noz-Bulnes, J.M.; Vermeij, M. New Coral Reefs-based Approaches for the Model Type Selection Problem: A Novel Method to Predict a Nation's Future Energy Demand. *Int. J. Bio-Inspired Comput.* **2017**, *10*, 145–158. [CrossRef]
61. Salcedo-Sanz, S.; Camacho-Gómez, C.; Mallol-Poyato, R.; Jiménez-Fernández, S.; Del Ser, J. A novel Coral Reefs Optimization algorithm with substrate layers for optimal battery scheduling optimization in micro-grids. *Soft Comput.* **2016**, *20*, 4287–4300. [CrossRef]
62. Sánchez-Montero, R.; Camacho-Gómez, C.; López-Espí, P.L.; Salcedo-Sanz, S. Optimal design of a planar textile antenna for Industrial Scientific Medical (ISM) 2.4 GHz Wireless Body Area Networks (WBAN) with the CRO-SL Algorithm. *Sensors* **2018**, *18*, 1982. [CrossRef] [PubMed]
63. Price, K.; Storn, R.; Lampinen, J. *Differential Evolution—A Practical Approach to Global Optimization*; Springer: Berlin, Germany, 2005.
64. Eiben, A.E.; Smith, J.E. *Introduction to Evolutionary Computing*; Springer: Berlin/Heidelberg, Germany, 2003.

Article

Global Stability of Delayed Ecosystem via Impulsive Differential Inequality and Minimax Principle

Ruofeng Rao [1,2]

[1] Department of Mathematics, Chengdu Normal University, Chengdu 611130, China; ruofengrao@163.com or ruofengrao@cdnu.edu.cn
[2] Institute of Financial Mathematics, Chengdu Normal University, Chengdu 611130, China

Abstract: This paper reports applying Minimax principle and impulsive differential inequality to derive the existence of multiple stationary solutions and the global stability of a positive stationary solution for a delayed feedback Gilpin–Ayala competition model with impulsive disturbance. The conclusion obtained in this paper reduces the conservatism of the algorithm compared with the known literature, for the impulsive disturbance is not limited to impulsive control.

Keywords: Minimax principle; linear approximation theory; ecosystem; steady state solution

Citation: Rao, R. Global Stability of Delayed Ecosystem via Impulsive Differential Inequality and Minimax Principle. *Mathematics* **2021**, *9*, 1943. https://doi.org/10.3390/math9161943

Academic Editor: Mustafa R.S. Kulenovic

Received: 11 July 2021
Accepted: 12 August 2021
Published: 14 August 2021

Publisher's Note: MDPI stays neutral with regard to jurisdictional claims in published maps and institutional affiliations.

1. Introduction

It is well known that Gilpin–Ayala competition model (GACM) has been hotly discussed (see in [1–7]) due to its importance in simulating two or more competing biological populations in nature. As diffusion is an essential characteristic of most biological populations, Ling Bai and Ke Wang began to investigate the global stability of reaction-diffusion Gilpin–Ayala ecosystem under Neumann zero boundary value in 2005 (see in [8]), and obtained good results. Actually, Neumann zero boundary value means that the populations do not migrate beyond the biosphere boundary. However, many animal populations are at the edge of the biosphere, where the population density is usually zero, which is not reflected by Neumann zero boundary value. Thus, the Dirichlet zero boundary value was considered in recent literature [6,7]. In recent years, linear impulsive control and nonlinear impulsive control technology are widely used in ordinary differential dynamical systems and infinite dimensional dynamical systems [9–16]. For example, in [14], event-triggered nonlinear impulsive control to stabilize damped wave equations was designed, and rapid exponential stabilization was achieved. However, in this paper, linear impulse is relatively simple and feasible in practical ecological management because ecological management is a natural system of impulse artificial intervention. Therefore, the linear impulsive control is considered in this paper. Note that impulse control is employed to make the GACM stable globally in [6,7], but this paper involves the impulsive disturbance, which is not limited to impulsive control. Minimax principle will be employed to derive the existence of multiple stationary solutions, which improve the method of Mountain Pass Lemma in [6]. On the other hand, impulsive disturbance is considered in this paper, not just impulse control. In fact, some impulse management measures other than impulsive control sometimes occur in ecological management due to accidents, such as releasing animals, hunting animals harmful to the population, and so on. These pulse measures mean that the pulse intensity is not necessarily less than 1 based on system stability.

2. Preparatory Knowledge

Consider the following reaction-diffusion Gilpin–Ayala competition model (RDGACM) with delayed feedback under Dirichlet boundary value

$$
\begin{cases}
\dfrac{\partial u_1}{\partial t} = d_1 \Delta u_1 + u_1(b_1 - a_{11}u_1^{\theta_1} - a_{12}u_2) - k_1(u_1 - u_1(t - \tau_1, x)), & t \geqslant 0,\ x \in \Omega, \\[2mm]
\dfrac{\partial u_2}{\partial t} = d_2 \Delta u_2 + u_2(b_2 - a_{21}u_1 - a_{22}u_2^{\theta_2}) - k_2(u_2 - u_2(t - \tau_2, x)), & t \geqslant 0,\ x \in \Omega, \\[2mm]
u_1(t, x) = u_2(t, x) = 0, & t \geqslant 0,\ x \in \partial\Omega,
\end{cases}
\tag{1}
$$

equipped with the initial value

$$
u_1(s, x) = \xi_1(x), \quad u_2(s, x) = \xi_2(x), \quad s \in [0, \tau_0], \quad x \in \Omega,
\tag{2}
$$

where Ω is a bounded domain in $\mathbb{R}^N (1 \leqslant N \leqslant 3)$ with smooth boundary $\partial\Omega$. Time delays $\tau_1, \tau_2 \in [0, \tau_0]$, $u_i(t, x)$ represents the population density of the ith population at time t and the spatial location x, $b_i > 0$ represents the birth rate of the population of the ith species, and $a_{ij} > 0$ represents the competition parameter between the species i and the species j. $d_i > 0$ represents the diffusion coefficient for the species i. Initial value function $\xi(x) = (\xi_1(x), \xi_2(x))^T$ is bounded and continuous.

Assume that

(A1) For $i = 1, 2$, setting $-1 < \theta_i < 4$, and $s^{2+\theta_i} \geqslant 0$ for all $s \in R^1$.

(A2) For $i = 1, 2$, there exist positive constants $M_i > 0$ such that

$$
0 \leqslant u_i \leqslant M_i.
$$

(A3) For $i = 1, 2$, $|\nabla u_i(t, x)|$ is bounded for all $x \in \Omega$.

Due to the limited natural resources, it is reasonable to assume in (A2) that each population density is limited. Besides, the limited natural resources imply that the boundedness assumption of (A3) is suitable to the real state of nature.

Remark 1. *(A1) expands greatly the allowable range of parameters θ_i, compared with the previous related literature (see, e.g., in [6,7]). For example, the harsh condition "$0 < \theta_i < 1$ with $\hat{\theta}_i$ being an even number, and $\check{\theta}_i$ being an odd number" is deleted.*

Lemma 1 (see, e.g., in [17]). *Let $J \in C^1(H_0^1(\Omega), \mathbb{R}^1)$. If there is an upper boundness of J in $H_0^1(\Omega)$, and J satisfies the (PS) condition, then $c_* = \sup\limits_{v \in H_0^1(\Omega)} J(v)$ is a critical value of J.*

Here, the $(PS)_c$ condition may be found in [18] (Definition 2). Actually, the (PS) condition is equivalent to the $(PS)_c$ condition. For convenience, the author describes the (PS) condition as follows:

Definition 1 ([17]). *Let ψ be a real C^1 functional defined on a Banach space X. If any sequence $\{u_n\}$ in X with $\|\psi'(u_n)\|_{X^*} \to 0$ and the bounded sequence $\{\psi(u_n)\}_{n=1}^{\infty}$ has a convergent subsequence in X, then ψ is called satisfying the (PS) condition.*

Lemma 2 ([7], Theorem 3.1). *Set $u^*(x) = (u_1^*(x), u_2^*(x))^T$. Suppose that the condition (A2) holds, and $0 < \theta_i < 1$ for $i = 1, 2$. Moreover, if there exists a positive constant $c_* > 0$ such that*

$$
0 \leqslant h(u^*(x)) \leqslant c_* D\mathfrak{H},
\tag{3}
$$

then there are at least a positive bounded equilibrium solution $u^*(x)$ for the RDAGCM (1), where $\mathfrak{H} = (1, 1)^T$, $h(u) = (h_1(u_1, u_2), h_2(u_1, u_2))^T$ with $u = (u_1, u_2)^T$ and

$$
h_1(u_1, u_2) = u_1(b_1 - a_{11}u_1^{\theta_1} - a_{12}u_2), \quad h_2(u_1, u_2) = u_2(b_2 - a_{21}u_1 - a_{22}u_2^{\theta_2}),
$$

$$
D = \begin{pmatrix} d_1 & 0 \\ 0 & d_2 \end{pmatrix} > 0.
$$

The conditions of Lemma 2 guarantee the existence of a positive stationary solution $(u_1^*(x), u_2^*(x))$ for the delayed feedback system (1). Set

$$\begin{cases} U_1 = u_1 - u_1^*(x) \\ U_2 = u_2 - u_2^*(x), \end{cases}$$

and the stationary solution $(u_1^*(x), u_2^*(x))$ of the system (1) corresponds to the zero solution $(0,0)^T$ of the following system:

$$\begin{cases} \dfrac{\partial U_1}{\partial t} = d_1 \Delta U_1 + b_1 U_1 - \Phi_1(U_1, U_2) - k_1[U_1 - U_1(t - \tau_1, x)], & t \geqslant 0,\, x \in \Omega, \\ \dfrac{\partial U_2}{\partial t} = d_2 \Delta U_2 + b_2 U_2 - \Phi_2(U_1, U_2) - k_2[U_2 - U_2(t - \tau_2, x)], & t \geqslant 0,\, x \in \Omega, \\ U_1(t,x) = U_2(t,x) = 0, & t \geqslant 0,\, x \in \partial\Omega, \end{cases}$$

or

$$\begin{cases} \dfrac{\partial U_1}{\partial t} = d_1 \Delta U_1 + (b_1 - k_1) U_1 - \Phi_1(U_1, U_2) + k_1 U_1(t - \tau_1, x), & t \geqslant 0,\, x \in \Omega, \\ \dfrac{\partial U_2}{\partial t} = d_2 \Delta U_2 + (b_2 - k_2) U_2 - \Phi_2(U_1, U_2) + k_2 U_2(t - \tau_2, x), & t \geqslant 0,\, x \in \Omega, \\ U_1(t,x) = U_2(t,x) = 0, & t \geqslant 0,\, x \in \partial\Omega, \end{cases} \tag{4}$$

where we denote $U = (U_1, U_2)^T$, and

$$\begin{aligned} \Phi_1(U) &= (U_1 + u_1^*(x))[a_{11}(U_1 + u_1^*(x))^{\theta_1} + a_{12}(U_2 + u_2^*(x))] - u_1^*(x)(a_{11}u_1^*(x)^{\theta_1} + a_{12}u_2^*(x)), \\ \Phi_2(U) &= (U_2 + u_2^*(x))[a_{21}(U_1 + u_1^*(x)) + a_{22}(U_2 + u_2^*(x))^{\theta_2}] - u_2^*(x)(a_{21}u_1^*(x) + a_{22}u_2^*(x)^{\theta_2}). \end{aligned} \tag{5}$$

The following system is the system (4) in form of vector-matrix:

$$\begin{cases} \dfrac{\partial U}{\partial t} = D\Delta U + (B - K)U - \Phi(U) + KU(t - \tau, x), & t \geqslant 0,\ x \in \Omega, \\ U(t, x) = 0, & t \geqslant 0,\, x \in \partial\Omega, \end{cases} \tag{6}$$

where $U = (U_1, U_2)^T$, $U(t - \tau, x) = (U(t - \tau_1, x), U(t - \tau_2, x))^T$, $\Phi(U) = (\Phi_1(U), \Phi_2(U))^T$ and

$$D = \begin{pmatrix} d_1 & 0 \\ 0 & d_2 \end{pmatrix}, \quad B = \begin{pmatrix} b_1 & 0 \\ 0 & b_2 \end{pmatrix}, \quad K = \begin{pmatrix} k_1 & 0 \\ 0 & k_2 \end{pmatrix}. \tag{7}$$

Considering the impulse disturbance on (6), one can get the following system:

$$\begin{cases} \dfrac{\partial U}{\partial t} = D\Delta U + (B - K)U - \Phi(U) + KU(t - \tau, x), & t \geqslant 0,\, t \neq t_k,\, x \in \Omega, \\ U(t_k^+, x) = A_k U(t_k^-, x), & k = 1, 2 \cdots \\ U(t, x) = 0, & t \geqslant 0,\, x \in \partial\Omega, \end{cases} \tag{8}$$

where $\{t_k\}_{k=1}^{\infty}$ is a sequence of fixed impulsive instants, satisfying $0 < t_1 < t_2 < \cdots < t_k < t_{k+1} < \cdots$ and $\lim\limits_{k \to \infty} t_k = +\infty$. Besides, $U_i(t_k^+, x) = U_i(t_k, x)$, $U_i(t_k^-, x) = \lim\limits_{t \to t_k^-} U_i(t, x)$ for all $i = 1, 2$, $k = 1, 2, \cdots$.

Definition 2. *$(u_1^*(x), u_2^*(x))^T$ is said to be globally exponentially stable under impulsive disturbances if the zero solution of the system (8) is globally exponentially stable.*

Lemma 3 (see [19]). *Consider the following differential inequality:*

$$\begin{cases} D^+ v(t) \leq -av(t) + b[v(t)]_\tau, & t \neq t_k \\ v(t_k) \leq a_k v(t_k^-) + b_k[v(t_k^-)]_\tau, \end{cases}$$

where $v(t) \geq 0$, $[v(t_k)]_\tau = \sup\limits_{t-\tau \leq s \leq t} v(s)$, $[v(t_k^-)]_\tau = \sup\limits_{t-\tau \leq s < t} v(s)$ and $v(t)$ is continuous except $t_k, k = 1, 2, \cdots$, where it has jump discontinuities. The sequence t_k satisfies $0 = t_0 < t_1 < t_2 < \cdots < t_k < t_{k+1} < \cdots$, and $\lim\limits_{k \to \infty} t_k = \infty$. Suppose that

   (1) $a > b \geq 0$;

   (2) $t_k - t_{k-1} > \delta\tau$, where $\delta > 1$, and there exist constants $\gamma > 0$, $M > 0$ such that

$$\rho_1 \rho_2 \cdots \rho_{k+1} e^{k\lambda\tau} \leqslant M e^{\gamma t_k}, \tag{9}$$

where $\rho_i = max\{1, a_i + b_i e^{\lambda\tau}\}$, $\lambda > 0$ is the unique solution of equation $\lambda = a - b e^{\lambda\tau}$; then

$$v(t) \leqslant M[v(0)]_\tau e^{-(\lambda-\gamma)t}.$$

In addition, if $\theta = \sup\limits_{k \in Z}\{1, a_k + b_k e^{\lambda\tau}\}$, then

$$v(t) \leqslant \theta[v(0)]_\tau e^{-(\lambda - \frac{\ln(\theta e^{\lambda\tau})}{\delta\tau})t}, \quad t \geq 0.$$

Notation 1. *Denote by λ_1 the first positive eigenvalue of the operator $-\Delta$ in the Sobolev space $H_0^1(\Omega)$ equipped with the norm $\|v\| = \sqrt{\int_\Omega |\nabla v|^2 dx}$ for any $v(x) \in H_0^1(\Omega)$. Denote by $E(\lambda_1)$ the eigenfunction space of λ_1. Denote by $\varphi_1(x) > 0$ the positive eigenfunction corresponding to $E(\lambda_1)$ with $\|\varphi_1(x)\| = 1$. Besides, I represents the identity matrix. Denote by $\lambda_{\max}(A)$ the maximum eigenvalue of symmetric matrix A, and by $\lambda_{\min}(A)$ the minimum eigenvalue of symmetric matrix A.*

3. Main Results

Theorem 1. *Suppose the conditions (A1)–(A3) and (3) hold, and if the following conditions are satisfied:*

$$b_1 < d_1 \lambda_1 \tag{10}$$

$$b_2 < d_2 \lambda_1 \tag{11}$$

then the system (1) owns multiple stationary solutions, including the positive solution $(u_1^*(x), u_2^*(x))^T$.

Proof. To complete the proof of Theorem 1, the author needs to do it step by step.

Step 1. Under the condition (10), there is at least a stationary solution $(\alpha_*(x), 0)$ for the system (1).

Let $(\alpha(x), 0)^T$ be a stationary solution of the system (1), satisfying

$$d_1\Delta\alpha(x) + \alpha(x)(b_1 - a_{11}\alpha(x)^{\theta_1} - a_{12} \cdot 0) = 0, \quad x \in \Omega; \quad \alpha(x)|_{\partial\Omega} = 0, \tag{12}$$

whose functional is

$$\Psi(\alpha) = \frac{1}{2}\int_\Omega |\nabla\alpha(x)|^2 dx - \frac{b_1}{2d_1}\int_\Omega |\alpha(x)|^2 dx + \frac{a_{11}}{(2+\theta_1)d_1}\int_\Omega \alpha(x)^{2+\theta_1} dx, \tag{13}$$

It is obvious that $\Psi(0) = 0$ and $\Psi \in C^1(H_0^1(\Omega), \mathbb{R}^1)$.

In fact, for example, when $N = 3$, the assumption (A1) yields that there exists a real number $c > 0$ big enough that

$$|\alpha(x)(b_1 - a_{11}\alpha(x)^{\theta_1} - a_{12} \cdot 0)| \leqslant c(1 + |\alpha(x)|^{1+\theta_1}) \leqslant c(1 + |\alpha(x)|^{2^*-1}),$$

where $2^* = \frac{2N}{N-2}$ is the Sobolev critical exponent in the case of $\Omega \subset R^N$. This means $\Psi \in C^1(H_0^1(\Omega), \mathbb{R}^1)$, and then a critical point of the functional Ψ is corresponding to the solution of the Equation (12).

Next, the author claim that Ψ satisfies the (PS) condition.

Indeed, if there exists a real number a and $\{\alpha_n\} \subset H_0^1(\Omega)$, satisfying $|\Psi(\alpha_n)| \leqslant a \in R^1$ and $\|\Psi'(\alpha_n)\|_{(H_0^1(\Omega))^*} \to 0$, it means that when n is big enough,

$$\Psi(\alpha_n) = \frac{1}{2}\int_\Omega |\nabla\alpha_n(x)|^2 dx - \frac{b_1}{2d_1}\int_\Omega |\alpha_n(x)|^2 dx + \frac{a_{11}}{(2+\theta_1)d_1}\int_\Omega \alpha_n(x)^{2+\theta_1} dx \leqslant a, \quad (14)$$

which together with (A1) and the poincare inequality means

$$\frac{1}{2}(1 - \frac{b_1}{d_1\lambda_1})\int_\Omega |\nabla\alpha_n(x)|^2 dx \leqslant a \tag{15}$$

Equation (15) implies the boundedness of $\{\alpha_n\}$ in the Sobolev space $H_0^1(\Omega)$. Moreover, Hilbert space $H_0^1(\Omega)$ yields that there exists a subsequence, say $\{\alpha_n\}$, such that $\alpha_n(x) \rightharpoonup \alpha(x)$ in $H_0^1(\Omega)$. Rellich Theorem means that $\alpha_n(x) \to \alpha(x)$ in $L^p(\Omega)$ with $1 \leqslant p < 2^*$. Therefore,

$$\int_\Omega |\alpha_n - \alpha|^2 dx \to 0, \quad \int_\Omega |\alpha_n - \alpha|^{2+\theta_1} dx \to 0, \quad n \to \infty.$$

Thus, as $n \to \infty$,

$$\|\alpha_n - \alpha\|^2 = \langle \Psi'(\alpha_n) - \Psi'(\alpha), \alpha_n - \alpha\rangle + \frac{b_1}{d_1}\int_\Omega |\alpha_n - \alpha|^2 dx - \frac{a_{11}}{d_1}\int_\Omega |\alpha_n - \alpha|^{2+\theta_1} dx \to 0,$$

which verifies that the (PS) condition is satisfied.

Next, the author claims that there is an upper boundedness for Ψ.

In fact, (A1) and (A2) yields

$$\Psi(\alpha) = \frac{1}{2}\int_\Omega |\nabla\alpha(x)|^2 dx - \frac{b_1}{2d_1}\int_\Omega |\alpha(x)|^2 dx + \frac{a_{11}}{(2+\theta_1)d_1}\int_\Omega \alpha(x)^{2+\theta_1} dx$$

$$\leqslant \frac{1}{2}\int_\Omega |\nabla\alpha(x)|^2 dx + \frac{a_{11}}{(2+\theta_1)d_1} M_1^{2+\theta_1} mes(\Omega),$$

which together with (A3) means that there exists an upper boundedness for Ψ.

According to Lemma 1, there exists $\alpha_*(x)$ such that

$$J(\alpha_*(x)) = \sup_{v \in H_0^1(\Omega)} J(v)$$

and $(\alpha_*(x), 0)^T$ is a stationary solution of the system (1).

Step 2. The author claims that the system (1) owns multiple stationary solutions, including the positive solution.

First, the condition (3) and Lemma 2 guarantee the existence of a positive stationary solution for the system (1). Second, zero solution $(0,0)^T$ is obviously another stationary solution for the system (1). Next, $(\alpha_*(x), 0)^T$ is the third stationary solution thanks to Step 1. In fact, the continuity of $\varphi_1(x)$ yields

$$J(\alpha_*(x)) = \sup_{v \in H_0^1(\Omega)} J(v) \geqslant J(\varphi_1) \geqslant \frac{a_{11}}{(2+\theta_1)d_1}\int_\Omega \varphi_1(x)^{2+\theta_1} dx > 0,$$

which means that $(\alpha_*(x), 0)^T$ is a nontrivial stationary solution for the system (1). Finally, one can similarly prove that there exists a nontrivial stationary solution $(0, \beta_*(x))^T$ for the system (1). $\square$

Theorem 2. *Suppose that all the conditions of Theorem 1 are satisfied. Assume, in addition,*

(B1) there exist three positive constants p_m, p_M, ε, and a positive definite diagonal matrix $P = diag(p_1, p_1) > 0$ such that the following LMI conditions hold:

$$2\lambda_1 PD - 2P(B - K) - p_M\Theta - \varepsilon PK > 0 \tag{16}$$

$$P < p_M I \tag{17}$$

$$p_M I < P \tag{18}$$

where

$$\Theta = \begin{pmatrix} 2\Big(a_{11}(1+\theta_1)(2M_1)^{\theta_1} + a_{12}M_2\Big) & a_{12}M_1 + a_{21}M_2 \\ * & 2\Big(a_{22}(1+\theta_2)(2M_2)^{\theta_2} + a_{21}M_1\Big) \end{pmatrix}$$

(B2) $a > b \geqslant 0$, where $a = \dfrac{\lambda_{\min}\Big(2\lambda_1 PD - 2P(B-K) - p_M\Theta - \varepsilon PK\Big)}{p_M}$, $b = \dfrac{\lambda_{\max}(K)}{\varepsilon}$

(B3) there exists a constant $\delta > 1$ such that $\inf_{k \in \mathbb{Z}}(t_k - t_{k-1}) > \delta\tau$ and $\lambda > \dfrac{\ln(\rho e^{\lambda\tau})}{\delta\tau}$, where $\rho = \sup\limits_{j \in \mathbb{Z}}\{1, a_j + b_j e^{\lambda\tau}\}$ with $a_j = \dfrac{\lambda_{\max}(A_j^T PA_j)}{p_m}$ and $b_j \equiv 0$, and $\lambda > 0$ is the unique solution of the equation $\lambda = a - be^{\lambda\tau}$.

then the zero solution of the system (8) is globally exponentially stable with convergence rate $\frac{1}{2}\Big(\lambda - \frac{\ln(\rho e^{\lambda\tau})}{\delta\tau}\Big)$, and $(u_1^*(x), u_2^*(x))^T$ is said to be globally exponentially stable under impulsive disturbances with convergence rate $\frac{1}{2}\Big(\lambda - \frac{\ln(\rho e^{\lambda\tau})}{\delta\tau}\Big)$.

Proof. Consider the following Lyapunov function:

$$V(t) = \int_\Omega U^T(t,x)PU(t,x)dx = \int_\Omega |U(t,x)|^T P|U(t,x)|dx$$

then for $t \geqslant 0, t \neq t_k$, the Poincare inequality yields

$$
\begin{aligned}
D^+ V =& 2\int_\Omega U^T P\Big(D\Delta U + (B - K)U - \Phi(U) + KU(t - \tau, x)\Big)dx \\
\leqslant& \int_\Omega U^T P\Big(-2\lambda_1 D + 2(B - K)\Big)Udx + \int_\Omega \Big(-2U^T P\Phi(U) + 2U^T PKU(t - \tau, x)\Big)dx \\
\leqslant& \int_\Omega |U|^T P\Big(-2\lambda_1 D + 2(B - K)\Big)|U|dx + \int_\Omega \Big(2|U|^T P|\Phi(U)| + 2|U|^T PK|U(t - \tau, x)|\Big)dx
\end{aligned} \tag{19}
$$

On the other hand, it follows from (5) that $\Phi_1(0,0) = 0 = \Phi_2(0,0)$, and

$$\Phi_1(0, U_2) = a_{12}u_1^*(x)(U_2 + u_2^*(x)) - a_{12}u_1^*(x)u_2^*(x) = a_{12}u_1^*(x)U_2 \tag{20}$$

and thus differential mean value theorem and (A2) yield

$$
\begin{aligned}
|\Phi_1(U)| = |\Phi_1(U) - \Phi_1(0)| \leqslant& |\Phi_1(U_1, U_2) - \Phi_1(0, U_2)| + |\Phi_1(0, U_2) - \Phi_1(0,0)| \\
\leqslant& \Big(a_{11}(1+\theta_1)(2M_1)^{\theta_1} + a_{12}M_2\Big)|U_1| + a_{12}M_1|U_2|.
\end{aligned} \tag{21}
$$

Similarly,

$$|\Phi_2(U)| \leqslant a_{21}M_2|U_1| + \Big(a_{22}(1+\theta_2)(2M_2)^{\theta_2} + a_{21}M_1\Big)|U_2| \tag{22}$$

Thus,

$$2|U|^T P|\Phi(U)| \leqslant p_M(2|U_1| \cdot |\Phi_1(U)| + 2|U_2| \cdot |\Phi_2(U)|) \tag{23}$$
$$\leqslant p_M|U|^T\Theta|U|$$

$$2|U|^T PK|U(t-\tau,x)| \leqslant \varepsilon|U|^T(PK)|U|) + \frac{1}{\varepsilon}\lambda_{\max}(K)|U(t-\tau,x)|^T P|U(t-\tau,x)| \tag{24}$$

Combining (19)–(24) results in

$$D^+V(t) \leqslant \int_\Omega |U|^T P\left(-2\lambda_1 D + 2(B-K)\right)|U|dx + \int_\Omega \left(2|U|^T P|\Phi(U)| + 2|U|^T PK|U(t-\tau,x)|\right)dx$$

$$\leqslant -\frac{\lambda_{\min}\left(2\lambda_1 PD - 2P(B-K) - p_M\Theta - \varepsilon PK\right)}{p_M}\int_\Omega |U|^T P|U|dx + \frac{\lambda_{\max}(K)}{\varepsilon}V(t-\tau) \tag{25}$$

$$\leqslant -av(t) + b[v(t)]_\tau, \quad t \neq t_k.$$

On the other hand, letting $\gamma = \frac{\ln(\rho e^{\lambda\tau})}{\delta\tau}$, one can conclude from Lemma 3 that

$$V(t) \leqslant (\rho^2 e^{\lambda\tau})[V(0)]_\tau e^{-(\lambda-\gamma)t}, \quad t \geqslant t_0, \tag{26}$$

or equivalently,

$$V(t) \leqslant (\rho^2 e^{\lambda\tau})[V(0)]_\tau e^{-(\lambda-\frac{\ln(\rho e^{\lambda\tau})}{\delta\tau})t}, \quad t \geqslant t_0, \tag{27}$$

Indeed,

$$V(t_k) = \int_\Omega U^T(t_k,x)PU(t_k,x)dx$$
$$\leqslant \frac{\lambda_{\max}(A_k^T PA_k)}{p_m}\int_\Omega U^T(t_k^-,x)PU(t_k^-,x)dx$$
$$= a_k V(t_k^-).$$

According the conditions (B1)–(B3), one can see it from Lemma 3 that (26) and (27) holds if the condition (9) is verified. In fact, in Lemma 3, let $M = \rho^2 e^{\lambda\tau}$, then

$$Me^{\gamma t_k} = (\rho^2 e^{\lambda\tau})e^{\gamma(t_k-t_0)}$$
$$\geqslant (\rho^2 e^{\lambda\tau})(\rho e^{\lambda\tau})^{k-1}$$
$$= (\rho^{k+1}e^{k\lambda\tau}),$$

which means that the condition (9) is satisfied, and then Lemma 3 makes (26) and (27) hold. Moreover, (27) yields

$$p_m\|U\|_{L^2(\Omega)} \leqslant V(t) \leqslant (\rho^2 e^{\lambda\tau})[V(0)]_\tau e^{-(\lambda-\frac{\ln(\rho e^{\lambda\tau})}{\delta\tau})t}$$
$$\leqslant (\rho^2 e^{\lambda\tau})p_M\|\xi(s,x) - u^*(x)\|_\tau^2 e^{-(\lambda-\frac{\ln(\rho e^{\lambda\tau})}{\delta\tau})t}, \quad t \geqslant t_0, \tag{28}$$

where $\|\xi(s,x) - u^*(x)\|_\tau^2 = \sup\limits_{s\in[-\tau,0]} \int_\Omega [\xi(s,x) - u^*(x)]^T[\xi(s,x) - u^*(x)]dx$ with $\xi = (\xi_1,\xi_2)^T$ and $u^* = (u_1^*, u_2^*)^T$. Obviously, (28) completes the proof. $\square$

Remark 2. *Theorem 2 offers a better stabilization criterion than the previous literature ([6,7]), which reduces the conservatism of the algorithm. In fact, in Theorem 2, the impulse condition $\lambda_{\min}A_k$ may not be smaller than 1, which implies that this paper deletes the harsh restrictions on small impulse of the related literature [6,7].*

Remark 3. *Compared with the previous related literature [6,7], Theorem 2 expands the range of the parameters θ_1, θ_2 from $(0,1)$ to $(-1,4)$.*

4. Numerical Examples

First, the following example shows the effectiveness of Theorem 1.

Example 1. *Let $\theta_1 = \frac{2}{3}$, $\theta_2 = \frac{4}{5}$, $b_i = 0.13 + 0.0001i$, $d_i = 0.1 + 0.0001i$, $i = 1, 2$, and $\Omega = (0,1) \times (0,1)$. Direct calculation yields that $\lambda_1 = 19.7392$ ([14] (Remark 14)), and $b_1 = 0.1301 < 0.1001 \times 19.7392 = d_1\lambda_1$ and $b_2 = 0.1302 < 0.1002 \times 19.7392 = d_2\lambda_1$. Furthermore, set $M_i = 2 + 0.1i$, $i = 1, 2$, and $a_{11} = 0.03, a_{12} = 0.02, a_{21} = 0.025, a_{22} = 0.03$, $k_1 = 0.15$, $k_2 = 0.12$. An accurate calculation can verify that the condition (3) is satisfied if letting $c_* = 100000$. Now, one can conclude from Theorem 1 that there is a positive stationary solution $(u_1^*(x), u_2^*(x))^T$ and other three stationary solutions for the ecosystem (1).*

Below, the feasibility of Theorem 2 need be verified, too.

Example 2. *All the data of Example 1 are employed in this example, then an accurate calculation yields that*

$$\Theta = \begin{pmatrix} 0.3483 & 0.0970 \\ 0.0970 & 0.4583 \end{pmatrix}$$

(B2) $a > b \geqslant 0$. Furthermore, using computer Matlab LMI toolbox to solve LMI condition (16)–(18) yields the following feasible data:

$$P = \begin{pmatrix} 0.9998 & 0 \\ 0 & 1.0013 \end{pmatrix}, \ \varepsilon = 0.9996, \ p_M = 1.0015, \ p_m = 0.9973.$$

Then, a direct calculation obtains $a = 3.3046, b = 0.1501$, and thus $a > b \geqslant 0$. Let $\tau = 0.5$, solving the equality $\lambda = a - be^{\lambda\tau}$ reaches $\lambda = 2.7199$. Set

$$A_j \equiv \begin{pmatrix} 1.0603 & 0 \\ 0 & 1.0783 \end{pmatrix}, \forall j \in \mathbb{Z}, \tag{29}$$

which together the above data derives that $a_j \equiv 1.1674$, and thus $\rho = 1.1674$. Set $\delta = 2$, then an immediate calculation yields $\lambda - \frac{\ln(\rho e^{\lambda\tau})}{\delta\tau} = 1.2052 > 0$, and $\frac{1}{2}[\lambda - \frac{\ln(\rho e^{\lambda\tau})}{\delta\tau}] = 0.6026$. According to Theorem 2, the zero solution of the system (8) is globally exponentially stable with convergence rate 60.26%.

Remark 4. *Example 2 verifies the advantages described in Remarks 2–3.*

5. Conclusions and Further Considerations

Compared with the known literature, this paper has double advantages in method and conclusion. On one hand, employing the Minimax principle and impulsive differential inequality improves the methods in [6,7]. For example, in deriving the existence of multiple stationary solutions of RDGACM, the methods involved in Minimax principle is more simpler than those in applying Mountain Pass Lemma of [6]. Besides, in stabilizing globally the ecosystem, utilizing the impulsive differential inequality makes the impulse range wider. Especially, an impulse range means that people can adjust and manage the ecosystem more flexibly.

For $v \in H_0^1(\Omega)$, the norm

$$\|v\| = \sqrt{\int_\Omega |\nabla v|^2 dx} \tag{30}$$

in this paper is simpler than the norm

$$\|v\| = \sqrt{\int_\Omega (|\nabla v|^2 + \frac{C}{D}v^2) dx} \tag{31}$$

in [18] (Statement 2). Now, with the help of such a simple norm (30), some further considerations are presented below.

In fact, in [18] (Statement 2), the following ordinary differential equation and its corresponding partial differential equation were considered:

$$\frac{dx(t)}{dt} = -Cx(t) + Af(x(t)) + Bf(x(t - \tau(t))) + J, \quad \text{and } x \in \mathbb{R}^1, \tag{32}$$

and its corresponding reaction-diffusion cellular neural networks

$$\begin{cases} \dfrac{\partial u(t,x)}{\partial t} = D\Delta u(t,x) - Cu(t,x) + Af(u(t,x)) + Bf(u(t - \tau(t), x)) + J, & \text{and } t \geqslant 0, x \in \Omega, \\ u(t,x) = 0, & x \in \partial\Omega, \end{cases} \tag{33}$$

where Ω is an open bounded domain in $\mathbb{R}^3$ with smooth boundary $\partial\Omega$, $D \in \mathbb{R}^1$ is the diffusion coefficient with $D > 0$, and C, A both are positive real numbers, $J = 0$, $B = 0$, the function f is defined as follows:

$$f(u) = \begin{cases} \dfrac{3D}{A}\mu_1 u^{\frac{1}{3}} + \dfrac{2D}{A}\mu_1, & u \leqslant -1; \\ \dfrac{D}{A}\mu_1 u, & u \in [-1, 1]; \\ \dfrac{3D}{A}\mu_1 u^{\frac{1}{3}} - \dfrac{2D}{A}\mu_1, & u \geqslant 1. \end{cases} \tag{34}$$

Here, we denote by μ_i the ith positive eigenvalue of the following eigenvalue problem:

$$\begin{cases} -\Delta u(x) + \dfrac{C}{D}u(x) = \mu u(x), x \in \Omega, \\ u(x) = 0, & u \in \partial\Omega, \end{cases} \tag{35}$$

Particularly in the case of $C = 0$, the norm (31) is just that of (30), and then $\mu_1 = \lambda_1 > 0$ is the first positive eigenvalue of the operator $-\Delta$ in $H_0^1(\Omega)$ with the norm (30). Thus, in the case of $C = 0$, the following theorem holds:

Theorem 3. *If zero solution is the global stable unique equilibrium point of the following ordinary differential system*

$$\frac{dx(t)}{dt} = Af(x(t)) + Bf(x(t - \tau(t))), \quad \text{and } x \in \mathbb{R}^1, \tag{36}$$

where

$$f(u) = \begin{cases} \dfrac{3D}{A}\lambda_1 u^{\frac{1}{3}} + \dfrac{2D}{A}\lambda_1, & u \leqslant -1; \\ \dfrac{D}{A}\lambda_1 u, & u \in [-1, 1]; \\ \dfrac{3D}{A}\lambda_1 u^{\frac{1}{3}} - \dfrac{2D}{A}\lambda_1, & u \geqslant 1, \end{cases} \tag{37}$$

then its corresponding reaction-diffusion system:

$$\begin{cases} \dfrac{\partial u(t,x)}{\partial t} = D\Delta u(t,x) + Af(u(t,x)) + Bf(u(t - \tau(t), x)), & \text{and } t \geqslant 0, x \in \Omega, \\ u(t,x) = 0, & x \in \partial\Omega, \end{cases} \tag{38}$$

owns zero solution and other stationary solutions which are at least two non-zero functions or infinitely many positive functions and negative functions.

Remark 5. *It follows from [18](Remark 11) that zero solution is actually the global stable unique equilibrium point of the ordinary differential system (36). In fact, according to the Introduction in [20], the function f defined by (37) satisfies the conditions [20] (Equation (7)) and [20] (Equation (8)), and thus the zero solution is actually the unique equilibrium point of the ordinary differential system (36). That is, there exists such an example that under the influence of diffusion, the unique equilibrium point of the ordinary differential system (36) with the Lipschitz activation function f can become at least three equilibrium points of its corresponding reaction-diffusion system (38).*

Now, in view of Theorem 3 and Remark 5, the author wants to know whether an example can be designed such that the global stable unique equilibrium point x^* of the ordinary differential system can become multiple equilibrium points $u_i^*(x)(i \in \Lambda)$ of its corresponding reaction-diffusion system under the influence of diffusion? Here, Λ is a finite index set or infinite index set. Furthermore, is the diffusion coefficient related to the number of the index set Λ? Is the smaller the diffusion coefficient, the fewer the number of the index set Λ? Moreover, if the diffusion coefficient is small enough, is the norm $\|u_i^*(x) - x^*\|_*$ is also small? Here, $\|\cdot\|_*$ may be $\|u_i^*(x) - x^*\|_* = \sup\limits_{x \in \Omega} |u_i^*(x) - x^*|$. All these problems are interesting.

Funding: The research was supported by the Application Basic Research Project of Science and Technology Department of Sichuan Province in China (No. 2020YJ0434).

Institutional Review Board Statement: Not applicable.

Informed Consent Statement: Not applicable.

Data Availability Statement: Not applicable.

Conflicts of Interest: The author declares no conflict of interest.

References

1. Vasilova, M.; Jovanovic, M. Stochastic Gilpin-Ayala competition model with infinite delay. *Appl. Math. Comput.* **2011**, *217*, 4944–4959. [CrossRef]
2. Lian, B.; Hu, S. Stochastic Delay Gilpin-Ayala Competition Models. *Stochastics Dyn.* **2006**, *6*, 561–576. [CrossRef]
3. Lian, B.; Hu, S. Asymptotic behaviour of the stochastic Gilpin-Ayala competition models. *J. Math. Anal. Appl.* **2008**, *339*, 419–428. [CrossRef]
4. Settati, A.; Lahrouz, A. On stochastic Gilpin-Ayala population model with Markovian switching. *Biosystems* **2015**, *130*, 17–27. [CrossRef] [PubMed]
5. Wu, R. Dynamics of stochastic hybrid Gilpin-Ayala system with impulsive perturbations. *J. Nonlinear Sci. Appl.* **2017**, *10*. [CrossRef]
6. Rao, R.; Yang, X.; Tang, R.; Zhang, Y.; Li, X. Impulsive stabilization and stability analysis for Gilpin-Ayala competition model involved in harmful species via LMI approach and variational methods. *Math. Comput. Simul.* **2021**, *188C*, 571–590. [CrossRef]
7. Rao, R.; Zhu, Q.; Shi, K. Input-to-State Stability for Impulsive Gilpin-Ayala Competition Model With Reaction Diffusion and Delayed Feedback. *IEEE Access* **2020**, *8*, 222625–222634. [CrossRef]
8. Bai, L.; Wang, K. Gilpin-Ayala model with spatial diffusion and its optimal harvesting policy. *Appl. Math. Comput.* **2005**, *171*, 531–546. [CrossRef]
9. Xue, Y.; Zhao, P. Input-to-State Stability and Stabilization of Nonlinear Impulsive Positive Systems. *Mathematics* **2021**, *9*, 1663. [CrossRef]
10. Rao, R. Global Stability of a Markovian Jumping Chaotic Financial System with Partially Unknown Transition Rates under Impulsive Control Involved in the Positive Interest Rate. *Mathematics* **2019**, *7*, 579. [CrossRef]
11. Yang, X.; Li, X.; Lu, J.; Cheng, Z. Synchronization of time-delayed complex networks with switching topology via hybrid actuator fault and impulsive effects control. *IEEE Trans. Cybern.* **2020**, *50*, 4043–4052. [CrossRef] [PubMed]
12. Tang, R.; Yang, X.; Wan, X.; Zou, Y.; Cheng, Z.; Fardoun, H.M. Finite-time synchronization of nonidentical BAM discontinuous fuzzy neural networks with delays and impulsive effects via non-chattering quantized control. *Commun. Nonlinear Sci. Numer. Simul.* **2019**, *79*, 104893. [CrossRef]
13. Yang, X.; Lu, J.; Ho, D.W.; Song, Q. Synchronization of uncertain hybrid switching and impulsive complex networks. *Appl. Math. Model.* **2018**, *59*, 379–392. [CrossRef]
14. Dlala, M.; Saud Almutairi, A. Rapid Exponential Stabilization of NonlinearWave Equation Derived from Brain Activity via Event-Triggered Impulsive Control. *Mathematics* **2021**, *9*, 516. [CrossRef]

15. Wang, X.; Rao, R.; Zhong, S. *p*th Moment Stability of a Stationary Solution for a Reaction Diffusion System with Distributed Delays. *Mathematics* **2020**, *8*, 200 [CrossRef]
16. Rao, R.; Huang, J.; Li, X. Stability analysis of nontrivial stationary solution of reaction-diffusion neural networks with time delays under Dirichlet zero boundary value. *Neurocomputing* **2021**, *445C*, 105–120. [CrossRef]
17. Willem, M. *Minimax Theorems*; Birhauser: Berlin, Germany, 1996.
18. Rao, R. Stability analysis of nontrivial stationary solution and constant equilibrium point of reaction-diffusion neural networks with time delays under Dirichlet zero boundary value. *Preprint* **2020**, 2020040277. [CrossRef]
19. Yue, D.; Xu, S.F.; Liu, Y.Q. Differential inequality with delay and impulse and its applications to design robust control. *Control Theory Appl.* **1999**, *16*, 519–524. (In Chinese)
20. Xu, D.; Yang, Z.; Huang, Y. Existence-uniqueness and continuation theorems for stochastic functional differential equations. *J. Differ. Equ.* **2008**, *245*, 1681–1703. [CrossRef]

 *mathematics*

MDPI

Article

Improved Rotor Flux and Torque Control Based on the Third-Order Sliding Mode Scheme Applied to the Asynchronous Generator for the Single-Rotor Wind Turbine

Habib Benbouhenni [1] and Nicu Bizon [2,3,4,*]

[1] Faculty of Engineering and Architecture, Department of Electrical & Electronics Engineering, Nisantasi University, Istanbul 34481742, Turkey; habib.benbouenni@nisantasi.edu.tr

[2] Faculty of Electronics, Communication and Computers, University of Pitesti, 110040 Pitesti, Romania

[3] Doctoral School, Polytechnic University of Bucharest, 313 Splaiul Independentei, 060042 Bucharest, Romania

[4] ICSI Energy, National Research and Development Institute for Cryogenic and Isotopic Technologies, 240050 Ramnicu Valcea, Romania

* Correspondence: nicu.bizon@upit.ro

Abstract: In this work, a third-order sliding mode controller-based direct flux and torque control (DFTC-TOSMC) for an asynchronous generator (AG) based single-rotor wind turbine (SRWT) is proposed. The traditional direct flux and torque control (DFTC) technology or direct torque control (DTC) with integral proportional (PI) regulator (DFTC-PI) has been widely used in asynchronous generators in recent years due to its higher efficiency compared with the traditional DFTC switching strategy. At the same time, one of its main disadvantages is the significant ripples of magnetic flux and torque that are produced by the classical PI regulator. In order to solve these drawbacks, this work was designed to improve the strategy by removing these regulators. The designed strategy was based on replacing the PI regulators with a TOSMC method that will have the same inputs as these regulators. The numerical simulation was carried out in MATLAB software, and the results obtained can evaluate the effectiveness of the designed strategy relative to the traditional strategy.

Keywords: asynchronous generator; single-rotor wind turbine; direct flux and torque control (DFTC); third-order sliding mode controller (TOSMC); integral proportional (PI) regulator; DFTC-PI control; DFTC-TOSMC strategy

Citation: Benbouhenni, H.; Bizon, N. Improved Rotor Flux and Torque Control Based on the Third-Order Sliding Mode Scheme Applied to the Asynchronous Generator for the Single-Rotor Wind Turbine. *Mathematics* **2021**, *9*, 2297. https://doi.org/10.3390/math9182297

Academic Editors: Francesc Pozo and Amir Mosavi

Received: 8 August 2021
Accepted: 13 September 2021
Published: 17 September 2021

Publisher's Note: MDPI stays neutral with regard to jurisdictional claims in published maps and institutional affiliations.

1. Introduction

The strategies of direct flux and torque control (DFTC) scheme or DTC of the asynchronous generator (AG) with constant switching frequency have become a focal point due to their easy design of the AC harmonic filter and power converter, and also due to the reduction in the ripples of the rotor flux and torque [1]. This work introduces a new technique for this technology. It is shown that the DFTC strategy with constant switching frequency is mainly achieved by using the PWM [2], SVM [3,4], DSVM-DFTC [5], and P-DFTC [6], respectively. There are many techniques in the literature that have been proposed to minimize the ripples of magnetic flux and torque [7–11]. However, the sliding mode control (SMC) technique has better dynamics and robustness compared to any other regulators [12]. It also has a better ability to reduce the ripples of torque and magnetic flux. Several works on the SMC technique for the control of an alternating current (AC) machine are available in the literature, which analyzes and discusses its disadvantages and advantages [13–19]. In [13], the SMC provided better results compared to the traditional proportional-integral (PI) controller.

Chattering at very high frequencies is defined as a shortcoming of SMC technology, which causes ripples in the motor. The high-end SMC technology is suitable for reducing this chattering phenomenon [14]. Many strategies like suboptimal, twisting and super twisting [15], terminal SMC [16], non-singular terminal SMC [17], fast integral terminal

SMC [18], and fast terminal SMC technique [19] are available in the references above-mentioned, but also in other works. These techniques are used to improve the performance of electric machines. There are several proposed techniques for controlling and reducing torque ripples, and these methods are divided into two main classes, namely, direct control and indirect control such as DFTC in the first class, and direct power control (DPC) and field-oriented control (FOC) in the second class. For the two methods in the second class, DPC and FOC, the active and reactive power are controlled. As is well-known, the torque is related to the active power and its reference value. In [20], the authors proposed using the virtual flux DPC control (VF-DPC) to minimize the electromagnetic torque of the AG-based wind power. This proposed strategy further minimizes torque ripple compared to the classic DPC method. On the other hand, the VF-DPC is easy to implement. In [21], a new DPC technique was proposed based on the terminal synergetic control theory to reduce ripples of rotor flux, current, and electromagnetic torque. This designed strategy was more robust compared to the traditional DPC strategy and other strategies such as the traditional DFTC and FOC control. A new FOC method was proposed in [22] to minimize the ripples of active power, current, rotor flux, and electromagnetic torque of the induction generator. This designed FOC strategy based on a hysteresis rotor current controller and experimental results showed the performance of the designed strategy. Another intelligent robust technique was designed in [23] to control and reduce the rotor flux and torque of the induction generator. The proposed method was a combination of two different methods. The first method was the SMC technique, where durability is its biggest advantage compared to its counterparts. Regarding the second method, it was based on fuzzy logic, where simplicity is the biggest advantage that distinguishes it compared to other methods. The obtained method was more robust, and the simulation results showed its effectiveness in reducing the value of harmonic distortion (THD) of the current compared to the traditional strategy. The second-order continuous SMC technique (SOCSMC) was proposed to improve the performances of the DFTC control of the induction generator [24]. The designed strategy minimizes the ripples of rotor flux, stator current, and electromagnetic torque compared to the traditional DFTC strategy with proportional-integral (PI) controllers. Although the designed strategy is simpler, more robust, and easier to implement, the THD value remains quite high. Additionally, it does not completely remove the torque ripples of the electric machine. DPC control with PI controllers (DPC-PI) reduces the ripples of electromagnetic torque, rotor flux, and stator current compared to traditional DPC and FOC strategies [25]. The experimental results showed a better performance obtained for the DPC-PI strategy, which is also easier to implement compared to traditional direct and indirect FOC strategies. In [26], the author combined two methods, different in principle, in order to obtain a more robust method. Thus, the SMC method was incorporated into the DTC method. One of the advantages of the resulting method is that it obtains much lower current ripples than in the classical method [26]. Moreover, the method obtained is very simple and can be easily accomplished.

Another robust strategy was proposed in [27] to minimize the ripple of electromagnetic torque of the induction generator-based dual-rotor wind power. This proposed method combines two different nonlinear methods: the SMC method, where chattering phenomenon is its biggest disadvantage compared to other nonlinear methods, and the synergetic control method, where simplicity is the biggest advantage that distinguishes it compared to other nonlinear methods. The resulting nonlinear strategy reduces the ripple of electromagnetic torque, stator current, and rotor flux compared to traditional direct FOC control and other strategies such as the DFTC, FOC, and SMC methods. However, the proposed nonlinear strategy is more robust and easier to implement and further reduces the chattering phenomenon compared to traditional SMC control. Using a research direction similar to the one in [27], the merger between the synergetic control and super twisting algorithm was proposed to reduce the ripple of electromagnetic torque of the AG-based dual-rotor wind turbine [28]. This proposed nonlinear strategy is more robust compared to traditional controllers such as the PI controller and SMC. Super Twisting algorithm (STA)-

based SOSM controllers have been proposed to control the AG-based wind power [29]. In order to show the effectiveness and superiority of the designed controller, the thermal exchange optimization (TEO) method was used. The integral sliding-mode DFTC method (ISM-DFTC) with space-vector modulation (SVM) for AG-based wind turbine conversion systems under unbalanced grid voltage was designed in [30]. This proposed DFTC method minimizes the torque ripples compared to the traditional DFTC strategy.

In this paper, a new high-order SMC technique was proposed and designed to improve the characteristics of the DFTC control and reduce the rotor flux, current, and torque ripples of the AG-based wind power. Compared to the classical SMC technique, the chattering phenomenon was reduced or eliminated. This proposed control technique was based on a super twisting algorithm (STA) applied for the third-order sliding mode controller (TOSMC) technique, called below as the DFTC-TOSMC method. In order to improve the performance of the conventional DFTC technique, the standard hysteresis comparators will be replaced by two TOSMC methods and the switching table by the SVM technique. The rotor flux and electromagnetic torque estimation block maintain the same shape as that established for classical DFTC, as described in [31,32]. In this DFTC control strategy, the rotor flux and torque are regulated by two proposed TOSMC regulators, while the SVM technique replaces the traditional switching table. The principle as well as the advantages and disadvantages of the DFTC-TOSMC method have been comparatively analyzed with other advanced control strategies proposed in the literature [10,20–29]. The main contributions of the proposed designed control scheme are to minimize the total harmonic distortion (THD) of current for an AG-based SRWP system, increases the robustness and stability of the controlled system, provides methodical and less-complicated techniques based on a novel SOSMC method to adjust the rotor voltage of DFIG, and reduced ripples of both rotor flux and electromagnetic torque.

The parameters used to observe the performance of the designed strategy are the total harmonic distortion (THD) for current, torque ripple, steady-state error, response time, and rotor flux undulations. The DFTC-PI structure shown in Figure 1 is the system considered in this paper as a reference to compare the improved performances of the proposed DFTC-TOSMC method.

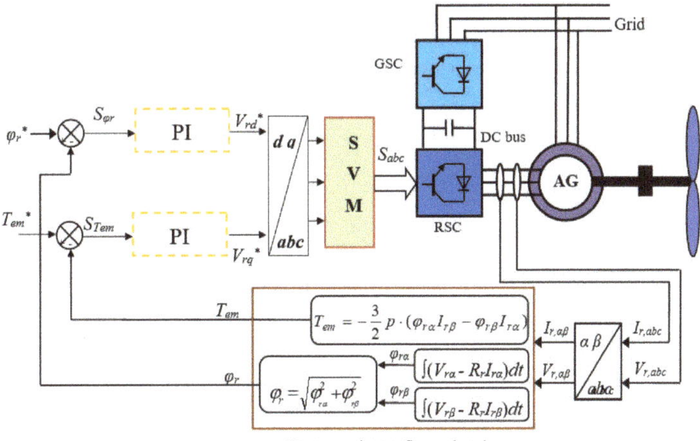

Torque and rotor flux estimation

Figure 1. Structure of the DFTC strategy with PI controllers.

In summary, the novelty and main findings of this paper are as follows:

- A new TOSMC method based on the DFTC method was designed to minimize ripples of both rotor flux and electromagnetic torque;

- Third-order sliding mode controllers reduces the tracking error for rotor flux and electromagnetic torque toward the references of AG-based SRWT systems; and
- The DFTC-TOSMC method with SVM strategy reduces harmonic distortion of the stator current and torque ripple of AG-based SRWT systems.

Thus, the rest of the paper is structured as follows. Section 2 presents models of single-rotor wind systems. The model of the AG is presented in Section 3 using Park transformations. The proposed TOSMC technique is presented in Section 4. DFTC-TOSMC control of the AG-based SRWP is presented in Section 5. Sections 6 and 7 present and discuss the results of the research carried out.

2. Single-Rotor Wind Power

Equation (1) expresses the power obtained from a wind turbine [33]:

$$P_t = \frac{1}{2}\rho R^2 C_p(\beta, \lambda) V^3 \tag{1}$$

where λ is the tip speed ration; R is the radius of the turbine (m); ρ is the air density (kg/m^3); V is the wind speed (m/s); β is the blade pitch angle (deg); and C_P is the power coefficient.

Equation (2) expresses the C_P of the wind turbine. The C_P is a nonlinear function [34]:

$$C_p = (0.5 - 0.167)(\beta - 2) \times \sin\left(\frac{\pi(\lambda - 0.1)}{18.5 - 0.3(\beta - 2)}\right) - 0.0018 \times (\beta - 3)(\beta - 2) \tag{2}$$

The λ is given by:

$$\lambda = \frac{R.\Omega_t}{V} \tag{3}$$

where Ω_t is the rotational speed of the SRWP.

3. The AG Model

The asynchronous generator is one of the most popular and widely used in the field of wind energy due to its low maintenance, reduced cost, robustness, efficiency, ease of control, minimum energy losses, and ability to work at a speed that varies by $\pm33\%$ around the synchronous speed [35]. On the other hand, this is evident in the number of papers published on AG, where several controls have been developed in order to improve the characteristics of this generator [36–40]. In order to obtain the mathematical form of the generator, the Park transform was used. The following equations represent the mathematical form of the generator [41,42]:

$$\begin{cases} V_{dr} = R_r I_{dr} + \frac{d}{dt}\Psi_{dr} - w_r\Psi_{qr} \\ V_{qr} = R_r I_{qr} + \frac{d}{dt}\Psi_{qr} + w_r\Psi_{dr} \\ V_{qs} = R_s I_{qs} + \frac{d}{dt}\Psi_{qs} + w_s\Psi_{ds} \\ V_{ds} = R_s I_{ds} + \frac{d}{dt}\Psi_{sd} - w_s\Psi_{qs} \end{cases} \tag{4}$$

$$\begin{cases} \Psi_{dr} = M I_{ds} + L_r I_{dr} \\ \Psi_{qr} = L_r I_{qr} + M I_{qs} \\ \Psi_{qs} = M I_{qr} + L_s I_{qs} \\ \Psi_{ds} = M I_{dr} + L_s I_{ds} \end{cases} \tag{5}$$

The electric machine consists of two main parts: the electrical part, and the mechanical part. The electrical part is represented in the equations of tension and flux, while the mechanical part of the electric machine is represented in the following equation:

$$T_e - T_r = J\frac{d\Omega}{dt} + f\Omega \tag{6}$$

Torque can be given by the following equation:

$$T_e = 1.5\, p\frac{M}{L_s}\left(-\Psi_{sd} I_{rq} + \Psi_{sq} I_{rd}\right) \tag{7}$$

4. Third-Order Sliding Mode Controller

There are many controllers proposed to regulate and reduce the torque of AC machines in the literature. Among all the techniques designed for the high-order SMC technique, the STA strategy is an exception, which only requires information on the nonlinear surface [43]. The proposed high-order SMC controller, named the third-order sliding mode controller (TOSMC), is an effective strategy for uncertain systems and overcomes the main drawbacks of the classical SMC technique described in the literature. TOSMC is a robust strategy and is an alternative to non-linear and linear strategies. In the STA strategy, the command input applies to the second-order derivative of the nonlinear surface, and reverses the SMC, which acts on the first derivative of the sliding surface. The proposed TOSMC technique is based on the STA algorithm. The control input of the proposed TOSMC technique uses the sum of three inputs, as defined below:

$$u(t) = u_1 + u_2 + u_3 \tag{8}$$

$$u_1(t) = \lambda_1 \sqrt{|S|} sign(S) \tag{9}$$

$$u_2(t) = \lambda_2 \int sign(S) dt \tag{10}$$

$$u_1(t) = \lambda_1 \sqrt{|S|} sign(S) \tag{11}$$

The control input of the proposed TOSMC method is obtained as Equation (12).

$$u(t) = \lambda_1 \sqrt{|S|} sign(S) + \lambda_2 \int sign(S) dt + \lambda_3 sign(S) \tag{12}$$

where S is the sliding surface.

The tuning constants λ_1, λ_2, and λ_3 were used to improve the performance of the TOSMC method. Therefore, this was the design process using TOSMC for the DFTC strategy. Figure 2 shows the structure of the TOSM controller for the DFTC strategy in wind power systems.

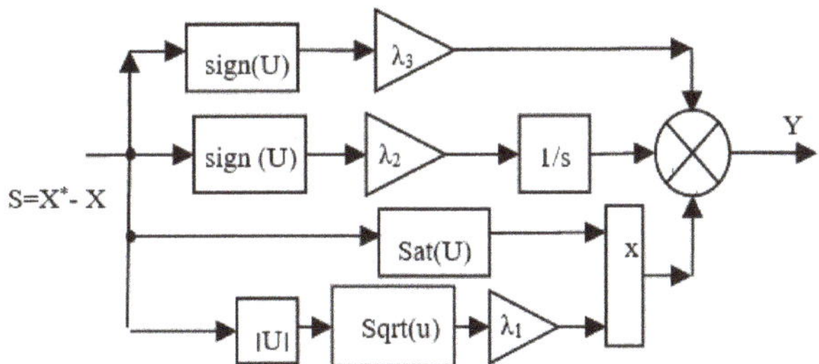

Figure 2. The command law structure of the proposed TOSM controller.

The stability condition is given by:

$$S \times \dot{S} < 0 \tag{13}$$

This proposed controller was used in this paper to reduce the THD of the current and ripples of the electromagnetic torque and rotor flux in the case of an AG-based SRWP system using the DFTC technique. Note that the inverter was controlled by the SVM strategy.

5. DFTC-TOSMC Control of the AG-Based SRWP

The traditional DFTC technique has been developed and investigated as a replacement for the classical FOC method in high-performance AC machine drives. DFTC is well-known for its robust strategy, simple algorithm, and fast-flux/torque response, which requires no modulation techniques, current control, or coordinate transformation [44]. This method has been applied to several electric machines such as induction motor [45], a brushless DC electric motor [46], interior permanent magnet synchronous motor [47], five-phase induction motor [48], brushless doubly-fed machine [49,50], permanent magnet synchronous motor (PMSM) [51], six-phase induction motor [52], and five-phase PMSM [53,54]. In [55], the DFTC control scheme reduced the electromagnetic torque, stator current, and rotor flux compared to the FOC method. The DFTC strategy was designed based on a model predictive controller [56]. This proposed DFTC is simpler and, in addition, reduces the torque ripple compared to the classical DFTC strategy. A DFTC method with a modified finite set model predictive technique was designed in [57]. Simplicity and durability are the two main advantages of this proposed method. A flexible switching table (FST) was designed for the DFTC method applied to PMSMs to enhance the dynamic performances and steady-state of the drive system [58]. The simulation results showed that the proposed method improved the efficiency of the electric machine.

Despite the many advantages that characterize the DFTC method, there are several problems that characterize it, for example, high ripples in rotor flux and torque, several current harmonics, and low-speed problems. Torque ripples represent the major problem of the traditional DFTC strategy, which can be very hurtful for the AG because of the use of hysteresis comparators and switching table or PI controllers [59]. Some solutions have been designed to avoid this disadvantage [60–65]. The essential idea was to replace the switching table and hysteresis comparators with intelligent techniques and at the same time conserve the essential performance of the traditional method.

In this section, a new DFTC control scheme was designed based on TOSMC techniques. In order to improve the performance of the classical DFTC strategy, the standard hysteresis comparators were replaced by two TOSMC controllers and the switching table by the traditional SVM strategy. The electromagnetic torque and rotor flux estimation block keep the same shape as that established for traditional DFTC with PI controllers, as described in [66,67]. In this proposed DFTC control strategy, the electromagnetic torque and rotor flux are regulated by two proposed TOSMC controllers, while the SVM technique replaces the switching table. However, this control by DFTC-TOSMC or DFTC-SVM-TOSMC has the advantages of vector control and conventional DFTC to overcome the problem of fluctuations in rotor flux and electromagnetic torque generated by the DFIG. TOSMC regulators and SVM techniques were used to obtain a fixed switching frequency and less pulsation of the rotor flux and torque.

This proposed strategy can be minimized more than the electromagnetic torque and rotor flux compared to traditional DFTC and strategies such as FOC, DPC control, and other control techniques. The DFTC-TOSMC principle is proposed to control the rotor flux and the torque of the AG-based SRWT systems. The electromagnetic torque is regulated utilizing the quadrature axis voltage V_{qr}^*, while the flux is regulated utilizing the direct axis voltage V_{dr}^*.

In this paper, we proposed the use of a new nonlinear controller (based on the TOSMC technique) to replace the conventional PI controllers.

The designed DFTC-TSOMC strategy is shown in Figure 3 and was designed to reduce the undulations of the torque and rotor flux of an AG, as presented below.

The estimation of the rotor flux can be done in different ways using the voltage model, and the rotor flux can be estimated by integrating from the rotor voltage equation.

$$Q_r = \int_0^t (V_r - R_r i_r)dt \tag{14}$$

In the reference $(\alpha\text{-}\beta)$, the components of the rotor flux are determined as follows:

$$\begin{cases} Q_{r\alpha} = \int_0^t (V_r - R_r i_{r\alpha})dt \\ Q_{r\beta} = \int_0^t (V_r - R_r i_{r\beta})dt \end{cases} \tag{15}$$

where $V_r = V_{r\alpha} + jV_{r\beta}$; $i_r = i_{r\alpha} + ji_{r\beta}$; $Q_r = Q_{r\alpha} + jQ_{r\beta}$.

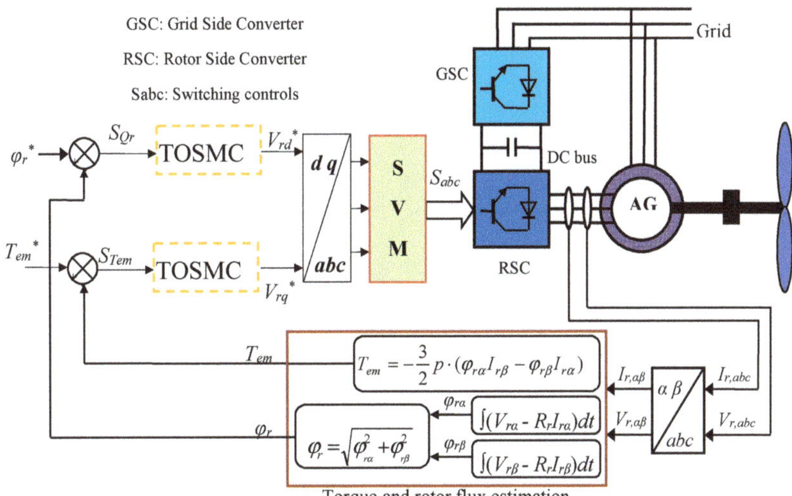

Figure 3. Bloc diagram of the AG with the DFTC-TOSMC method.

From these two equations, the modulus of the rotor flux and the angle θr result is as follows:

$$|Q_r| = \sqrt{\left(Q_{r\beta}^2 + Q_{r\alpha}^2\right)} \tag{16}$$

$$\theta r = artg\frac{Q_{r\beta}}{Q_{r\alpha}} \tag{17}$$

The errors of the flux and electromagnetic torque are shown in Equations (18) and (19).

$$S_{Tem} = T_{em}^* - T_{em} \tag{18}$$

$$S_{Qr} = Q_r^* - Q_r \tag{19}$$

where the surfaces are the flux magnitude error $S_{Qr} = Q_r{}^* - Q_r$ and the electromagnetic torque error $S_{Tem} = T_{em}{}^* - T_{em}$.

The errors shown in Equations (18) and (19) were used as input to the TOSMC techniques. Electromagnetic torque and rotor flux TOSMC regulators were used to respectively influence the $V_{dr}{}^*$ and $V_{qr}{}^*$ as in Equations (20) and (21):

$$V_{dr}^* = \lambda_1\sqrt{|S_{Qr}|}sign(S_{Qr}) + \lambda_2\int sign(S_{Qr}).dt + \lambda_3 sign\left(S_{Qr}\right) \tag{20}$$

$$V_{qr}^* = \lambda_1\sqrt{|S_{Tem}|}.sign(S_{Tem}) + \lambda_2\int sign(S_{Tem})dt + \lambda_3 sign(S_{Tem}) \tag{21}$$

The TOSMC controller structure for the torque and flux of the DFTC strategy are presented in Figures 4 and 5, respectively.

This proposed controller was applied for a DFTC strategy based on the TOSMC technique to obtain a minimum torque ripple and to minimize the chattering phenomenon.

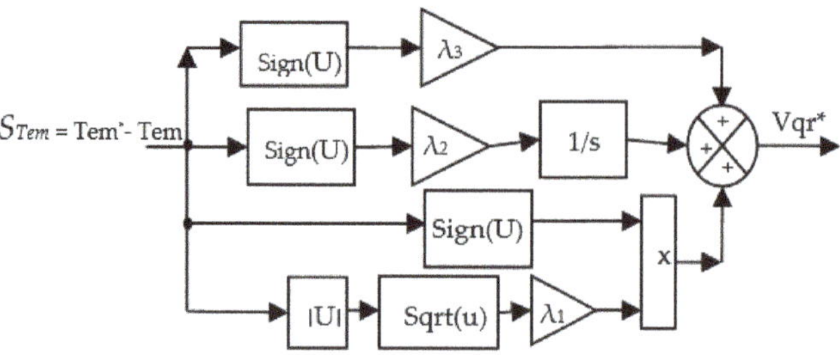

Figure 4. Proposed TOSMC torque controller.

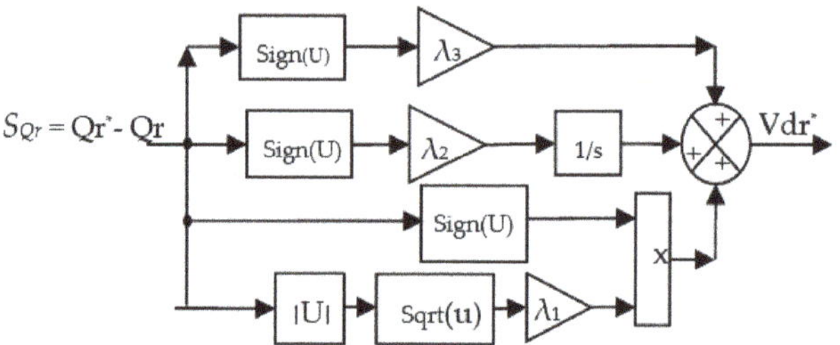

Figure 5. Proposed TOSMC flux controller.

6. Analysis of the Simulation Results

This work aimed to reduce the flux and torque ripples of an asynchronous generator. The latter operated at nominal speed. The values of the electric machine elements are shown in Table A1 (see Appendix A). A generator with a power of 1.5 megawatts was used, operating under a voltage of 380 V, and the frequency of the network was 50 Hz. The two DFTC techniques, DFTC-PI and DFTC-TOSMC, were studied, simulated, and compared in terms of torque ripple, reference tracking, THD value of the current, and rotor flux ripple.

The results obtained by using the MATLAB/Simulink® software are shown in Figures 6–10. The Simulink diagrams presented above and built-in MATLAB functions were run on a personal computer with an Intel® Core™ i9-9900K processor. Looking at Figures 8 and 9, it is worth noting that the rotor flux and electromagnetic torque for the designed DFTC techniques followed their reference values almost perfectly.

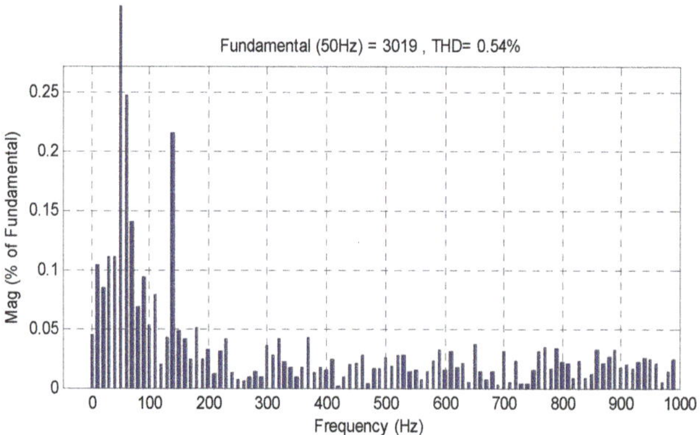

Figure 6. THD value of the stator current (DFTC-PI).

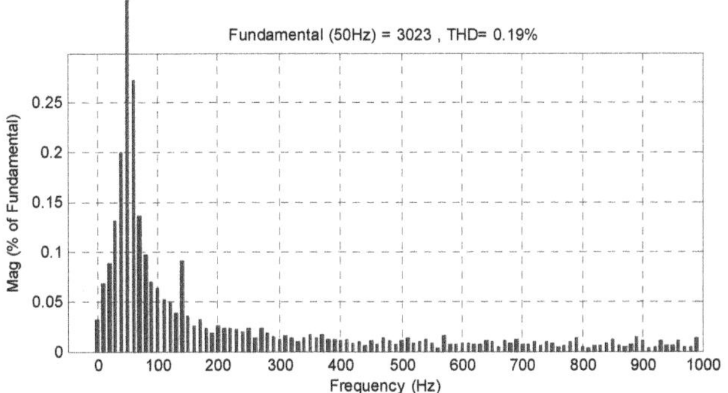

Figure 7. THD value of the stator current (DFTC-TOSMC).

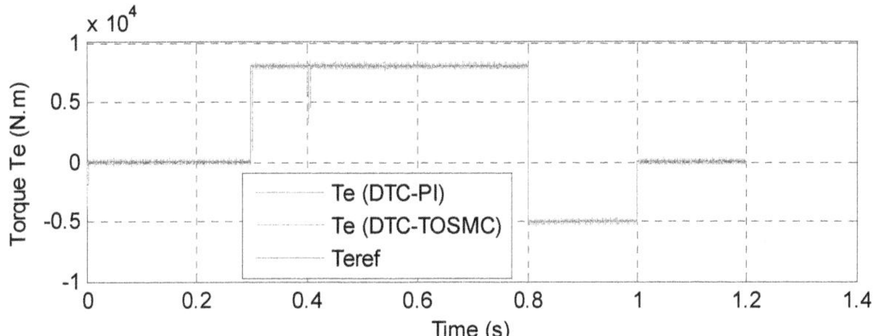

Figure 8. Electromagnetic torque.

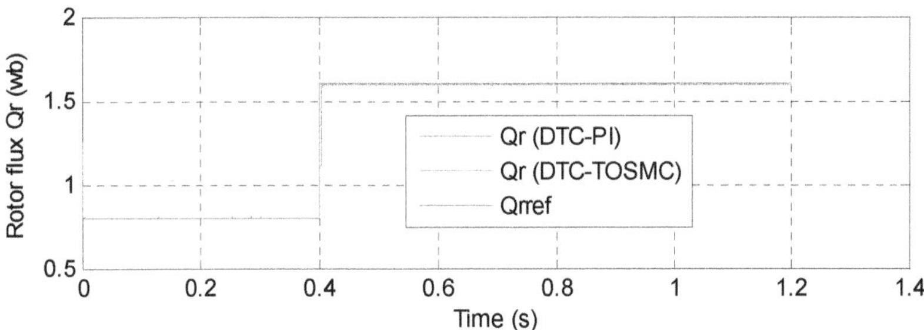

Figure 9. Rotor flux.

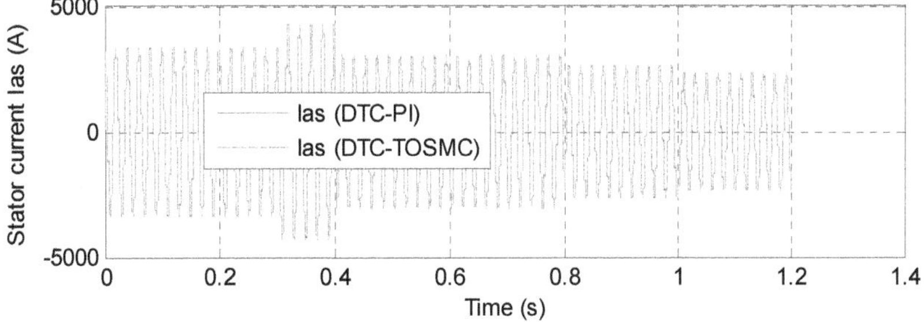

Figure 10. Stator current.

Figure 10 shows the stator current of the designed DFTC strategies and it can be seen that the current was correlated with the torque and flux reference values.

Figures 6 and 7 show the THD value of the stator current of the designed DFTC techniques. It is worth noting that the THD value was lower for DFTC-TOSMC (0.19%) when compared to DFTC-PI (0.54%).

The zoom in the torque, flux, and current is shown in Figures 11–13, respectively. The DFTC-TOSMC technique minimized the undulations in torque, flux, and current compared to the DFTC-PI technique.

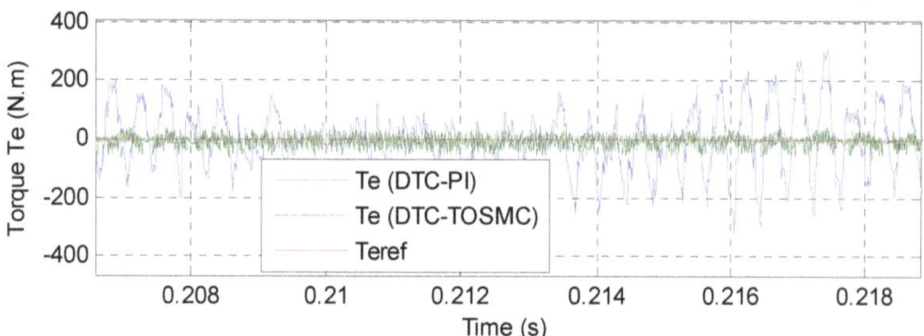

Figure 11. Zoom in the torque.

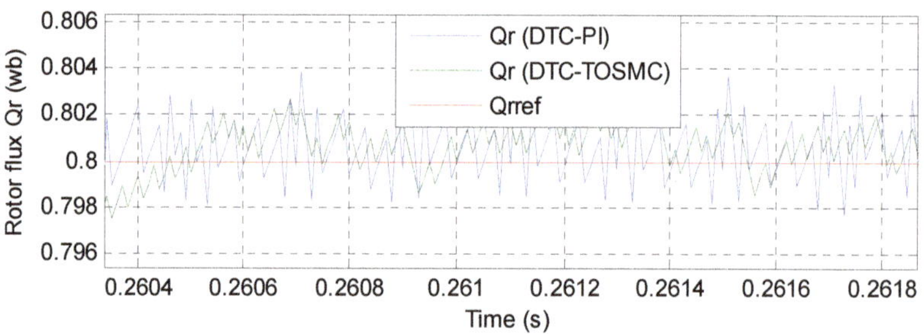

Figure 12. Zoom in the rotor flux.

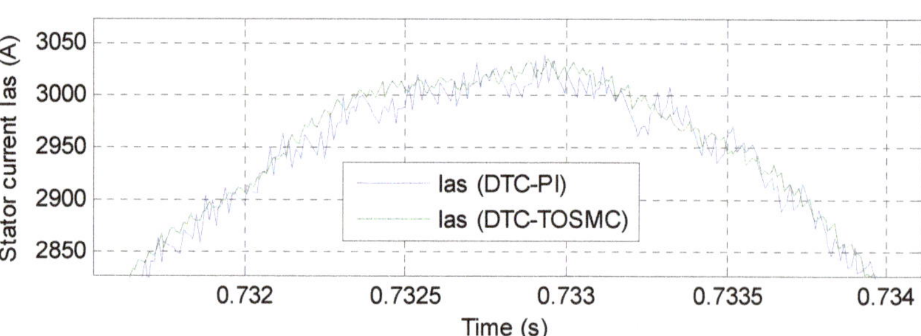

Figure 13. Zoom in the stator current.

7. Discussion

Based on the above results, it can be said that the DFTC-TOSMC strategy has proven its effectiveness in minimizing undulations and the chattering phenomenon, in addition to keeping the other advantages of the DFTC-PI technique. This proposed strategy minimized the THD value of stator current compared to other strategies (see Table 1).

Table 1. Comparison of the THD values obtained from the proposed method with values from several published methods.

Reference	Strategy	THD (%)
Ref. [20]	DPC	4.88
	VF-DPC	4.19
Ref. [21]	DPC-TSC	0.25
Ref. [10]	PI controller	0.77
	STA-SOSMC controller	0.28
Ref. [22]	FOC	3.70
Ref. [23]	Fuzzy SMC control	3.05
Ref. [24]	DFTC-SOCSMC	0.98
Ref. [25]	DPC-IP	0.43
Ref. [26]	DFTC	1.45
Ref. [27]	Direct FOC with synergetic sliding mode controller	0.50
Ref. [32]	Two-level DFTC method	9.87
	Three-level DFTC method	1.52
Ref. [36]	DFTC method	7.54
	DFTC method with genetic algorithm	4.80
Ref. [66]	Traditional DFTC strategy	6.70
	Fuzzy DFTC technique	2.04
Ref. [68]	FOC with Type 2 fuzzy logic controller (FOC-T2FLC)	1.14
	FOC with neuro-fuzzy controller (FOC-NFC)	0.78
Ref. [69]	ISMC	9.71
	MRSMC	3.14
Ref. [70]	DPC control with intelligent metaheuristics	4.05
Proposed strategy	DFTC-TOSM	0.19

The FOC-T2FLC strategy [68] is used as a reference strategy in the same class as the FOC-NFC strategy. The multi-resonant sliding mode controller (MRSMC) and the integral sliding mode controller (ISMC) have been proposed for the DFIG-based wind system in unbalanced and harmonic grid conditions [69].

Table 2 presents a brief comparative study using the simulation results of Figures 6–13. It is clear that the designed DFTC technique based on TOSMC controllers was more robust than the traditional one using the PI controller, except for the dynamic response, which was faster in TOSMC than PI. The analytical reason that proves that the overshoot is very small in the designed DFTC technique using TOSMC is the absence of zero in the transfer function of this one. On the other hand, the designed DFTC technique based on TOSMC controllers improved the rise time, THD, torque and flux tracking, transient performance, quality of stator current, sensitivity to a parameter change, and settling time compared to the DFTC with PI controllers.

Table 2. Comparison of the results obtained from the proposed method with the classical method.

Criteria	Control	
	DFTC-PI	**DFTC-TOSMC**
Dynamic response (s)	Medium	Fast
Settling time (ms)	High	Medium
Overshoot (%)	Remarkable $\approx 22\%$	Neglected near $\approx 1.5\%$
Torque and flux tracking	Good	Excellent
Sensitivity to parameter change	High	Medium
Rise Time (s)	High	Medium
THD (%)	0.54	0.19
Simplicity of converter and filter design	Simple	Simple
Torque: ripple (N.m)	Around 500	Around 60
Simplicity of calculations	Simple	Simple
Rotor flux: ripple (wb)	Around 0.006	Around 0.004
Improvement of transient performance	Good	Excellent
Reduce torque and flux ripples	Acceptable	Excellent
Quality of stator current	Acceptable	Excellent

8. Conclusions

The paper addressed a third-order sliding mode control-based STA technique for a DFTC technique used in wind power. An SVM technique was used for controlling the inverter of AG-based SRWP systems. The mathematical design of the proposed TOSMC technique was discussed in detail for the DFTC technique. The controller was applied both on the torque and flux to regulate the direct and quadrature rotor voltage and also to minimize the undulations in stator current, electromagnetic torque, and rotor flux of the AG. The proposed strategy minimized the THD value of stator current compared to traditional DFTC, FOC, DPC, FSMC, and DFTC-SOCSMC methods (see Table 1). The proposed DFTC technique has improved the robustness of the traditional DFTC method, increasing its performances in transient and dynamic conditions in terms of efficiency, rapidity, overshoot, rise time, and stability. It was observed that this designed DFTC technique is robust with less steady-state error and less settling time compared to a traditional PI controller (for more information, see Table 2). On the other hand, this proposed strategy is a simple structure, no dynamic coordinate transforms are needed, no PI current controllers, and the switching frequency of the transistors is constant. At higher speeds, the proposed technique is not sensitive to any generator parameters. Good tracking capabilities of the desired variable, very fast steady-state reaching speed, robust dynamic nature of the controller, and also the elimination of chattering problem in SMC were realized. Zoom has been shown to compare and highlight its performance. This controller can be an alternative to STA. This proposed controller can be applied to direct power control and a field-oriented control scheme. A comparison was undertaken concerning the PI controller in terms of ripple, tracking, and output current THD for use of this proposed controller for the DFTC technique. Indeed, this proposed DFTC technique deserves attention because it solves the problem of high ripples torque and flux for wind turbines.

The current research work is limited given that the wind speed was fixed. Furthermore, the designed DFTC control scheme investigated a high voltage dip condition. Robustness enhancement of the AG-SRWP system under the previous concerns will be carried out in future papers. This will be implemented through interactions among AGs with various strategies such as neural algorithm, fractional-order PI, and a type 2 fuzzy logic controller.

Therefore, in summary, the main findings of this research are as follows:

- Reduces the electromagnetic torque and rotor flux;
- Simple control was proposed;

- Minimization of the total harmonic distortion of stator current by 64.81%; and
- A new nonlinear controller was presented and confirmed with numerical simulation.

The paper can be extended with fuzzy-TOSMC controllers (FTOSMC) to obtain zero settling time, minimum torque ripple, and zero steady-state error. DPC-based TOSMC controllers can also be taken up as an extension of this paper.

Author Contributions: Conceptualization, H.B.; Methodology, H.B.; Software, H.B.; Validation, N.B.; Investigation, H.B.; Resources, H.B. and N.B.; Data curation, N.B.; Writing—original draft preparation, H.B.; Supervision, N.B.; Project administration, N.B.; Formal analysis: N.B.; Funding acquisition: N.B.; Visualization: N.B.; Writing—review and editing: N.B. and H.B. All authors have read and agreed to the published version of the manuscript.

Funding: There is no funding available for this.

Institutional Review Board Statement: Not applicable.

Informed Consent Statement: Not applicable.

Data Availability Statement: Not applicable.

Conflicts of Interest: The authors declare no conflict of interest.

List of Symbols

$\phi_r, \phi_r{}^*$	Actual and reference rotor flux
V_s, I_s	Vectors of the stator voltage and current
$Vr_{a,b,c}, Ir_{a,b,c}$	Rotor voltage and current in abc frame
$V_{\alpha,\beta}, I_{\alpha,\beta}$	Voltage and current in $\alpha\beta$ frame
$T_e, T_e{}^*$	Actual and reference torques
ω_n, ω_r	Nominal and rotor speeds
R_s, R_r	Stator and rotor resistances
$\phi_{\alpha s}, \phi_{\beta s}$	Stator flux components in $\alpha\beta$ frame
θr	Rotor flux angle
Ki, Kp	Integral and proportional gains
Lr, Ls, Lm	Rotor, stator and mutual inductances
p	Generator pole pairs
Wb	Weber (unit)
Hz	Hertz (unit)
Mw	Migawatt (Unit)
mH	Millihenry (unit)
N.m	Newton-meter (Unit)

List of Acronyms

DTC	Direct torque control
PI	Proportional integral
DPC	Direct power control
SMC	Sliding mode control
DFTC	Direct flux and torque control
THD	Total harmonic distortion
SOCSMC	Second-order continuous sliding mode control
FOC	Field oriented control
FSMC	Fuzzy sliding mode control
SVM	Space vector modulation
IP	Integral-proportional
AG	Asynchronous generator
TOSMC	Third-order sliding mode controller
STA	Super twisting algorithm
THD	Total harmonic distortion
ISM	Integral sliding mode.
MRSMC	Multi-resonant-based sliding mode controller
ISMC	Integral sliding mode controller.

Appendix A

Table A1. The AG parameters [22,27,71].

P_{SRWT}	1.5 MW
Pn	1.5 MW
Rs	0.012 Ω
Ls	0.0137 H
Lm	0.0135 H
Rr	0.021 Ω
Lr	0.0136 H
fr	0.0024 Nm·s/rad
V_n	380 V
p	2
Ω	150 rad/s
F	50 Hz

References

1. Wang, F.; Zhang, Z.; Mei, X.; Rodríguez, J.; Kennel, R. Advanced Control Strategies of Induction Machine: Field Oriented Control, Direct Torque Control and Model Predictive Control. *Energies* **2018**, *11*, 120. [CrossRef]
2. Kim, S.J.; Kim, J.-W.; Park, B.-G.; Lee, D.-H. A Novel Predictive Direct Torque Control Using an Optimized PWM Approach. *IEEE Trans. Ind. Appl.* **2021**, *57*, 2537–2546. [CrossRef]
3. Reddy, C.U.; Prabhakar, K.K.; Singh, A.K.; Kumar, P. Speed Estimation Technique Using Modified Stator Current Error-Based MRAS for Direct Torque Controlled Induction Motor Drives. *IEEE J. Emerg. Sel. Top. Power Electron.* **2020**, *8*, 1223–1235. [CrossRef]
4. Abosh, A.H.; Zhu, Z.Q.; Ren, Y. Reduction of Torque and Flux Ripples in Space Vector Modulation-Based Direct Torque Control of Asymmetric Permanent Magnet Synchronous Machine. *IEEE Trans. Power Electron.* **2017**, *32*, 2976–2986. [CrossRef]
5. Khoucha, F.; Khoudir, M.; Kheloui, A.; Benbouzid, M. Electric Vehicle Induction Motor DSVM-DTC with Torque Ripple Minimization. *Int. Rev. Electr. Eng.* **2009**, *4*, 501–508.
6. Mossa, M.A.; Duc Do, T.; Saad Al-Sumaiti, A.; Quynh, N.V.; Diab, A.A.Z. Effective Model Predictive Voltage Control for a Sensorless Doubly Fed Induction Generator. *IEEE Can. J. Electr. Comput. Eng.* **2021**, *44*, 50–64. [CrossRef]
7. Alsofyani, I.M.; Lee, K.-B. Evaluation of Direct Torque Control with a Constant-Frequency Torque Regulator under Various Discrete Interleaving Carriers. *Electronics* **2019**, *8*, 820. [CrossRef]
8. Wu, C.; Zhou, D.; Blaabjerg, F. Direct Power Magnitude Control of DFIG-DC System without Orientation Control. *IEEE Trans. Ind. Electron.* **2021**, *68*, 1365–1373. [CrossRef]
9. Do, T.D.; Choi, H.H.; Jung, J. Nonlinear Optimal DTC Design and Stability Analysis for Interior Permanent Magnet Synchronous Motor Drives. *IEEE/ASME Trans. Mechatron.* **2015**, *20*, 2716–2725. [CrossRef]
10. Mazen Alhato, M.; Bouallègue, S.; Rezk, H. Modeling and Performance Improvement of Direct Power Control of Doubly-Fed Induction Generator Based Wind Turbine through Second-Order Sliding Mode Control Approach. *Mathematics* **2020**, *8*, 2012. [CrossRef]
11. Xiong, C. A Fault-Tolerant FOC Strategy for Five-Phase SPMSM with Minimum Torque Ripples in the Full Torque Operation Range Under Double-Phase Open-Circuit Fault. *IEEE Trans. Ind. Electron.* **2020**, *67*, 9059–9072. [CrossRef]
12. Venkatesh, N.; Satish, K.G. An Enhanced Exponential Reaching Law Based Sliding Mode Control Strategy for a Three Phase UPS System. *Serb. J. Electr. Eng.* **2020**, *17*, 313–336. [CrossRef]
13. Velasco, J.; Calvo, I.; Barambones, O.; Venegas, P.; Napole, C. Experimental Validation of a Sliding Mode Control for a Stewart Platform Used in Aerospace Inspection Applications. *Mathematics* **2020**, *8*, 2051. [CrossRef]
14. Alanis, A.Y.; Munoz-Gomez, G.; Rivera, J. Nested High Order Sliding Mode Controller with Back-EMF Sliding Mode Observer for a Brushless Direct Current Motor. *Electronics* **2020**, *9*, 1041. [CrossRef]
15. Gao, P.; Zhang, G.; Lv, X. Model-Free Hybrid Control with Intelligent Proportional Integral and Super-Twisting Sliding Mode Control of PMSM Drives. *Electronics* **2020**, *9*, 1427. [CrossRef]
16. Zhang, C.; Lin, Z.; Yang, S.X.; He, J. Total-Amount Synchronous Control Based on Terminal Sliding-Mode Control. *IEEE Access* **2017**, *5*, 5436–5444. [CrossRef]
17. Haibo, L.; Heping, W.; Junlei, S. Attitude control for QTR using exponential nonsingular terminal sliding mode control. *J. Syst. Eng. Electr.* **2019**, *30*, 191–200. [CrossRef]
18. Lin, X.; Shi, X.; Li, S. Adaptive Tracking Control for Spacecraft Formation Flying System via Modified Fast Integral Terminal Sliding Mode Surface. *IEEE Access* **2020**, *8*, 198357–198367. [CrossRef]
19. Wang, Z.; Li, S.; Li, Q. Discrete-Time Fast Terminal Sliding Mode Control Design for DC–DC Buck Converters with Mismatched Disturbances. *IEEE Trans. Ind. Inform.* **2020**, *16*, 1204–1213. [CrossRef]
20. Yusoff, N.A.; Razali, A.M.; Karim, K.A.; Sutikno, T.; Jidin, A. A Concept of Virtual-Flux Direct Power Control of Three-Phase AC-DC Converter. *Int. J. Power Electron. Drive Syst.* **2017**, *8*, 1776–1784. [CrossRef]

21. Amrane, F.; Chaiba, A.; Babes, B.E.; Mekhilef, S. Design and implementation of high performance field oriented control for grid-connected doubly fed induction generator via hysteresis rotor current controller. *Rev. Roum. Sci. Techn.-Electrotechn. Energ.* **2016**, *61*, 319–324.
22. Benbouhenni, H.; Bizon, N. Terminal Synergetic Control for Direct Active and Reactive Powers in Asynchronous Generator-Based Dual-Rotor Wind Power Systems. *Electronics* **2021**, *10*, 1880. [CrossRef]
23. Boudjema, Z.; Meroufel, A.; Djerriri, Y.; Bounadja, E. Fuzzy sliding mode control of a doubly fed induction generator for energy conversion. *Carpath. J. Electron. Comput. Eng.* **2013**, *6*, 7–14.
24. Boudjema, Z.; Taleb, R.; Djerriri, Y.; Yahdou, A. A novel direct torque control using second order continuous sliding mode of a doubly fed induction generator for a wind energy conversion system. *Turk. J. Electr. Eng. Comput. Sci.* **2017**, *25*, 965–975. [CrossRef]
25. Fayssal, A.; Bruno, F.; Azeddine, C. Experimental investigation of efficient and simple wind-turbine based on DFIG-direct power control using LCL-filter for stand-alone mode. *ISA Trans.* **2021**, 1–34, in press. [CrossRef]
26. Shehata, E.G. Sliding mode direct power control of RSC for DFIGs driven by variable speed wind turbines. *Alex. Eng. J.* **2015**, *54*, 1067–1075. [CrossRef]
27. Benbouhenni, H.; Bizon, N. A Synergetic Sliding Mode Controller Applied to Direct Field-Oriented Control of Induction Generator-Based Variable Speed Dual-Rotor Wind Turbines. *Energies* **2021**, *14*, 4437. [CrossRef]
28. Habib, B.; Lemdani, S. Combining synergetic control and super twisting algorithm to reduce the active power undulations of doubly fed induction generator for dual-rotor wind turbine system. *Electr. Eng. Electromech.* **2021**, *3*, 8–17. [CrossRef]
29. Alhato, M.M.; Ibrahim, M.N.; Rezk, H.; Bouallègue, S. An Enhanced DC-Link Voltage Response for Wind-Driven Doubly Fed Induction Generator Using Adaptive Fuzzy Extended State Observer and Sliding Mode Control. *Mathematics* **2021**, *9*, 963. [CrossRef]
30. Chen, S.Z.; Cheung, N.C.; Chung Wong, K.; Wu, J. Integral Sliding-Mode Direct Torque Control of Doubly-Fed Induction Generators Under Unbalanced Grid Voltage. *IEEE Trans. Energy Convers.* **2010**, *25*, 356–368. [CrossRef]
31. Klarmann, K.; Thielmann, M.; Schumacher, W. Comparison of Hysteresis Based PWM Schemes $\Delta\Sigma$-PWM and Direct Torque Control. *Appl. Sci.* **2021**, *11*, 2293. [CrossRef]
32. Najib, E.; Aziz, D.; Abdelaziz, E.; Mohammed, T.; Youness, E.; Khalid, M.; Badre, B. Direct torque control of doubly fed induction motor using three-level NPC inverter. *Protect. Control Mod. Power Syst.* **2019**, *4*, 1–9. [CrossRef]
33. Cortajarena, J.A.; Barambones, O.; Alkorta, P.; Cortajarena, J. Grid Frequency and Amplitude Control Using DFIG Wind Turbines in a Smart Grid. *Mathematics* **2021**, *9*, 143. [CrossRef]
34. Yin, M. Turbine Stability-Constrained Available Wind Power of Variable Speed Wind Turbines for Active Power Control. *IEEE Trans. Power Syst.* **2017**, *32*, 2487–2488. [CrossRef]
35. Hemeyine, A.V.; Abbou, A.; Bakouri, A.; Mokhlis, M.; El Moustapha, S.M.o.M. A Robust Interval Type-2 Fuzzy Logic Controller for Variable Speed Wind Turbines Based on a Doubly Fed Induction Generator. *Inventions* **2021**, *6*, 21. [CrossRef]
36. Mahfoud, S.; Derouich, A.; EL Ouanjli, N.; EL Mahfoud, M.; Taoussi, M. A New Strategy-Based PID Controller Optimized by Genetic Algorithm for DTC of the Doubly Fed Induction Motor. *Systems* **2021**, *9*, 37. [CrossRef]
37. Nguyen, T.-T. A Rotor-Sync Signal-Based Control System of a Doubly-Fed Induction Generator in the Shaft Generation of a Ship. *Processes* **2019**, *7*, 188. [CrossRef]
38. Pan, L.; Zhu, Z.; Xiong, Y.; Shao, J. Integral Sliding Mode Control for Maximum Power Point Tracking in DFIG Based Floating Offshore Wind Turbine and Power to Gas. *Processes* **2021**, *9*, 1016. [CrossRef]
39. Yousefi-Talouki, A.; Zalzar, S.; Pouresmaeil, E. Direct Power Control of Matrix Converter-Fed DFIG with Fixed Switching Frequency. *Sustainability* **2019**, *11*, 2604. [CrossRef]
40. Ma, Y.; Liu, J.; Liu, H.; Zhao, S. Active-Reactive Additional Damping Control of a Doubly-Fed Induction Generator Based on Active Disturbance Rejection Control. *Energies* **2018**, *11*, 1314. [CrossRef]
41. Giaourakis, D.G.; Safacas, A.N. Effect of Short-Circuit Faults in the Back-to-Back Power Electronic Converter and Rotor Terminals on the Operational Behavior of the Doubly-Fed Induction Generator Wind Energy Conversion System. *Machines* **2015**, *3*, 2–26. [CrossRef]
42. Abdelrahem, M.; Hackl, C.M.; Kennel, R. Limited-Position Set Model-Reference Adaptive Observer for Control of DFIGs without Mechanical Sensors. *Machines* **2020**, *8*, 72. [CrossRef]
43. Benbouhenni, H.; Boudjema, Z.; Belaidi, A. Direct power control with NSTSM algorithm for DFIG using SVPWM technique. *Iran. J. Electr. Electron. Eng.* **2021**, *17*, 1–11.
44. Farajpour, Y.; Alzayed, M.; Chaoui, H.; Kelouwani, S. A Novel Switching Table for a Modified Three-Level Inverter-Fed DTC Drive with Torque and Flux Ripple Minimization. *Energies* **2020**, *13*, 4646. [CrossRef]
45. Hakami, S.S.; Mohd Alsofyani, I.; Lee, K.-B. Low-Speed Performance Improvement of Direct Torque Control for Induction Motor Drives Fed by Three-Level NPC Inverter. *Electronics* **2020**, *9*, 77. [CrossRef]
46. Coballes-Pantoja, J.; Gómez-Fuentes, R.; Noriega, J.R.; García-Delgado, L.A. Parallel Loop Control for Torque and Angular Velocity of BLDC Motors with DTC Commutation. *Electronics* **2020**, *9*, 279. [CrossRef]
47. Hakami, S.S.; Lee, K.-B. Four-Level Hysteresis-Based DTC for Torque Capability Improvement of IPMSM Fed by Three-Level NPC Inverter. *Electronics* **2020**, *9*, 1558. [CrossRef]

48. Mossa, M.A.; Echeikh, H.; Diab, A.A.Z.; Haes Alhelou, H.; Siano, P. Comparative Study of Hysteresis Controller, Resonant Controller and Direct Torque Control of Five-Phase IM under Open-Phase Fault Operation. *Energies* **2021**, *14*, 1317. [CrossRef]

49. Xia, C.; Hou, X. Study on the Static Load Capacity and Synthetic Vector Direct Torque Control of Brushless Doubly Fed Machines. *Energies* **2016**, *9*, 966. [CrossRef]

50. Xia, C.; Hou, X.; Chen, F. Flux-Angle-Difference Feedback Control for the Brushless Doubly Fed Machine. *Energies* **2018**, *11*, 71. [CrossRef]

51. Song, Q.; Li, Y.; Jia, C. A Novel Direct Torque Control Method Based on Asymmetric Boundary Layer Sliding Mode Control for PMSM. *Energies* **2018**, *11*, 657. [CrossRef]

52. Heidari, H.; Rassõlkin, A.; Vaimann, T.; Kallaste, A.; Taheri, A.; Holakooie, M.H.; Belahcen, A. A Novel Vector Control Strategy for a Six-Phase Induction Motor with Low Torque Ripples and Harmonic Currents. *Energies* **2019**, *12*, 1102. [CrossRef]

53. Mehedi, F.; Habib, B.; Nezli, L.; Boudana, D. Feedforward neural network-DTC of multi-phase permanent magnet synchronous motor using five-phase neural space vector pulse width modulation strategy. *J. Eur. Syst. Autom.* **2021**, *54*, 345–354. [CrossRef]

54. Mehedi, F.; Yahdou, A.; Djilali, A.B.; Benbouhenni, H. Direct torque fuzzy controlled drive for multi-phase IPMSM based on SVM technique. *J. Eur. Syst. Autom.* **2020**, *53*, 259–266.

55. Ortega-García, L.E.; Rodriguez-Sotelo, D.; Nuñez-Perez, J.C.; Sandoval-Ibarra, Y.; Perez-Pinal, F.J. DSP-HIL Comparison between IM Drive Control Strategies. *Electronics* **2021**, *10*, 921. [CrossRef]

56. Yao, J.; Wang, M.; Li, Z.; Jia, Y. Research on Model Predictive Control for Automobile Active Tilt Based on Active Suspension. *Energies* **2021**, *14*, 671. [CrossRef]

57. Bao, G.; Qi, W.; He, T. Direct Torque Control of PMSM with Modified Finite Set Model Predictive Control. *Energies* **2020**, *13*, 234. [CrossRef]

58. Nasr, A.; Gu, C.; Bozhko, S.; Gerada, C. Performance Enhancement of Direct Torque-Controlled Permanent Magnet Synchronous Motor with a Flexible Switching Table. *Energies* **2020**, *13*, 1907. [CrossRef]

59. Li, Y.; Hang, L.; Li, G.; Guo, Y.; Zou, Y.; Chen, J.; Li, J.; Zhunag, J.; Li, S. An improved DTC controller for DFIG-based wind generation system. In Proceedings of the 2016 IEEE 8th International Power Electronics and Motion Control Conference (IPEMC-ECCE Asia), Hefei, China, 22–26 May 2016; pp. 1423–1426. [CrossRef]

60. Habib, B.; Zinelaabidine, B. Two-level DTC based on ANN controller of DFIG using 7-level hysteresis command to reduce flux ripple comparing with traditional command. In Proceedings of the 2018 International Conference on Applied Smart Systems (ICASS), Medea, Algeria, 24–25 November 2018; pp. 1–8. [CrossRef]

61. Boussekra, F.; Makouf, A. Sensorless speed control of IPMSM using sliding mode observer based on active flux concept. *Model. Meas. Control A* **2020**, *93*, 1–9. [CrossRef]

62. Djafar, D.; Belhamdi, S. Speed control of induction motor with broken bars using sliding mode control (SMC) based to on type-2 fuzzy logic controller (T2FLC). *Adv. Model. Anal. C* **2018**, *73*, 197–201. [CrossRef]

63. Habib, B.; Rachid, T.; Faycal, C. *Improvement of DTC with 24 Sectors of Induction Motor by Using a Three-Level Inverter and Intelligent Hysteresis Controllers*; Springer International Publishing AG: Berlin, Germany, 2018; Volume 35, pp. 99–107.

64. Arbi, J.; Ghorbal, M.J.; Slama-Belkhodja, I.; Charaabi, L. Direct Virtual Torque Control for Doubly Fed Induction Generator Grid Connection. *IEEE Trans. Ind. Electron.* **2009**, *56*, 4163–4173. [CrossRef]

65. Mondal, S.; Kastha, D. Input Reactive Power Controller with a Novel Active Damping Strategy for a Matrix Converter Fed Direct Torque Controlled DFIG for Wind Power Generation. *IEEE J. Emerg. Sel. Top. Power Electron.* **2020**, *8*, 3700–3711. [CrossRef]

66. Ayrira, W.; Ourahoua, M.; El Hassounia, B.; Haddi, A. Direct torque control improvement of a variable speed DFIG based on a fuzzy inference system. *Math. Comput. Simul.* **2020**, *167*, 308–324. [CrossRef]

67. Bounadja, E.; Djahbar, A.; Mahmoudi, M.O.; Matallah, M. Direct torque control of saturated doubly-fed induction generator using high order sliding mode controllers. *Int. J. Adv. Comput. Sci. Appl.* **2016**, *7*, 55–61. [CrossRef]

68. Amrane, F.; Chaiba, A. A novel direct power control for grid-connected doubly fed induction generator based on hybrid artificial intelligent control with space vector modulation. *Rev. Roum. Sci. Techn.-Electrotechn. Energ.* **2016**, *61*, 263–268.

69. Quan, Y.; Hang, L.; He, Y.; Zhang, Y. Multi-Resonant-Based Sliding Mode Control of DFIG-Based Wind System under Unbalanced and Harmonic Network Conditions. *Appl. Sci.* **2019**, *9*, 1124. [CrossRef]

70. Alhato, M.M.; Bouallègue, S. Direct Power Control Optimization for Doubly Fed Induction Generator Based Wind Turbine Systems. *Math. Comput. Appl.* **2019**, *24*, 77. [CrossRef]

71. Benbouhenni, H. Intelligent super twisting high order sliding mode controller of dual-rotor wind power systems with direct attack based on doubly-fed induction generators. *J. Electr. Eng. Electron. Control. Comput. Sci.* **2021**, *7*, 1–8. Available online: https://jeeeccs.net/index.php/journal/article/view/219 (accessed on 1 September 2021).

Article

Finite-Time Passivity Analysis of Neutral-Type Neural Networks with Mixed Time-Varying Delays

Issaraporn Khonchaiyaphum [1], Nayika Samorn [2], Thongchai Botmart [1] and Kanit Mukdasai [1,*]

[1] Department of Mathematics, Faculty of Science, Khon Kaen University, Khon Kaen 40002, Thailand; k.issaraporn@kkumail.com (I.K.); thongbo@kku.ac.th (T.B.)
[2] Faculty of Agriculture and Technology, Nakhon Phanom University, Nakhon Phanom 48000, Thailand; nayika@npu.ac.th
* Correspondence: kanit@kku.ac.th

Abstract: This research study investigates the issue of finite-time passivity analysis of neutral-type neural networks with mixed time-varying delays. The time-varying delays are distributed, discrete and neutral in that the upper bounds for the delays are available. We are investigating the creation of sufficient conditions for finite boundness, finite-time stability and finite-time passivity, which has never been performed before. First, we create a new Lyapunov–Krasovskii functional, Peng–Park's integral inequality, descriptor model transformation and zero equation use, and then we use Wirtinger's integral inequality technique. New finite-time stability necessary conditions are constructed in terms of linear matrix inequalities in order to guarantee finite-time stability for the system. Finally, numerical examples are presented to demonstrate the result's effectiveness. Moreover, our proposed criteria are less conservative than prior studies in terms of larger time-delay bounds.

Keywords: neural networks; finite-time passivity; linear matrix inequality; distributed delay; neutral system

Citation: Khonchaiyaphum, I.; Samorn, N.; Botmart, T.; Mukdasai, K. Finite-Time Passivity Analysis of Neutral-Type Neural Networks with Mixed Time-Varying Delays. *Mathematics* **2021**, *9*, 3321. https://doi.org/10.3390/math9243321

Academic Editors: Quanxin Zhu and Ezequiel López-Rubio

Received: 2 November 2021
Accepted: 15 December 2021
Published: 20 December 2021

1. Introduction

Neural networks have been intensively explored in recent decades due to their vast range of applications in a variety of fields, including signal processing, associative memories, learning ability and so on [1–10]. In the study of real systems, time-delay phenomena are unavoidable. Many interesting neural networks, such as Hopfield neural networks, cellular neural networks, Cohen-Grossberg neural networks and bidirectional associative memory neural networks frequently exhibit time delays. In addition, time delays are well recognized as a source of instability and poor performance [11]. Accordingly, stability analysis of delayed neural networks has become a topic of significant theoretical and practical relevance (see [12–15]), and many important discoveries have been reported on this subject. In recent years, T-S fuzzy delayed neural networks with Markovian jumping parameters using sampled-data control have been presented by Syed Ali et al. [16]. The global stability analysis of fractional-order fuzzy BAM neural networks with time delay and impulsive effects was considered in [17].

Furthermore, conventional neural network models are often unable to accurately represent the qualities of a neural reaction process due to the complex dynamic features of neural cells in the real world. It is only natural for systems to store information about the derivative of a previous state in order to better characterize and analyze the dynamics of such complicated brain responses. Neutral neural networks and neutral-type neural networks are the names given to this new type of neural network. Several academics [18–23] have studied neutral-type neural networks with time-varying delays in recent years. In 2018 [24], the authors investigated improved results on passivity analysis of neutral-type neural networks with mixed time-varying delays. In particular, a type of time-varying delay known as distributed delay occurs in networked-based systems and has received a lot of academic interest because of its significance in digital control systems [25]. Then, this system has all three types of delays: discrete delay, neutral delay and distributed delay. As a result, the neutral delay in neural networks has recently been reported, as well as some stability analysis results for neutral-type neural networks with mixed time-varying delays.

The passive theory [26] is a useful tool for analyzing system stability, and it can deal with systems based solely on the input–output dynamics' general features. The passive theory has been used in engineering applications such as in high-integrity and safety-critical systems. Krasovskii and Lidskii proposed this family of linear systems in 1961 [27]. Researchers have been looking at the passivity of neural networks with delays since then. Many studies have been performed on stability in recent years, including Lyapunov stability, asymptotic stability, uniform stability, eventually uniformly bounded stability and exponential stability, all of which are concerned with the behavior of systems over an indefinite time span. Most actual neural systems, on the other hand, only operate over finite-time intervals. Finite-time passivity is obviously vital and vital for investigating finite-time stabilization of neural networks as a useful tool for analyzing system stability.

This topic has piqued the curiosity of researchers [28–35]. They deal with by Jensen's and Coppel's inequality in [28], which is concerned with the problem of finite-time stability of continuous time delay systems. The authors used an unique control protocol based on the Lyapunov theory and inequality technology to examine the finite-time stabilization of delayed neural networks in [29]. Rajavel et al. [30] solves the problem of finite-time non-fragile passivity control for neural networks with time-varying delay using the Lyapunov–Krasovskii functional technique. Researchers used a new Lyapunov–Krasovskii function with triple and four integral terms to examine finite-time passive filtering for a class of neutral time-delayed systems in [31]. The free-weighting matrix approach and Wirtinger's double integral inequality were used to demonstrate finite-time stability of neutral-type neural networks with random time-varying delays in [32]. Syed Ali et al. [33] studied finite-time passivity for neutral-type neural networks with time-varying delays using the auxiliary integral inequality. Ali et al. [34] explored popular topics including the finite-time H_∞ boundedness of discrete-time neural networks and norm-bounded disturbances with time-varying delay. In 2021, Phanlert et al. [35] has been researching a finite-time non-neutral system. Based on the above research, there are many different methods for stability analysis. Our research will make stability stronger. However, no results on finite-time passivity analysis of neutral-type neural networks with mixed time-varying delays latency have been reported to the best of the authors' knowledge. This is the driving force behind our current investigation.

As a result of the foregoing, we investigate three types of finite passivity in neural networks and provide matching criteria for judging network properties using Lyapunov functional theory and inequality technology. The following are the primary contributions of this paper:

(i)   We examine a system with mixed time-varying delays in this study. Furthermore, because time-varying delays are distributed, discrete and neutral, the upper bounds for the delays are known.

(ii)   We then used the theorems to derive finite-time boundedness, finite-stability and finite-time passivity requirements.

(iii)   By using Peng-integral Park's inequality, model transformation, zero equation and subsequently Wirtinger-based integral inequality approach, some of the simplest LMI-based criteria have been developed.

(iv)   Several cases have been examined to ensure that the primary theorem and its corollaries are accurate.

The following is a breakdown of the paper's structure. Section 2 introduces the network under consideration and offers some definitions, propositions and lemmas. In Section 3, three types of finite-time passivity of the neural network are introduced, and finite-time stability is achieved. In Section 4, several useful outcomes are observed. In

Section 4, five numerical examples are presented to demonstrate the usefulness of our proposed results. Finally, in Section 5, we bring this study to a close.

2. Preliminaries

We begin by explaining various notations and lemmas that will be used throughout the study. R denotes the set of all real numbers; R^n denotes the n-dimensional space; $R^{m \times n}$ denotes the set of all $m \times n$ real matrices ; A^T denotes the transpose of the matrix A; A is symmetric if $A = A^T$; $\lambda(A)$ denotes the set of all eigenvalues of A; and $\lambda_{max}(A)$ and $\lambda_{min}(A)$ represent the maximum and minimum eigenvalues of the matrix A, respectively. $*$ represents the elements below the main diagonal of the symmetric matrices; $\mathbf{diag}\{.\}$ stands for the diagonal matrix.

Consider the study of finite-time passivity analysis of neutral-type neural networks with mixed time-varying delays of the following form:

$$\left.\begin{array}{rcl} \dot{\xi}(t) - G_c \dot{\xi}(t - \tau(t)) & = & -A\xi(t) + G_b f(\xi(t)) + G_d f(\xi(t - \mu(t))) + H\kappa(t) \\ & & + G_e \int_{t-\rho(t)}^t f(\xi(s))ds, \\ z(t) & = & G_1 f(\xi(t)) + G_2 \kappa(t), \quad t \in R^+ \\ \xi(t) & = & \phi(t), \quad t \in [-\bar{h}, 0], \end{array}\right\} \tag{1}$$

where $\xi(t) = [\xi_1(t), \xi_2(t), \ldots, \xi_n(t)]^T \in R^n$ is the neural state vector, $z(t)$ is the output vector of neuron network, and $\kappa(t)$ is the exogenous disturbance input vector belongs to $L_2[0, \infty)$. $A = \mathbf{diag}\{a_1, a_2, \ldots, a_n\} > 0$ is a diagonal matrix with $a_i > 0, i = 1, 2, \ldots, n$. Matrices G_b, G_d and G_e are the interconnection matrices representing the weight coefficients of the neurons. Matrices G_1, G_2, H and G_c are known real constant matrices with appropriate dimensions. $f(\xi(t)) = [f_1(\xi_1(t)), f_2(\xi_2(t)), \ldots, f_n(\xi_n(t))]^T \in R^n$ is the neuron activation function, and $\phi(t) \in C[[-\bar{h}, 0], R^n]$ denotes the initial function. $\mu(t)$ is the discrete time-varying delay, $\rho(t)$ is the distributed time-varying delay, $\tau(t)$ is neutral delay and $\bar{h} = \mathbf{max}\{\mu_M, \rho_M, \tau_M\}$.

The variables $\mu(t)$, $\rho(t)$ and $\tau(t)$ represent the mixed delays of the model in (1) and satisfy the following:

$$\begin{array}{ll} 0 \le \mu(t) \le \mu_M, & 0 \le \dot{\mu}(t) \le \mu_d, \\ 0 \le \rho(t) \le \rho_M, & 0 \le \dot{\rho}(t) \le \rho_d, \\ 0 \le \tau(t) \le \tau_M, & 0 \le \dot{\tau}(t) \le \tau_d, \end{array} \tag{2}$$

where $\mu_M, \mu_d, \rho_M, \rho_d, \tau_M$ and τ_d are positive real constants.

Assumtion 1. *The activation function f is continuous and the exist real constants F_i^- and F_i^+ such that the following is the case:*

$$F_i^- \le \frac{f_i(c_1) - f_i(c_2)}{c_1 - c_2} \le F_i^+, \tag{3}$$

for all $c_1 \ne c_2$, and $f_i = [f_1, f_2, \ldots, f_n]^T$ for any $i \in \{1, 2, \ldots, n\}$ satisfies $f_i(0) = 0$. For the sake of presentation convenience, in the following, we denote $F_1 = \mathbf{diag}(F_1^- F_1^+, F_2^- F_2^+, \ldots, F_n^- F_n^+)$ and $F_2 = \mathbf{diag}(\frac{F_1^- + F_1^+}{2}, \frac{F_2^- + F_2^+}{2}, \ldots, \frac{F_n^- + F_n^+}{2})$.

Assumtion 2. *In the case of a positive parameter δ, $\kappa(t)$ is a time-varying external disturbance that satisfies the following.*

$$\int_0^{T_f} \kappa^T(t)\kappa(t)dt \ge \delta, \quad \delta > 0. \tag{4}$$

Definition 1 ((Finite-time boundedness) [36,37]). *For a positive constant of T, system (1) is finite-time bounded with respect to $(g_1,\ g_2,\ T_f,\ P_1,\ \delta)$ if there exist constants $g_2 > g_1 > 0$ such that the following is the case:*

$$\sup_{-\mu_M \leq t_0 \leq 0} \{\xi^T(t_0)P_1\xi(t_0),\ \dot{\xi}^T(t_0)P_1\dot{\xi}(t_0)\} \geq g_1 \implies \xi^T(t)P_1\xi(t) \geq g_2,\ for\ t \in [0,\ T_f],$$

for a given positive constant T_f, and P_1 is a positive definite matrix.

Definition 2 ((Finite-time stability) [36,37]). *System (1) with $\kappa(t) = 0$ is said to be finite-time stable with respect to $(g_1,\ g_2,\ T_f,\ P_1)$ if there exist constants $g_2 > g_1 > 0$ such that the following is the case:*

$$\sup_{-\mu_M \leq t_0 \leq 0} \{\xi^T(t_0)P_1\xi(t_0),\ \dot{\xi}^T(t_0)P_1\dot{\xi}(t_0)\} \geq g_1 \implies \xi^T(t)P_1\xi(t) \geq g_2,\ for\ t \in [0,\ T_f],$$

for a given positive constant T_f, and P_1 is a positive definite matrix.

Definition 3 ((Finite-time passivity) [37]). *System (1) is said to be a finite-time passive with with a prescribed dissipation performance level $\gamma > 0$, if the following relations hold:*

(a) *For any external disturbances $\kappa(t)$, system (1) is finite-time bounded;*

(b) *For a given positive scalar $\gamma > 0$,   the following relationship holds under a zero initial condition.*

$$\int_0^{T_f} \kappa^T(t)z(t)dt \geq \gamma \int_0^{T_f} \kappa^T(t)\kappa(t)dt.$$

Lemma 1 ((Jensen's Inequality) [38]). *For each positive definite symmetric matrix P_7, positive real constant μ_M and vector function $\dot{\xi} : [-\mu_M, 0] \to R^n$ such that the following integral is well defined, then the following is obtained.*

$$-\mu_M \int_{-\mu_M}^0 \dot{\xi}^T(s+t)P_7\dot{\xi}(s+t)ds \leq -\left(\int_{-\mu_M}^0 \dot{\xi}(s+t)ds\right)^T P_7 \left(\int_{-\mu_M}^0 \dot{\xi}(s+t)ds\right).$$

Lemma 2 ((Wirtinger-based integral inequality) [39]). *For any matrix $P_{12} > 0$,   the following inequality holds for all continuously differentiable function $\dot{\xi} : [\alpha, \beta] \to R^n$*

$$-(\beta - \alpha)\int_\alpha^\beta \dot{\xi}^T(s)P_{12}\dot{\xi}(s)ds \leq \kappa^T \begin{bmatrix} -4P_{12} & -2P_{12} & 6P_{12} \\ * & -4P_{12} & 6P_{12} \\ * & * & -12P_{12} \end{bmatrix} \kappa,$$

where $\kappa = [\xi^T(\beta),\ \xi^T(\alpha),\ \frac{1}{\beta-\alpha}\int_\alpha^\beta \xi^T(s)ds]^T$.

Lemma 3 ((Peng-Park's integral inequality) [40,41]). *For any matrix of the following:* $\begin{bmatrix} P_{13} & S \\ * & P_{13} \end{bmatrix} \geq 0,\ 0 < \mu(t) < \mu_M$ *is satisfied by positive constants μ_M and $\mu(t)$, and $\dot{\xi} :$ $[-\mu_M, 0] \to R^n$ is a vector function that verifies the integrations in question are correctly specified. We then have the following:*

$$-\mu_M \int_{t-\mu_M}^t \dot{\xi}^T(s)P_{13}\dot{\xi}(s)ds \leq \Psi^T \begin{bmatrix} -P_{13} & P_{13} - S & S \\ * & -2P_{13} + S + S^T & P_{13} - S \\ * & * & -P_{13} \end{bmatrix} \Psi,$$

where $\Psi = [\xi^T(t),\ \xi^T(t - \mu(t)),\ \xi^T(t - \mu_M)]^T$ and $\Theta = \begin{bmatrix} -P_{13} & P_{13} - S & S \\ * & -2P_{13} + S + S^T & P_{13} - S \\ * & * & -P_{13} \end{bmatrix}$.

Lemma 4 ([42]). *The following inequality applies to a positive matrix P_{10}*

$$-\frac{(\alpha-\beta)^2}{2}\int_\beta^\alpha\int_s^\alpha \xi^T(u)P_{10}\xi(u)duds \leq -\Big(\int_\beta^\alpha\int_s^\alpha \xi(u)duds\Big)^T P_{10}\Big(\int_\beta^t\int_s^\alpha \xi(u)duds\Big).$$

Lemma 5 ([43]). *$P_6 \in R^{n\times n}$ is a constant symmetric positive definite matrix. For any constant symmetric positive definite matrix $P_6 \in R^{n\times n}$, $\mu(t)$ is a discrete time-varying delay with (2), vector function $\xi : [-\mu_M, 0] \to R^n$ such that the integrations concerned are well defined, then the following is the case.*

$$-\mu_M \int_{-\mu_M}^0 \xi^T(s)P_6\xi(s)ds \;\leq\; -\int_{-\mu(t)}^0 \xi^T(s)ds P_6 \int_{-\mu(t)}^0 \xi(s)ds$$

$$-\int_{-\mu_M}^{-\mu(t)} \xi^T(s)ds P_6 \int_{-\mu_M}^{-\mu(t)} \xi(s)ds.$$

Lemma 6 ([43]). *For any constant matrices R_7, R_8, $R_9 \in R^{n\times n}$, $R_7 \geq 0$, $R_9 > 0$, $\begin{bmatrix} R_7 & R_8 \\ * & R_9 \end{bmatrix} \geq 0$, $\mu(t)$ is a discrete time-varying delay with (2) and vector function $\dot\xi : [-\mu_M,\, 0] \to R^n$ such that the following integration is well defined:*

$$-\mu_M \int_{t-\mu_M}^t \begin{bmatrix} \xi(s) \\ \dot\xi(s) \end{bmatrix}^T \begin{bmatrix} R_7 & R_8 \\ * & R_9 \end{bmatrix} \begin{bmatrix} \xi(s) \\ \dot\xi(s) \end{bmatrix} ds \leq Y^T \Pi Y,$$

where $Y^T = \begin{bmatrix} \xi(t) & \xi(t-\mu(t)) & \xi(t-\mu_M) & \int_{t-\mu(t)}^t \xi(s)ds & \int_{t-\mu_M}^{t-\mu(t)} \xi(s)ds \end{bmatrix}$.

and the following is the case $\Pi = \begin{bmatrix} -R_9 & R_9 & 0 & -R_8^T & 0 \\ * & -R_9-R_9^T & R_9 & R_8^T & -R_8^T \\ * & * & -R_9 & 0 & R_8^T \\ * & * & * & -R_7 & 0 \\ * & * & * & * & -R_7 \end{bmatrix}$.

Lemma 7 ([43]). *Let $\xi(t) \in R^n$ be a vector-valued function with first-order continuous-derivative entries. For any constant matrices P_5, $\widehat{M}_i \in R^{n\times n}$, then the following integral inequality holds, $i = 1, 2, \ldots, 5$ and $\mu(t)$ is a discrete time-varying delay with (2):*

$$-\int_{t-\mu_M}^t \dot\xi^T(s)P_5\dot\xi(s)ds \;\leq\; \Gamma^T \begin{bmatrix} \widehat{M}_1+\widehat{M}_1^T & -\widehat{M}1^T+\widehat{M}_2 & 0 \\ * & \widehat{M}_1+\widehat{M}_1^T-\widehat{M}_2-\widehat{M}_2^T & -\widehat{M}_1^T+\widehat{M}_2 \\ * & * & -\widehat{M}_2-\widehat{M}_2^T \end{bmatrix} \Gamma$$

$$+\mu_M\Gamma^T \begin{bmatrix} \widehat{M}_3 & \widehat{M}_4 & 0 \\ * & \widehat{M}_3+\widehat{M}_5 & \widehat{M}_4 \\ * & * & \widehat{M}_5 \end{bmatrix}\Gamma,$$

where $\Gamma = \begin{bmatrix} \xi(t) \\ \xi(t-\mu(t)) \\ \xi(t-\mu_M) \end{bmatrix}$, $\begin{bmatrix} P_5 & \widehat{M}_1 & \widehat{M}_2 \\ * & \widehat{M}_3 & \widehat{M}_4 \\ * & * & \widehat{M}_5 \end{bmatrix} \geq 0.$

Lemma 8 ([44]). *For a positive definite matrix $P_8, P_9 > 0$ and any continuously differentiable function $\dot\xi : [a,\, b] \to R^n$, the following inequality holds:*

$$\int_a^b \dot\xi^T(s)P_5\dot\xi(s)ds \;\geq\; \frac{1}{b-a}\Theta_1^T P_8 \Theta_1 + \frac{3}{b-a}\Theta_2^T P_8 \Theta_2 + \frac{5}{b-a}\Theta_3^T P_8 \Theta_3,$$

$$\int_a^b \int_u^b \dot\xi^T(s)P_5\dot\xi(s)dsdu \;\geq\; 2\Theta_4^T P_9 \Theta_4 + 4\Theta_5^T P_9 \Theta_5 + 6\Theta_6^T P_9 \Theta_6,$$

where the following is the case.

$$\Theta_1 = \xi(a) - \xi(b),$$

$$\Theta_2 = \xi(a) + \xi(b) - \frac{2}{b-a}\int_a^b \xi(s)ds,$$

$$\Theta_3 = \xi(a) - \xi(b) + \frac{6}{b-a}\int_a^b \xi(s)ds - \frac{12}{(b-a)^2}\int_a^b\int_u^b \xi(s)dsdu,$$

$$\Theta_4 = \xi(b) - \frac{1}{b-a}\int_a^b \xi(s)ds,$$

$$\Theta_5 = \xi(b) + \frac{2}{b-a}\int_a^b \xi(s)ds - \frac{6}{(b-a)^2}\int_a^b\int_u^b \xi(s)dsdu,$$

$$\Theta_6 = \xi(b) - \frac{3}{b-a}\int_a^b \xi(s)ds + \frac{24}{(b-a)^2}\int_a^b\int_u^b \xi(s)dsdu$$

$$- \frac{60}{(b-a)^3}\int_a^b\int_u^b\int_s^b \xi(r)drdsdu.$$

3. Main Results

3.1. Finite-Time Boundedness Analysis

The following finite-time boundedness analysis of neutral-type neural networks with mixed time-varying delays is discussed in this subsection.

$$\left.\begin{aligned}
\dot{\xi}(t) - G_c\dot{\xi}(t-\tau(t)) &= -A\xi(t) + G_b f(\xi(t)) + G_d f(\xi(t-\mu(t))) + H\kappa(t) \\
&\quad + G_e \int_{t-\rho(t)}^t f(\xi(s))ds, \\
\xi(t) &= \phi(t), \quad t \in [-\bar{h}, 0].
\end{aligned}\right\} \tag{5}$$

In the first subsection, we look at system (5) with (2) that uses new criteria for systems introduced via the LMIs approach.

$$\sum = \left[\Pi_{(i,j)}\right]_{23\times23}. \tag{6}$$

For future reference, we introduce the following notations in the Appendix A.

Theorem 1. *For* $\| C \| < 1$, *system* (5) *is finite-time bounded if there exist positive definite matrices* P_i, R_j, $i = 1, 2, 3, \ldots, 16$, $j = 1, 2, 3, \ldots, 9$ *any appropriate matrices* S, P_{13}, R_8, Q_k, $R_7 \geq 0$, Z_l, $l = 1, 2$ *and* O_e, $e = 1, 2, 3, \ldots, 8$, $\begin{bmatrix} R_{n+3n} & R_{2+3n} \\ R_{2+3n}^T & R_{3+3n} \end{bmatrix} \geq 0$, $\begin{bmatrix} P_{13} & S \\ * & P_{13} \end{bmatrix} \geq 0$ *where* $n = 0, 1, 2$, $k = 1, 2, \ldots, 14$, *positive diagonal matrices* H_p, W_p, $p = 1, 2$ *and positive real constants* μ_M, ρ_M, μ_d, τ_M, τ_d, δ, α, g_1, g_2, T *such that the following symmetric linear matrix inequality holds:*

$$\begin{bmatrix} P_5 & M_1 & M_2 \\ * & M_3 & M_4 \\ * & * & M_5 \end{bmatrix} \geq 0, \tag{7}$$

$$\sum < 0, \tag{8}$$

$$\lambda_1 g_2 e^{-\alpha T} > \Lambda g_1 + \delta(1 - e^{-\alpha T}). \tag{9}$$

For future reference, we introduce the following notations in Appendix A. Then, λ_i, $i = 1, 2, \ldots, 31$ *in system* (9) *is defined in Remark 1.*

Proof. First, we show that system (5) is the finite-time bounded analysis. As a result, we consider system (5) to satisfy the following.

$$\dot{\xi}(t) = -A\xi(t) + G_b f(\xi(t)) + G_c\dot{\xi}(t-\tau(t)) + G_d f(\xi(t-\mu(t))) + H\kappa(t) \tag{10}$$
$$+G_e \int_{t-\rho(t)}^{t} f(\xi(s))ds.$$

We can rewrite system (10) to the following system:

$$\dot{\xi}(t) = y(t), \tag{11}$$
$$0 = -y(t) - A\xi(t) + G_b f(\xi(t)) + G_c\dot{\xi}(t-\tau(t)) + G_d f(\xi(t-\mu(t))) + H\kappa(t)$$
$$+G_e \int_{t-\rho(t)}^{t} f(\xi(s))ds, \tag{12}$$

by using the model transformation approach. Construct a Lyapunov–Krasovskii functional candidate for system (10)–(12) of the following form:

$$V(t) = \sum_{i=1}^{10} V_i(t), \tag{13}$$

where the following is the case:

$$V_1(t) = \xi^T(t)P_1\xi(t) + 2\sum_{i=1}^{N} w_{i1} \int_{0}^{\xi_i(t)} (f_i(s) - F^-s)ds,$$

$$V_2(t) = \zeta^T(t)GP_2\zeta(t) + 2\sum_{i=1}^{N} w_{i2} \int_{0}^{\xi_i(t)} (F^+s - f_i(s))ds,$$

$$V_3(t) = \int_{t-\mu_M}^{t} \xi^T(s)P_3\xi(s)ds$$
$$+ \int_{t-\mu(t)}^{t} \begin{bmatrix} \xi(s) \\ f(\xi(s)) \end{bmatrix}^T \begin{bmatrix} R_1 & R_2 \\ * & R_3 \end{bmatrix} \begin{bmatrix} \xi(s) \\ f(\xi(s)) \end{bmatrix} ds$$
$$+ \int_{t-\mu_M}^{t} \begin{bmatrix} \xi(s) \\ f(\xi(s)) \end{bmatrix}^T \begin{bmatrix} R_4 & R_5 \\ * & R_6 \end{bmatrix} \begin{bmatrix} \xi(s) \\ f(\xi(s)) \end{bmatrix} ds,$$

$$V_4(t) = \mu_M \int_{-\mu_M}^{0} \int_{t+s}^{t} \begin{bmatrix} \xi(\theta) \\ y(\theta) \end{bmatrix}^T \begin{bmatrix} R_7 & R_8 \\ * & R_9 \end{bmatrix} \begin{bmatrix} \xi(\theta) \\ y(\theta) \end{bmatrix} d\theta ds,$$

$$V_5(t) = \mu_M \int_{-\mu_M}^{0} \int_{t+s}^{t} \xi^T(\theta)P_4\xi(\theta)d\theta ds$$
$$+ \int_{-\mu_M}^{0} \int_{t+s}^{t} y^T(\theta)P_5 y(\theta)d\theta ds,$$

$$V_6(t) = \mu_M \int_{-\mu_M}^{0} \int_{t+s}^{t} y^T(\theta)P_6 y(\theta)d\theta ds$$
$$+\mu_M \int_{-\mu_M}^{0} \int_{t+s}^{t} \dot{\xi}^T(\theta)P_7\dot{\xi}(\theta)d\theta ds,$$

$$V_7(t) = \mu_M \int_{-\mu_M}^{0} \int_{t+s}^{t} y^T(\theta)S_1 y(\theta)d\theta ds$$
$$+\mu_M \int_{-\mu_M}^{0} \int_{\lambda}^{0} \int_{t+s}^{t} y^T(\theta)S_2 y(\theta)d\theta ds d\lambda,$$

$$V_8(t) = \frac{(\mu_M)^2}{2} \int_{-\mu_M}^{0} \int_{\lambda}^{0} \int_{t+s}^{t} \xi^T(\theta)P_{10}\xi(\theta)d\theta ds d\lambda$$
$$+\frac{(\mu_M)^2}{2} \int_{-\mu_M}^{0} \int_{\lambda}^{0} \int_{t+s}^{t} y^T(\theta)P_{11} y(\theta)d\theta ds d\lambda,$$

$$V_9(t) = \mu_M \int_{-\mu_M}^{0} \int_{t+s}^{t} y^T(\theta) P_{12} y(\theta) d\theta ds$$
$$+ \mu_M \int_{-\mu_M}^{0} \int_{t+s}^{t} y^T(\theta) P_{13} y(\theta) d\theta ds,$$

$$V_{10}(t) = \int_{t-\tau(t)}^{t} \dot{\xi}^T(s) P_{14} \dot{\xi}(s) ds + \tau_M \int_{t-\tau_M}^{t} \dot{\xi}^T(s) P_{15} \dot{\xi}(s) ds,$$

$$V_{11}(t) = \rho_M \int_{-\rho_M}^{0} \int_{t+s}^{t} f(\xi(\theta))^T P_{16} f(\xi(\theta)) d\theta ds,$$

where $G = \begin{bmatrix} I & 0 & 0 & 0 \\ 0 & 0 & 0 & 0 \\ 0 & 0 & 0 & 0 \\ 0 & 0 & 0 & 0 \end{bmatrix}$, $\zeta^T(t) = \begin{bmatrix} \xi(t) \\ \int_{t-\mu(t)}^{t} y(s)ds \\ \int_{t-\mu_M}^{t-d(t)} y(s)ds \\ y(t) \end{bmatrix}^T$.

Along the trajectory of system (10)–(12), the time derivative of $V(t)$ is equivalent to the following.

$$\dot{V}(t) = \sum_{i=1}^{10} \dot{V}_i(t). \tag{14}$$

The time derivative of $V_1(t)$ is then computed as the following.

$$\dot{V}_1(t) = 2\xi^T(t) P_1 \Big[-A\xi(t) + G_b f(\xi(t)) + G_c \dot{\xi}(t-\tau(t)) + G_d f(\xi(t-\mu(t)))$$
$$+ G_e \int_{t-\rho(t)}^{t} f(\xi(s))ds + H\kappa(t) \Big] + 2f^T(\xi(t)) W_1 \dot{\xi}(t) - \xi^T(t) W_1 F_1 \dot{\xi}(t).$$

Taking the derivative of $V_2(t)$ along any system solution trajectory, we have the following.

$$\dot{V}_2(t) = 2\xi^T(t) P_2 \Big[-A\xi(t) + G_b f(\xi(t)) + G_c \dot{\xi}(t-\tau(t)) + G_d f(\xi(t-\mu(t)))$$
$$+ G_e \int_{t-\rho(t)}^{t} f(\xi(s))ds + H\kappa(t) \Big] + 2\dot{\xi}^T(t) Q_{13}^T [\dot{\xi}(t) - y(t)]$$
$$+ 2y^T(t) Q_{14}^T [\dot{\xi}(t) - y(t)] + \xi^T(t) W_2 F_2 \dot{\xi}(t) - 2f^T(\xi(t)) W_2 \dot{\xi}(t)$$
$$= 2\xi^T(t) P_2 [-A\xi(t) + G_b f(\xi(t)) + G_c \dot{\xi}(t-\tau(t)) + G_d f(\xi(t-\mu(t)))$$
$$+ G_e \int_{t-\rho(t)}^{t} f(\xi(s))ds + H\kappa(t)] + 2\dot{\xi}^T(t) Q_{13}^T [\dot{\xi}(t) - y(t)]$$
$$+ 2y^T(t) Q_{14}^T [\dot{\xi}(t) - y(t)] + 2 \Big[\xi^T(t) Q_1^T + \int_{t-\mu(t)}^{t} y^T(s)ds Q_4^T$$
$$+ \int_{t-\mu_M}^{t-\mu(t)} y^T(s)ds Q_7^T + y^T(t) Q_{10}^T \Big] [-y(t) - A\xi(t) + G_b f(\xi(t))$$
$$+ G_c \dot{\xi}(t-\tau(t)) + G_d f(\xi(t-\mu(t))) + G_e \int_{t-\rho(t)}^{t} f(\xi(s))ds$$
$$+ H\kappa(t)] + 2 \Big[\xi^T(t) Q_2^T + \int_{t-\mu(t)}^{t} y^T(s)ds Q_5^T + \int_{t-\mu_M}^{t-\mu(t)} y^T(s)ds Q_8^T$$
$$+ y^T(t) Q_{11}^T \Big] \times \Big[\xi(t) - \xi(t-\mu(t)) - \int_{t-\mu(t)}^{t} y(s)ds \Big]$$
$$+ 2 \Big[\xi^T(t) Q_3^T + \int_{t-\mu(t)}^{t} y^T(s)ds Q_6^T + \int_{t-\mu_M}^{t-\mu(t)} y^T(s)ds Q_9^T + y^T(t) Q_{12}^T \Big]$$
$$\times \Big[\xi(t-\mu(t)) - \xi(t-\mu_M) - \int_{t-\mu_M}^{t-\mu(t)} y(s)ds \Big]$$
$$+ \xi^T(t) W_2 F_2 \dot{\xi}(t) - 2f^T(\xi(t)) W_2 \dot{\xi}(t).$$

For $V_3(t)$ and $\dot{\mu}(t) \leq \mu_d$, we now have the following.

$$
\begin{aligned}
\dot{V}_3(t) &= \xi^T(t)P_3\xi(t) - \xi^T(t-\mu_M)P_3\xi(t-\mu_M) \\
&\quad + \begin{bmatrix} \xi(t) \\ f(\xi(t)) \end{bmatrix}^T \begin{bmatrix} R_1 & R_2 \\ R_2^T & R_3 \end{bmatrix} \begin{bmatrix} \xi(t) \\ f(\xi(t)) \end{bmatrix} \\
&\quad - (1-\dot{\mu}(t)) \begin{bmatrix} \xi(t-\mu(t)) \\ f(\xi(t-\mu(t))) \end{bmatrix}^T \begin{bmatrix} R_1 & R_2 \\ R_2^T & R_3 \end{bmatrix} \begin{bmatrix} \xi(t-\mu(t)) \\ f(\xi(t-\mu(t))) \end{bmatrix} \\
&\quad + \begin{bmatrix} \xi(t) \\ f(\xi(t)) \end{bmatrix}^T \begin{bmatrix} R_4 & R_5 \\ R_5^T & R_6 \end{bmatrix} \begin{bmatrix} \xi(t) \\ f(\xi(t)) \end{bmatrix} \\
&\quad - \begin{bmatrix} \xi(t-\mu_M) \\ f(\xi(t-\mu_M)) \end{bmatrix}^T \begin{bmatrix} R_4 & R_5 \\ R_5^T & R_6 \end{bmatrix} \begin{bmatrix} \xi(t-\mu_M) \\ f(\xi(t-\mu_M)) \end{bmatrix} \\
&\leq \xi^T(t)P_3\xi(t) - \xi^T(t-\mu_M)P_3\xi(t-\mu_M) + \begin{bmatrix} \xi(t) \\ f(\xi(t)) \end{bmatrix}^T \begin{bmatrix} R_1 & R_2 \\ R_2^T & R_3 \end{bmatrix} \begin{bmatrix} \xi(t) \\ f(\xi(t)) \end{bmatrix} \\
&\quad - \begin{bmatrix} \xi(t-\mu(t)) \\ f(\xi(t-\mu(t))) \end{bmatrix}^T \begin{bmatrix} R_1 & R_2 \\ R_2^T & R_3 \end{bmatrix} \begin{bmatrix} \xi(t-\mu(t)) \\ f(\xi(t-\mu(t))) \end{bmatrix} \\
&\quad + \mu_d \begin{bmatrix} \xi(t-\mu(t)) \\ f(\xi(t-\mu(t))) \end{bmatrix}^T \begin{bmatrix} R_1 & R_2 \\ R_2^T & R_3 \end{bmatrix} \begin{bmatrix} \xi(t-\mu(t)) \\ f(\xi(t-\mu(t))) \end{bmatrix} \\
&\quad + \begin{bmatrix} \xi(t) \\ f(\xi(t)) \end{bmatrix}^T \begin{bmatrix} R_4 & R_5 \\ R_5^T & R_6 \end{bmatrix} \begin{bmatrix} \xi(t) \\ f(\xi(t)) \end{bmatrix} \\
&\quad - \begin{bmatrix} \xi(t-\mu_M) \\ f(\xi(t-\mu_M)) \end{bmatrix}^T \begin{bmatrix} R_4 & R_5 \\ R_5^T & R_6 \end{bmatrix} \begin{bmatrix} \xi(t-\mu_M) \\ f(\xi(t-\mu_M)) \end{bmatrix}.
\end{aligned}
$$

It is from Lemma 6 that we have the following.

$$
\begin{aligned}
\dot{V}_4(t) &= \mu_M^2 \begin{bmatrix} \xi(t) \\ y(t) \end{bmatrix}^T \begin{bmatrix} R_7 & R_8 \\ R_8^T & R_9 \end{bmatrix} \begin{bmatrix} \xi(t) \\ y(t) \end{bmatrix} - \mu_M \int_{t-\mu_M}^{t} \begin{bmatrix} \xi(s) \\ y(s) \end{bmatrix}^T \begin{bmatrix} R_7 & R_8 \\ R_8^T & R_9 \end{bmatrix} \begin{bmatrix} \xi(s) \\ y(s) \end{bmatrix} ds \\
&\leq \mu_M^2 \begin{bmatrix} \xi(t) \\ y(t) \end{bmatrix}^T \begin{bmatrix} R_7 & R_8 \\ R_8^T & R_9 \end{bmatrix} \begin{bmatrix} \xi(t) \\ y(t) \end{bmatrix} \\
&\quad + \begin{bmatrix} \xi(t) \\ \xi(t-\mu(t)) \\ \xi(t-\mu_M) \\ \int_{t-\mu(t)}^{t} \xi(s)ds \\ \int_{t-\mu_M}^{t-\mu(t)} \xi(s)ds \end{bmatrix}^T \Pi \begin{bmatrix} \xi(t) \\ \xi(t-\mu(t)) \\ \xi(t-\mu_M) \\ \int_{t-\mu(t)}^{t} \xi(s)ds \\ \int_{t-\mu_M}^{t-\mu(t)} \xi(s)ds \end{bmatrix}
\end{aligned}
$$

where

$$
\Pi = \begin{bmatrix}
-R_9 & R_9 & 0 & -R_8^T & 0 \\
R_9^T & -R_9 - R_9^T & R_9 & R_8^T & -R_8^T \\
0 & R_9^T & -R_9 & 0 & R_8^T \\
-R_9 & R_8 & 0 & -R_7 & 0 \\
0 & -R_8 & R_8 & 0 & -R_7
\end{bmatrix}.
$$

Using Lemmas 5 and 7, $V_5(t)$ is computed as follows:

$$
\begin{aligned}
\dot{V}_5(t) \;=\;& \mu_M^2 \xi^T(t) P_4 \xi(t) - \mu_M \int_{t-\mu_M}^{t} \xi^T(s) P_4 \xi(s) ds + \mu_M y^T(t) P_5 y(t) \\
& - \int_{t-\mu_M}^{t} \dot{\xi}^T(s) P_5 \dot{\xi}(s) ds \\
\leq\;& \mu_M^2 \xi^T(t) P_4 \xi(t) + \mu_M y^T(t) P_5 y(t) - \int_{t-\mu(t)}^{t} \xi^T(s) ds P_4 \int_{t-\mu(t)}^{t} \xi(s) ds \\
& - \int_{t-\mu_M}^{t-\mu(t)} \xi^T(s) ds P_4 \int_{t-\mu_M}^{t-\mu(t)} \xi(s) ds \\
& + \begin{bmatrix} \xi(t) \\ \xi(t-\mu(t)) \\ \xi(t-\mu_M) \end{bmatrix}^T \Theta \begin{bmatrix} \xi(t) \\ \xi(t-\mu(t)) \\ \xi(t-\mu_M) \end{bmatrix} \\
& + \mu_M \begin{bmatrix} \xi(t) \\ \xi(t-\mu(t)) \\ \xi(t-\mu_M) \end{bmatrix}^T \begin{bmatrix} \widehat{M}_3 & \widehat{M}_4 & 0 \\ \widehat{M}_4^T & \widehat{M}_3 + \widehat{M}_5 & \widehat{M}_4 \\ 0 & \widehat{M}_4^T & \widehat{M}_5 \end{bmatrix} \begin{bmatrix} \xi(t) \\ \xi(t-\mu(t)) \\ \xi(t-\mu_M) \end{bmatrix}
\end{aligned}
$$

where the following is the case.

$$
\Theta = \begin{bmatrix} \widehat{M}_1 + \widehat{M}_1^T & -\widehat{M}_1^T + \widehat{M}_2 & 0 \\ -\widehat{M}_1 + \widehat{M}_2^T & \widehat{M}_1 + \widehat{M}_1^T - \widehat{M}_2 - \widehat{M}_2^T & -\widehat{M}_1^T + \widehat{M}_2 \\ 0 & -\widehat{M}_1 + \widehat{M}_2^T & -\widehat{M}_2 - \widehat{M}_2^T \end{bmatrix}.
$$

Using Lemma 1 (Jensen's Inequality), we have the following.

$$
\begin{aligned}
\dot{V}_6(t) \;\leq\;& \mu_M^2 y^T(t) P_6 y(t) - \int_{t-\mu_M}^{t} y^T(s) ds P_6 \int_{t-\mu_M}^{t} y(s) ds \\
& + \mu_M^2 \dot{\xi}^T(t) P_7 \dot{\xi}(t) - \int_{t-\mu_M}^{t} \dot{\xi}^T(s) ds P_7 \int_{t-\mu_M}^{t} \dot{\xi}(s) ds \\
\leq\;& \mu_M^2 y^T(t) P_6 y(t) + \mu_M^2 \dot{\xi}^T(t) P_7 \dot{\xi}(t) \\
& - \left[\int_{t-\mu(t)}^{t} y^T(s) ds + \int_{t-\mu_M}^{t-\mu(t)} y^T(s) ds \right] P_6 \left[\int_{t-\mu(t)}^{t} y^T(s) ds + \int_{t-\mu_M}^{t-\mu(t)} y^T(s) ds \right] \\
& - \left[\int_{t-\mu(t)}^{t} \dot{\xi}^T(s) ds + \int_{t-\mu_M}^{t-\mu(t)} \dot{\xi}^T(s) ds \right] P_7 \left[\int_{t-\mu(t)}^{t} \dot{\xi}^T(s) ds + \int_{t-\mu_M}^{t-\mu(t)} \dot{\xi}^T(s) ds \right].
\end{aligned}
$$

Using Lemma 8 to confront $\dot{V}_7(t)$, we can obtain the following:

$$
\begin{aligned}
\dot{V}_7(t) \;=\;& \mu_M^2 y(t) P_8 y(t) - \mu_M \int_{t-\mu_M}^{t} \dot{\xi}^T(s) P_8 \dot{\xi}(s) ds \\
& + \frac{\mu_M^2}{2} y^T(t) P_9 y(t) - \int_{t-\mu_M}^{t} \int_{u}^{t} \dot{\xi}^T(\lambda) P_{11} \dot{\xi}(\lambda) d\lambda du \\
\leq\;& \mu_M^2 y(t) P_8 y(t) + \frac{\mu_M^2}{2} y^T(t) P_9 y(t) \\
& - [\Theta_1^T P_8 \Theta_1 + 3\Theta_2^T P_8 \Theta_2 + 5\Theta_3^T P_8 \Theta_3] - [2\Theta_4^T P_9 \Theta_4 + 4\Theta_5^T P_9 \Theta_5 + 6\Theta_6^T P_9 \Theta_6],
\end{aligned}
$$

where the following is the case.

$$
\begin{aligned}
\Theta_1 &= \xi(t - \mu_M) - \xi(t), \\
\Theta_2 &= \xi(t - \mu_M) + \xi(t) - \frac{2}{\mu_M}\int_{t-\mu_M}^{t}\xi(s)ds, \\
\Theta_3 &= \xi(t - \mu_M) - \xi(t) + \frac{6}{\mu_M}\int_{t-\mu_M}^{t}\xi(s)ds - \frac{12}{\mu_M^2}\int_{t-\mu_M}^{t}\int_{u}^{t}\xi(s)dsdu, \\
\Theta_4 &= \xi(t) - \frac{1}{\mu_M}\int_{t-\mu_M}^{t}\xi(s)ds, \\
\Theta_5 &= \xi(t) + \frac{2}{\mu_M}\int_{t-\mu_M}^{t}\xi(s)ds - \frac{6}{\mu_M^2}\int_{t-\mu_M}^{t}\int_{u}^{t}\xi(s)dsdu, \\
\Theta_6 &= \xi(t) - \frac{3}{\mu_M}\int_{t-\mu_M}^{t}\xi(s)ds + \frac{24}{\mu_M^2}\int_{t-\mu_M}^{t}\int_{u}^{t}\xi^{T}(s)dsdu \\
&\quad - \frac{60}{\mu_M^3}\int_{t-\mu_M}^{t}\int_{u}^{t}\int_{s}^{t}\xi(s)drdsdu.
\end{aligned}
$$

According to Lemma 4, we can obtain $\dot{V}_8(t)$ by performing the following.

$$
\begin{aligned}
\dot{V}_8(t) &\leq \frac{\mu_M^4}{4}\xi^{T}(t)P_{10}\xi(t) - \frac{\mu_M^2}{2}\int_{t-\mu_M}^{t}\int_{u}^{t}\xi^{T}(\lambda)P_{10}\xi(\lambda)d\lambda du \\
&\quad + \frac{\mu_M^4}{2}y^{T}(t)P_{11}y(t) - \mu_M^2\int_{t-\mu_M}^{t}\int_{u}^{t}\dot{\xi}^{T}(\lambda)P_{11}\dot{\xi}(\lambda)d\lambda du \\
&\leq \frac{\mu_M^4}{4}\xi^{T}(t)P_{10}\xi(t) - \int_{t-\mu_M}^{t}\int_{u}^{t}\xi^{T}(\lambda)d\lambda du P_{10}\int_{t-\mu_M}^{t}\int_{u}^{t}\xi(\lambda)d\lambda du \\
&\quad + \frac{\mu_M^4}{2}y^{T}(t)P_{11}y(t) - 2\int_{t-\mu_M}^{t}\int_{u}^{t}\dot{\xi}^{T}(\lambda)d\lambda du P_{11}\int_{t-\mu_M}^{t}\int_{u}^{t}\dot{\xi}(\lambda)d\lambda du \\
&\leq \frac{\mu_M^4}{4}\xi^{T}(t)P_{10}\xi(t) - \int_{t-\mu_M}^{t}\int_{u}^{t}\xi^{T}(\lambda)d\lambda du P_{10}\int_{t-\mu_M}^{t}\int_{u}^{t}\xi(\lambda)d\lambda du \\
&\quad + \frac{\mu_M^4}{2}y^{T}(t)P_{11}y(t) - 2\mu_M^2\xi^{T}(t)P_{11}\xi(t) + 2\mu_M\xi^{T}(t)P_{11}\int_{t-\mu_M}^{t}\xi^{T}(u)du \\
&\quad + 2\mu_M\int_{t-\mu_M}^{t}\xi^{T}(u)du P_{11}\xi(t) - 2\int_{t-\mu_M}^{t}\xi^{T}(u)du P_{11}\int_{t-\mu_M}^{t}\xi^{T}(u)du.
\end{aligned}
$$

Using Lemmas 2 and 3, an upper bound of $V_9(t)$ can be obtained as follows.

$$
\begin{aligned}
\dot{V}_9(t) &\leq \mu_M^2 y^{T}(t)P_{12}y(t) + \mu_M^2 y^{T}(t)P_{13}y(t) \\
&\quad + \begin{bmatrix}\xi(t) \\ \xi(t-\mu_M) \\ \frac{1}{\mu_M}\int_{t-\mu_M}^{t}\xi(s)ds\end{bmatrix}^{T}\begin{bmatrix} -4P_{12} & -2P_{12} & 6P_{12} \\ -2P_{12}^{T} & -4P_{12} & 6P_{12} \\ 6P_{12}^{T} & 6P_{12}^{T} & -12P_{12}\end{bmatrix}\begin{bmatrix}\xi(t) \\ \xi(t-\mu_M) \\ \frac{1}{\mu_M}\int_{t-\mu_M}^{t}\xi(s)ds\end{bmatrix} \\
&\quad + \begin{bmatrix}\xi(t) \\ \xi(t-\mu(t)) \\ \xi(t-\mu_M)\end{bmatrix}^{T}\begin{bmatrix} -P_{13} & P_{13}-S & S \\ P_{13}^{T}-S^{T} & -2P_{13}+S+S^{T} & P_{13}-S \\ S^{T} & P_{13}^{T}-S^{T} & -P_{13}\end{bmatrix}\begin{bmatrix}\xi(t) \\ \xi(t-\mu(t)) \\ \xi(t-\mu_M)\end{bmatrix}.
\end{aligned}
$$

Taking the time derivative of $V_{10}(t)$, we have the following.

$$
\begin{aligned}
\dot{V}_{10}(t) &\leq \dot{\xi}^{T}(t)P_{14}\dot{\xi}(t) - (1-\dot{\tau}(t))\dot{\xi}^{T}(t-\tau(t))P_{14}\dot{\xi}(t-\tau(t)) + \dot{\xi}^{T}(t)P_{15}\dot{\xi}(t) \\
&\quad - \tau_M\dot{\xi}^{T}(t-\tau_M)P_{15}\dot{\xi}(t-\tau_M) \\
&\leq \dot{\xi}^{T}(t)P_{14}\dot{\xi}(t) - (1-\tau_d)\dot{\xi}^{T}(t-\tau(t))P_{14}\dot{\xi}(t-\tau(t)) + \tau_M\dot{\xi}^{T}(t)P_{15}\dot{\xi}(t) \\
&\quad - \tau_M\dot{\xi}^{T}(t-\tau_M)P_{15}\dot{\xi}(t-\tau_M).
\end{aligned}
$$

Calculating $\dot{V}_{11}(t)$ yields the following.

$$
\begin{aligned}
\dot{V}_{11}(t) &= \rho_M^2 f^T(\xi(t))P_{16}f(\xi(t)) - \rho_M \int_{t-\rho_M}^t f^T(\xi(s))ds P_{16}f(\xi(s))ds \\
&\leq \rho_M^2 f^T(\xi(t))P_{16}f(\xi(t)) - \int_{t-\rho(t)}^t f^T(\xi(s))ds P_{16} \int_{t-\rho(t)}^t f^T(\xi(s))ds.
\end{aligned}
$$

From (3), for any positive diagonal matrices $H_1 > 0$, $H_2 > 0$, the following is obtained.

$$
\begin{bmatrix} \xi(t) \\ f(\xi(t)) \end{bmatrix}^T \begin{bmatrix} -F_1 H_1 & F_2 H_1 \\ F_2^T H_1^T & -H_1 \end{bmatrix} \begin{bmatrix} \xi(t) \\ f(\xi(t)) \end{bmatrix} \geq 0, \tag{15}
$$

$$
\begin{bmatrix} \xi(t-\mu(t)) \\ f(\xi(t-\mu(t))) \end{bmatrix}^T \begin{bmatrix} -F_1 H_2 & F_2 H_2 \\ F_2^T H_2^T & -H_2 \end{bmatrix} \begin{bmatrix} \xi(t-\mu(t)) \\ f(\xi(t-\mu(t))) \end{bmatrix} \geq 0. \tag{16}
$$

Furthermore, for any real matrices Z_i, $i = 1, 2$ and O_j, $j = 1, 2, 3, \ldots, 8$ of compatible dimensions, we obtain

$$
2 \int_{t-\mu(t)}^t \dot{\xi}(s)ds Z_1^T \left[\xi(t) - \xi(t-\mu(t)) - \int_{t-\mu(t)}^t \dot{\xi}(s)ds \right] = 0, \tag{17}
$$

$$
2 \int_{t-\mu_M}^{t-\mu(t)} \dot{\xi}(s)ds Z_2^T \left[\xi(t-\mu(t)) - \xi(t-\mu_M) - \int_{t-\mu_M}^{t-\mu(t)} \dot{\xi}(s)ds \right] = 0, \tag{18}
$$

$$
\begin{aligned}
&2\left[\xi^T(t)O_1^T + \xi^T(t)O_2^T + f(\xi(t))O_3^T + f(\xi(t-\mu(t)))O_4^T \right]\left[-\dot{\xi}(t) - A\xi(t) \right. \\
&+ G_b f(\xi(t)) + G_c \dot{\xi}(t-\tau(t)) + G_d f(\xi(t-\mu(t))) + G_e \int_{t-\rho(t)}^t f(\xi(s))ds \\
&\left. + H\kappa(t) \right] 2\left[y^T(t)O_5^T + \xi^T(t)O_6^T + f(\xi(t))O_7^T + f(\xi(t-\mu(t)))O_8^T \right] \\
&\times \left[-y(t) - A\xi(t) + G_b f(\xi(t)) + G_c \dot{\xi}(t-\tau(t)) + G_d f(\xi(t-\mu(t))) \right. \\
&\left. + G_e \int_{t-\rho(t)}^t f(\xi(s))ds + H\kappa(t) \right] = 0. \tag{19}
\end{aligned}
$$

Based on (14)–(19), it is clear that the following is observed:

$$
\eta^T(t) \sum \eta(t) < 0, \tag{20}
$$

where the following is the case.

$$
\begin{aligned}
\eta(t) = &\left[\xi(t),\ y(t),\ f(\xi(t)),\ f(\xi(t-\mu(t))),\ \xi(t-\mu(t)),\ \xi(t-\mu_M), \right. \\
&\int_{t-\mu(t)}^t y(s)ds,\ \int_{t-\mu_M}^{t-\mu(t)} y(s)ds,\ f(\xi(t-\mu_M)),\ \int_{t-\mu(t)}^t \xi(s)ds,\ \int_{t-\mu_M}^{t-\mu(t)} \xi(s)ds, \\
&\dot{\xi}(t),\ \frac{1}{\mu_M}\int_{t-\mu_M}^t \xi(s)ds,\ \frac{1}{\mu_M^2}\int_{t-\mu_M}^t \int_{t-\mu_M}^t \xi(s)ds,\ \frac{1}{\mu_M^3}\int_{t-\mu_M}^t \int_{t-\mu_M}^t \int_{t-\mu_M}^t \xi(s)ds, \\
&\int_{t-\mu_M}^t \xi(u)du,\ \int_{-\mu_M}^t \int_u^t \xi(\lambda)d\lambda du,\ \int_{t-\mu(t)}^t \dot{\xi}(s)ds,\ \int_{t-\mu_M}^{t-\mu(t)} \dot{\xi}(s)ds,\ \dot{\xi}(t-\tau_M), \\
&\left. \dot{\xi}(t-\tau(t)),\ \int_{t-\rho(t)}^t f(\xi(s))ds,\ \kappa(t) \right].
\end{aligned}
$$

Then, $\alpha > 0$ and we are able to obtain the following.

$$
\dot{V}(t) - \alpha V(t) - \alpha \kappa^T(t)\kappa(t) \leq \zeta^T(t) \sum \zeta(t). \tag{21}
$$

By multiplying the above inequality by $e^{\alpha t}$, we can obtain the following.

$$\frac{d}{dt}[e^{-\alpha t}V(t)] \le \alpha e^{-\alpha t}\kappa^T(t)\kappa(t). \tag{22}$$

Integrating the two sides of the inequality (22) from 0 to t, with $t \in [0, T]$, we have obtained the following.

$$V(t) \le e^{\alpha t}V(0) + \alpha e^{\alpha t}\int_0^t e^{-\alpha s}\kappa^T(t)\kappa(t)ds. \tag{23}$$

They include the following.

$$
\begin{aligned}
V(0) \;=\; & \xi^T(0)P_1\xi(0) + 2\sum_{i=1}^N k_i \int_0^{\xi_i(0)}(f_i(s) - F^- s)ds + \zeta^T(0)GP_2\zeta(0) \\
& + 2\sum_{i=1}^N w_i \int_0^{\xi_i(0)}(F^+ s - f_i(s))ds + \int_{0-\mu_M}^0 \xi^T(s)P_3\xi(s)ds \\
& + \int_{\mu(t)}^0 \begin{bmatrix} \xi(s) \\ f(\xi(s)) \end{bmatrix}^T \begin{bmatrix} R_1 & R_2 \\ * & R_3 \end{bmatrix} \begin{bmatrix} \xi(s) \\ f(\xi(s)) \end{bmatrix} ds \\
& + \int_{-\mu_M}^0 \begin{bmatrix} \xi(s) \\ f(\xi(s)) \end{bmatrix}^T \begin{bmatrix} R_4 & R_5 \\ * & R_6 \end{bmatrix} \begin{bmatrix} \xi(s) \\ f(\xi(s)) \end{bmatrix} ds \\
& + \mu_M \int_{-\mu_M}^0 \int_s^0 \begin{bmatrix} \xi(\theta) \\ y(\theta) \end{bmatrix}^T \begin{bmatrix} R_7 & R_8 \\ * & R_9 \end{bmatrix} \begin{bmatrix} \xi(\theta) \\ y(\theta) \end{bmatrix} d\theta ds \\
& + \mu_M \int_{-\mu_M}^0 \int_s^0 \xi^T(\theta)P_4\xi(\theta)d\theta ds \\
& + \int_{-\mu_M}^0 \int_s^0 y^T(\theta)P_5 y(\theta)d\theta ds + \mu_M \int_{-\mu_M}^0 \int_s^0 y^T(\theta)P_6 y(\theta)d\theta ds \\
& + \mu_M \int_{-\mu_M}^0 \int_s^0 y^T(\theta)P_7 y(\theta)d\theta ds + \mu_M \int_{-\mu_M}^0 \int_s^0 y^T(\theta)P_8 y(\theta)d\theta ds \\
& + \mu_M \int_{-\mu_M}^0 \int_\lambda^0 \int_s^0 \xi^T(\theta)P_9\xi(\theta)d\theta ds d\lambda \\
& + \frac{(\mu_M)^2}{2}\int_{-\mu_M}^0 \int_\lambda^0 \int_s^0 \xi^T(\theta)P_{10}\xi(\theta)d\theta ds d\lambda \\
& + \frac{(\mu_M)^2}{2}\int_{-\mu_M}^0 \int_\lambda^0 \int_s^0 y^T(\theta)P_{11}y(\theta)d\theta ds d\lambda \\
& + \mu_M \int_{-\mu_M}^0 \int_s^0 y^T(\theta)P_{12}y(\theta)d\theta ds \\
& + \mu_M \int_{-\mu_M}^0 \int_s^0 y^T(\theta)P_{13}y(\theta)d\theta ds + \int_{\tau(t)}^0 \xi^T(s)P_{14}\xi(s)ds \\
& + \tau_M \int_{-\tau_M}^0 \xi^T(s)P_{15}\xi(s)ds + \rho_M \int_{-\rho_M}^0 \int_s^0 f(\xi(\theta))^T P_{16}f(\xi(\theta))d\theta ds.
\end{aligned}
$$

Note that $\widetilde{P} = L^{-\frac{1}{2}} P_i L^{-\frac{1}{2}}; i = 1, 2, 3, \ldots, 13$, $\widetilde{R} = L^{-\frac{1}{2}} R_i L^{-\frac{1}{2}}; i = 1, 2, 3, \ldots, 9$ and the following relationship can be found.

$$
\begin{aligned}
V(0) \;=\; & \xi^T(0) L^{\frac{1}{2}} \widetilde{P}_1 L^{\frac{1}{2}} \xi(0) + 2\widetilde{W}_1 f(\xi^T(0)) + \xi^T(0) L^{\frac{1}{2}} G \widetilde{P}_2 L^{\frac{1}{2}} \zeta(0) 2\widetilde{W}_2 f(\xi^T(0)) \\
& + \int_{0-\mu_M}^{0} \xi^T(s) L^{\frac{1}{2}} \widetilde{P}_3 L^{\frac{1}{2}} \xi(s)\, ds \\
& + \int_{\mu(t)}^{0} \begin{bmatrix} \xi(s) \\ f(\xi(s)) \end{bmatrix}^T \begin{bmatrix} L^{\frac{1}{2}} \widetilde{R}_1 L^{\frac{1}{2}} & L^{\frac{1}{2}} \widetilde{R}_2 L^{\frac{1}{2}} \\ L^{\frac{1}{2}} \widetilde{R}_2^T L^{\frac{1}{2}} & L^{\frac{1}{2}} \widetilde{R}_3 L^{\frac{1}{2}} \end{bmatrix} \begin{bmatrix} \xi(s) \\ f(\xi(s)) \end{bmatrix} ds \\
& + \int_{-\mu_M}^{0} \begin{bmatrix} \xi(s) \\ f(\xi(s)) \end{bmatrix}^T \begin{bmatrix} L^{\frac{1}{2}} \widetilde{R}_4 L^{\frac{1}{2}} & L^{\frac{1}{2}} \widetilde{R}_5 L^{\frac{1}{2}} \\ L^{\frac{1}{2}} \widetilde{R}_5^T L^{\frac{1}{2}} & L^{\frac{1}{2}} \widetilde{R}_6 L^{\frac{1}{2}} \end{bmatrix} \begin{bmatrix} \xi(s) \\ f(\xi(s)) \end{bmatrix} ds \\
& + \mu_M \int_{-\mu_M}^{0} \int_{s}^{0} \begin{bmatrix} \xi(\theta) \\ y(\theta) \end{bmatrix}^T \begin{bmatrix} L^{\frac{1}{2}} \widetilde{R}_7 L^{\frac{1}{2}} & L^{\frac{1}{2}} \widetilde{R}_8 L^{\frac{1}{2}} \\ L^{\frac{1}{2}} \widetilde{R}_8^T L^{\frac{1}{2}} & L^{\frac{1}{2}} \widetilde{R}_9 L^{\frac{1}{2}} \end{bmatrix} \begin{bmatrix} \xi(\theta) \\ y(\theta) \end{bmatrix} d\theta\, ds \\
& + \mu_M \int_{-\mu_M}^{0} \int_{s}^{0} \xi^T(\theta) L^{\frac{1}{2}} \widetilde{P}_4 L^{\frac{1}{2}} \xi(\theta)\, d\theta\, ds \\
& + \int_{-\mu_M}^{0} \int_{s}^{0} y^T(\theta) L^{\frac{1}{2}} \widetilde{P}_5 L^{\frac{1}{2}} y(\theta)\, d\theta\, ds + \mu_M \int_{-\mu_M}^{0} \int_{s}^{0} y^T(\theta) L^{\frac{1}{2}} \widetilde{P}_6 L^{\frac{1}{2}} y(\theta)\, d\theta\, ds \\
& + \mu_M \int_{-\mu_M}^{0} \int_{s}^{0} y^T(\theta) L^{\frac{1}{2}} \widetilde{P}_7 L^{\frac{1}{2}} y(\theta)\, d\theta\, ds + \mu_M \int_{-\mu_M}^{0} \int_{s}^{0} y^T(\theta) L^{\frac{1}{2}} \widetilde{P}_8 L^{\frac{1}{2}} y(\theta)\, d\theta\, ds \\
& + \mu_M \int_{-\mu_M}^{0} \int_{\lambda}^{0} \int_{s}^{0} \xi^T(\theta) L^{\frac{1}{2}} \widetilde{P}_9 L^{\frac{1}{2}} \xi(\theta)\, d\theta\, ds\, d\lambda \\
& + \frac{(\mu_M)^2}{2} \int_{-\mu_M}^{0} \int_{\lambda}^{0} \int_{s}^{0} \xi^T(\theta) L^{\frac{1}{2}} \widetilde{P}_{10} L^{\frac{1}{2}} \xi(\theta)\, d\theta\, ds\, d\lambda \\
& + \frac{(\mu_M)^2}{2} \int_{-\mu_M}^{0} \int_{\lambda}^{0} \int_{s}^{0} y^T(\theta) L^{\frac{1}{2}} \widetilde{P}_{11} L^{\frac{1}{2}} y(\theta)\, d\theta\, ds\, d\lambda \\
& + \mu_M \int_{-\mu_M}^{0} \int_{s}^{0} y^T(\theta) L^{\frac{1}{2}} \widetilde{P}_{12} L^{\frac{1}{2}} y(\theta)\, d\theta\, ds \\
& + \mu_M \int_{-\mu_M}^{0} \int_{s}^{0} y^T(\theta) L^{\frac{1}{2}} \widetilde{P}_{13} L^{\frac{1}{2}} y(\theta)\, d\theta\, ds + \int_{\tau(t)}^{0} \xi^T(s) L^{\frac{1}{2}} \widetilde{P}_{14} L^{\frac{1}{2}} \xi(s)\, ds \\
& + \tau_M \int_{-\tau_M}^{0} \xi^T(s) L^{\frac{1}{2}} \widetilde{P}_{15} L^{\frac{1}{2}} \xi(s)\, ds + \rho_M \int_{-\rho_M}^{0} \int_{s}^{0} f(\xi(\theta))^T L^{\frac{1}{2}} \widetilde{P}_{16} L^{\frac{1}{2}} f(\xi(\theta))\, d\theta\, ds, \\
\leq\; & [\lambda_{\max}(\widetilde{P}_1 + \widetilde{P}_2) + 2\lambda_{\max}(K + W) + \mu_M \lambda_{\max}(\widetilde{P}_3 + \widetilde{R}_1 + \widetilde{R}_2 + \widetilde{R}_2^T + \widetilde{R}_3 + \widetilde{R}_4 \\
& + \widetilde{R}_5 + \widetilde{R}_5^T + \widetilde{R}_6) + \frac{\mu_M^3}{2} \lambda_{max}(\widetilde{P}_4 + \widetilde{P}_5 + \widetilde{P}_6 + \widetilde{P}_7 + \widetilde{R}_7 + \widetilde{R}_8 + \widetilde{R}_8^T + \widetilde{R}_9 + \widetilde{P}_8 \\
& + \widetilde{P}_{12} + \widetilde{P}_{13}) + \frac{\mu_M^5}{12} \lambda_{max}(\widetilde{P}_9 + \widetilde{P}_{10}) + \tau_M \lambda_{max}(\widetilde{P}_{14}) + \tau_M^2 \lambda_{max}(\widetilde{P}_{15}) \\
& + \frac{\rho_M^3}{2} \lambda_{max}(\widetilde{P}_{16})] \times \sup_{-\mu_M \leq t_0 \leq 0} \{\xi^T(t_0) L \xi(t_0), \dot{\xi}^T(t_0) L \dot{\xi}(t_0)\}, \\
\leq\; & \Lambda g_1.
\end{aligned}
$$

We have the following:

$$
\begin{aligned}
e^{\alpha t} V(0) + \alpha e^{\alpha t} \int_0^t e^{-\alpha s} \kappa^T(t) \kappa(t)\, ds \;\leq\;& e^{\alpha t} \Lambda g_1 + \alpha e^{\alpha t} \int_0^t e^{-\alpha s} \kappa^T(s) \kappa(s)\, ds, \\
\leq\;& e^{\alpha T} \Lambda g_1 + e^{\alpha T} \delta(1 - e^{-\alpha T}), \\
\leq\;& e^{\alpha T} [\Lambda g_1 + \delta(1 - e^{-\alpha T})],
\end{aligned}
\tag{24}
$$

where the following is the case.

$$\begin{aligned}
\Lambda \;=\; & \lambda_2 + \lambda_4 + 2(\lambda_3 + \lambda_5) + \mu_M \lambda_{\max}(\lambda_6 + \lambda_7 + \lambda_8 + \lambda_9 + \lambda_{10} + \lambda_{11} + \lambda_{12} \\
& + \lambda_{13} + \lambda_{14}) + \frac{\mu_M^2}{2}\lambda_{20} + \frac{\mu_M^3}{2}(\lambda_{15} + \lambda_{16} + \lambda_{17} + \lambda_{18} + \lambda_{19} + \lambda_{21} + \lambda_{22} \\
& + \lambda_{23} + \lambda_{27} + \lambda_{28}) + \frac{\mu_M^4}{6}\lambda_{24} + \frac{\mu_M^5}{12}(\lambda_{25} + \lambda_{26}) + \tau_M(\lambda_{29}) + \tau_M^2(\lambda_{30}) \\
& + \frac{\rho_M^3}{2}(\lambda_{31}).
\end{aligned} \tag{25}$$

On the other hand, the following condition holds.

$$V(t) \geq \xi^T(t) P_1 \xi(t) \geq \lambda_{min}(\widetilde{P}_1)\xi^T(t)L\xi(t) = \lambda_1 \xi^T(t)L\xi(t). \tag{26}$$

From Equations (24) and (27), we obtain the following.

$$\xi^T(t)L\xi(t) \leq \frac{e^{\alpha T}\left[\Lambda g_1 + \delta(1 - e^{-\alpha T})\right]}{\lambda_1}. \tag{27}$$

Condition $[\lambda_1 g_2 e^{-\alpha T} > \Lambda g_1 + \delta(1 - e^{-\alpha T})]$ indicates that for $\forall t \in [0, T]$, $\xi^T(t)L\xi(t) < g_2$. From Definition 2, system (5) is finite-time bounded with regard to $(g_1,\ g_2,\ T,\ L,\ \delta)$. The proof is now finished. $\square$

Remark 1. *Condition (9) is not a standard form of LMIs. In order to verify that this condition is equivalent to the relation of LMIs, let $\lambda_i,\ i = 1, 2, 3, \ldots, 31$ be some positive scalars with the following.*

$$\begin{array}{llll}
\lambda_1 I \leq \widetilde{P}_1 \leq \lambda_2 I, & 0 \leq \widetilde{W}_1 \leq \lambda_3 I, & 0 \leq \widetilde{P}_2 \leq \lambda_4 I, & 0 \leq \widetilde{W}_2 \leq \lambda_5 I, \\
0 \leq \widetilde{P}_3 \leq \lambda_6 I, & 0 \leq \widetilde{R}_1 \leq \lambda_7 I, & 0 \leq \widetilde{R}_2 \leq \lambda_8 I, & 0 \leq \widetilde{R}_2^T \leq \lambda_9 I, \\
0 \leq \widetilde{R}_3 \leq \lambda_{10} I, & 0 \leq \widetilde{R}_4 \leq \lambda_{11} I, & 0 \leq \widetilde{R}_5 \leq \lambda_{12} I, & 0 \leq \widetilde{R}_5^T \leq \lambda_{13} I, \\
0 \leq \widetilde{R}_6 \leq \lambda_{14} I, & 0 \leq \widetilde{R}_7 \leq \lambda_{15} I, & 0 \leq \widetilde{R}_8 \leq \lambda_{16} I, & 0 \leq \widetilde{R}_8^T \leq \lambda_{17} I, \\
0 \leq \widetilde{R}_9 \leq \lambda_{18} I, & 0 \leq \widetilde{P}_4 \leq \lambda_{19} I, & 0 \leq \widetilde{P}_5 \leq \lambda_{20} I, & 0 \leq \widetilde{P}_6 \leq \lambda_{21} I, \\
0 \leq \widetilde{P}_7 \leq \lambda_{22} I, & 0 \leq \widetilde{P}_8 \leq \lambda_{23} I, & 0 \leq \widetilde{P}_9 \leq \lambda_{24} I, & 0 \leq \widetilde{P}_{10} \leq \lambda_{25} I, \\
0 \leq \widetilde{P}_{11} \leq \lambda_{26} I, & 0 \leq \widetilde{P}_{12} \leq \lambda_{27} I, & 0 \leq \widetilde{P}_{13} \leq \lambda_{28} I. & 0 \leq \widetilde{P}_{14} \leq \lambda_{29} I, \\
0 \leq \widetilde{P}_{15} \leq \lambda_{30} I. & 0 \leq \widetilde{P}_{16} \leq \lambda_{31} I.
\end{array}$$

Consider the following.

$$\begin{array}{llllllll}
\lambda_1 & = & \lambda_{min}(\widetilde{P}_1), & \lambda_2 & = & \lambda_{max}(\widetilde{P}_1), & \lambda_3 & = & \lambda_{max}(\widetilde{W}_1), & \lambda_4 & = & \lambda_{max}(\widetilde{P}_2), \\
\lambda_5 & = & \lambda_{max}(\widetilde{W}_2), & \lambda_6 & = & \lambda_{max}(\widetilde{P}_3), & \lambda_7 & = & \lambda_{max}(\widetilde{R}_1), & \lambda_8 & = & \lambda_{max}(\widetilde{R}_2), \\
\lambda_9 & = & \lambda_{max}(\widetilde{R}_2^T), & \lambda_{10} & = & \lambda_{max}(\widetilde{R}_3), & \lambda_{11} & = & \lambda_{max}(\widetilde{R}_4), & \lambda_{12} & = & \lambda_{max}(\widetilde{R}_5), \\
\lambda_{13} & = & \lambda_{max}(\widetilde{R}_5^T), & \lambda_{14} & = & \lambda_{max}(\widetilde{R}_6), & \lambda_{15} & = & \lambda_{max}(\widetilde{R}_7), & \lambda_{16} & = & \lambda_{max}(\widetilde{R}_8), \\
\lambda_{17} & = & \lambda_{max}(\widetilde{R}_8^T), & \lambda_{18} & = & \lambda_{max}(\widetilde{R}_9), & \lambda_{19} & = & \lambda_{max}(\widetilde{P}_4), & \lambda_{20} & = & \lambda_{max}(\widetilde{P}_5), \\
\lambda_{21} & = & \lambda_{max}(\widetilde{P}_6), & \lambda_{22} & = & \lambda_{max}(\widetilde{P}_7), & \lambda_{23} & = & \lambda_{max}(\widetilde{P}_8), & \lambda_{24} & = & \lambda_{max}(\widetilde{P}_9), \\
\lambda_{25} & = & \lambda_{max}(\widetilde{P}_{10}), & \lambda_{26} & = & \lambda_{max}(\widetilde{P}_{11}), & \lambda_{27} & = & \lambda_{max}(\widetilde{P}_{12}), & \lambda_{28} & = & \lambda_{max}(\widetilde{P}_{13}), \\
\lambda_{29} & = & \lambda_{max}(\widetilde{P}_{14}), & \lambda_{30} & = & \lambda_{max}(\widetilde{P}_{15}), & \lambda_{31} & = & \lambda_{max}(\widetilde{P}_{16}).
\end{array}$$

3.2. Finite-Time Stability Analysis

Remark 2. *If there is an external disruption $\kappa(t) = 0$, system (5) changes into the following.*

$$\left.\begin{aligned}
\dot{\xi}(t) - G_c\dot{\xi}(t - \tau(t)) &= -A\xi(t) + G_b f(\xi(t)) + G_d f(\xi(t - \mu(t))) \\
&\quad + G_e \int_{t-\rho(t)}^{t} f(\xi(s))ds, \\
\xi(t) &= \phi(t), \qquad t \in [-\bar{h}, 0].
\end{aligned}\right\} \tag{28}$$

By (8), we provide additional notation for finite-time stability analysis for (28).

$$\widetilde{\Sigma} = \left[\Pi_{(i,j)}\right]_{22 \times 22}. \tag{29}$$

We obtain that $\Pi_{(1,1)} - \Pi_{(22,22)}$ is the same as in Theorem 1. Then, we define the following:

$$
\begin{aligned}
\widetilde{\eta}(t) = \Big[& \xi(t),\ y(t),\ f(\xi(t)),\ f(\xi(t-\mu(t))),\ \xi(t-\mu(t)),\ \xi(t-\mu_M), \\
& \int_{t-\mu(t)}^{t} y(s)ds,\ \int_{t-\mu_M}^{t-\mu(t)} y(s)ds,\ f(\xi(t-\mu_M)),\ \int_{t-\mu(t)}^{t} \xi(s)ds,\ \int_{t-\mu_M}^{t-\mu(t)} \xi(s)ds, \\
& \dot{\xi}(t),\ \frac{1}{\mu_M}\int_{t-\mu_M}^{t} \xi(s)ds,\ \frac{1}{\mu_M^2}\int_{t-\mu_M}^{t}\int_{t-\mu_M}^{t} \xi(s)ds,\ \frac{1}{\mu_M^3}\int_{t-\mu_M}^{t}\int_{t-\mu_M}^{t}\int_{t-\mu_M}^{t} \xi(s)ds, \\
& \int_{t-\mu_M}^{t} \xi(u)du,\ \int_{-\mu_M}^{t}\int_{u}^{t} \xi(\lambda)d\lambda du,\ \int_{t-\mu(t)}^{t} \dot{\xi}(s)ds,\ \int_{t-\mu_M}^{t-\mu(t)} \dot{\xi}(s)ds, \\
& \dot{\xi}(t-\tau(t)),\ \int_{t-\rho(t)}^{t} f(\xi(s))ds \Big],
\end{aligned}
$$

and construct a new theorem that follows Corollary 1.

Corollary 1. *For $\| C \| < 1$, system (28) with $\kappa(t) = 0$ is finite-time stable if there exist positive symmetric matrices $P_i,\ R_j,\ i = 1, 2, 3, \ldots, 16,\ j = 1, 2, 3, \ldots, 9$ any appropriate matrices $S,\ P_{13},\ R_8,\ Q_k,\ R_7 \geq 0, Z_l,\ l = 1, 2$ and $O_e,\ e = 1, 2, 3, \ldots, 8,\ \begin{bmatrix} R_{n+3n} & R_{2+3n} \\ R_{2+3n}^T & R_{3+3n} \end{bmatrix} \geq 0,\ \begin{bmatrix} P_{13} & S \\ * & P_{13} \end{bmatrix} \geq 0,$ where $n = 0, 1, 2,\ k = 1, 2, \ldots, 14,$ positive diagonal matrices are $H_p, W_p,$ $p = 1, 2$ and positive real constants are $\mu_M,\ \rho_M,\ \mu_d,\ \tau_M,\ \tau_d,\ \alpha,\ g_1,\ g_2,\ T$ such that the following symmetric linear matrix inequality holds:*

$$
\begin{bmatrix} P_5 & M_1 & M_2 \\ * & M_3 & M_4 \\ * & * & M_5 \end{bmatrix} \geq 0, \tag{30}
$$

$$
\widetilde{\Sigma} < 0, \tag{31}
$$

$$
\lambda_1 g_2 e^{-\alpha T} > \Lambda g_1, \tag{32}
$$

where $\kappa(t) = 0$ as described in Theorem 1.

Proof. Since the proof is identical to that of Theorem 1, it is excluded from this section. $\square$

3.3. Finite-Time Passivity Analysis

This section discusses the topic of finite-time passivity analysis investigated for the following system.

$$
\left.
\begin{aligned}
\dot{\xi}(t) - G_c \dot{\xi}(t-\tau(t)) &= -A\xi(t) + G_b f(\xi(t)) + G_d f(\xi(t-\mu(t))) + H\kappa(t) \\
&\quad + G_e \int_{t-\rho(t)}^{t} f(\xi(s))ds, \\
z(t) &= G_1 f(\xi(t)) + G_2 \kappa(t), \quad t \in R^+ \\
\xi(t) &= \phi(t), \quad t \in [-\bar{h}, 0].
\end{aligned}
\right\} \tag{33}
$$

Theorem 2. *For $\| C \| < 1$, system (43) is finite-time passivity if there exist positive symmetric matrices $P_i,\ R_j,\ G_t,\ i = 1, 2, 3, \ldots, 16,\ j = 1, 2, 3, \ldots, 9,\ t = 1, 2$ any appropriate matrices $S,\ P_{13},\ R_8,\ Q_k,\ R_7 \geq 0, Z_l,\ l = 1, 2$ and $O_e,\ e = 1, 2, 3, \ldots, 8,\ \begin{bmatrix} R_{n+3n} & R_{2+3n} \\ R_{2+3n}^T & R_{3+3n} \end{bmatrix} \geq 0,\ \begin{bmatrix} P_{13} & S \\ * & P_{13} \end{bmatrix} \geq 0,$ where $n = 0, 1, 2,\ k = 1, 2, \ldots, 14,$ positive diagonal matrices are $H_p, W_p,$*

$p = 1, 2$ *and positive real constants are* μ_M, ρ_M, μ_d, τ_M, τ_d, α, δ, β, g_1, g_2, T *such that the following symmetric linear matrix inequality holds:*

$$\begin{bmatrix} P_5 & M_1 & M_2 \\ * & M_3 & M_4 \\ * & * & M_5 \end{bmatrix} \geq 0, \tag{34}$$

$$\widehat{\Sigma} = \left[\widehat{\Pi}_{(i,j)} \right]_{23 \times 23} < 0, \tag{35}$$

$$\lambda_1 g_2 e^{-\alpha T} > \Lambda g_1 + \delta(1 - e^{-\alpha T}), \tag{36}$$

where $\widehat{\Pi}_{(i,j)} = \Pi_{(i,j)}$, $i, j = 1, 2, \ldots, 23$ *except* $\widehat{\Pi}_{4,19} = \Pi_{3,23} - G_1^T, \widehat{\Pi}_{23,3} = \Pi_{19,4} - G_1, \widehat{\Pi}_{23,23} = -\beta I - G_2^T - G_2$.

Proof. The following function is defined using the same Lyapunov–Krasovskii function as Theorem 1.

$$\dot{V}(t) - [\alpha V(t) + 2\kappa^T(t)z(t) - \beta\kappa^T(t)\kappa(t)] \leq \eta^T(t)\widehat{\Sigma}\eta(t). \tag{37}$$

$\widehat{\Sigma}$ is show in (35), and then the following is the case.

$$\dot{V}(t) - \alpha V(t) \leq 2\kappa^T(t)z(t) - \beta\kappa^T(t)\kappa(t). \tag{38}$$

Then, multiplying (38) by $e^{-\alpha T}$ and integrating it between 0 and T, we can obtain the following:

$$\begin{aligned} V(t)e^{-\alpha T} &\leq 2\int_0^T e^{-\alpha t}\kappa^T(t)z(t)dt - \beta\int_0^T e^{-\alpha t}\kappa^T(t)\kappa(t)dt, \\ &\leq 2\int_0^T \kappa^T(t)z(t)dt - \beta e^{-\alpha T}\int_0^T \kappa^T(t)\kappa(t)dt, \end{aligned}$$

which implies the following.

$$V(t) \leq 2e^{\alpha T}\int_0^T \kappa^T(t)z(t)dt - \beta\int_0^T \kappa^T(t)\kappa(t)dt. \tag{39}$$

Due to $V(t) \geq 0$, it is reasonable to obtain it from (39) and the following:

$$\int_0^T \kappa^T(t)z(t)dt \geq \gamma\int_0^T \kappa^T(t)\kappa(t), \tag{40}$$

where $\gamma = \frac{\beta e^{-\alpha T}}{2}$. As a result, we may infer that system (33) is finite-time passive. This completes the proof. $\square$

Remark 3. *When* $E = 0$, $C = 0$ *and* $H = 0$ *system* (5) *changes to delayed neural network, the following is the case.*

$$\dot{\xi}(t) = -A\xi(t) + G_b f(\xi(t)) + G_a f(\xi(t - \mu(t))). \tag{41}$$

By (8), *we consider system* (41) *without finite-time stability condition and same proof line of Theorem* 1. *Moreover, the system is said to be asymptotically stable:*

$$\bar{\Sigma} = \left[\bar{\Pi}_{(i,j)} \right]_{19 \times 19}, \tag{42}$$

where $\bar{\Pi}_{12,12} = \Pi_{12,12} - P_{14} - \tau_M P_{15}$, $\bar{\Pi}_{4,4} = \Pi_{3,3} - \rho_M^2 P_{16}$, *and the parameters are as defined in Theorem* 1. *Then, we define the following.*

$$\bar{\eta}(t) = \Big[\xi(t),\, y(t),\, f(\xi(t)),\, f(\xi(t-\mu(t))),\, \xi(t-\mu(t)),\, \xi(t-\mu_M),\, \int_{t-\mu(t)}^{t} y(s)ds,$$

$$\int_{t-\mu_M}^{t-\mu(t)} y(s)ds,\, f(\xi(t-\mu_M)),\, \int_{t-\mu(t)}^{t} \xi(s)ds,\, \int_{t-\mu_M}^{t-\mu(t)} \xi(s)ds,\, \dot{\xi}(t),\, \frac{1}{\mu_M}\int_{t-\mu_M}^{t} \xi(s)ds,$$

$$\frac{1}{\mu_M^2}\int_{t-\mu_M}^{t}\int_{t-\mu_M}^{t} \xi(s)ds,\, \frac{1}{\mu_M^3}\int_{t-\mu_M}^{t}\int_{t-\mu_M}^{t}\int_{t-\mu_M}^{t} \xi(s)ds,\, \int_{t-\mu_M}^{t} \xi(u)du,$$

$$\int_{-\mu_M}^{t}\int_{u}^{t} \xi(\lambda)d\lambda du,\, \int_{t-\mu(t)}^{t} \dot{\xi}(s)ds,\, \int_{t-\mu_M}^{t-\mu(t)} \dot{\xi}(s)ds\Big].$$

4. Numerical Examples

Simulation examples are provided in this part to show the feasibility and efficiency of theoretic solutions. Five examples are given in this part to demonstrate the key theoretical conclusions that have been offered.

Example 1. *Consider the following matrix parameters for the neutral-type neural networks:*

$$\dot{\xi}(t) - G_c\dot{\xi}(t-\tau(t)) = -A\xi(t) + G_b f(\xi(t)) + G_d f(\xi(t-\mu(t))) + H\kappa(t)$$
$$+ G_e\int_{t-\rho(t)}^{t} f(\xi(s))ds,$$

with the following.

$$A = \begin{bmatrix} 3.6 & 0 \\ 0 & 3.6 \end{bmatrix},\quad G_b = \begin{bmatrix} -0.34 & 0 \\ -0.1 & -0.1 \end{bmatrix},\quad G_d = \begin{bmatrix} 0.1 & 0.2 \\ -0.15 & -0.18 \end{bmatrix},$$

$$G_c = \begin{bmatrix} -0.5 & 0 \\ 0.2 & 0.5 \end{bmatrix},\quad H = \begin{bmatrix} 0.41 & 0.5 \\ 0.69 & -0.31 \end{bmatrix}.$$

Let the following be the case:

$$\begin{array}{llll} \tau_M = 0.2, & g_1 = 0.4, & T = 6, & \rho_M = 0.1, \\ \alpha = 0.10, & \delta = 0.005, & \mu_d = 0.5, & \tau_d = 0.2, \end{array}$$

and $\mu_M = 1.3, F_1 = \mathbf{diag}\{0,0\}, F_2 = \mathbf{diag}\{1,1\}$. *Using the MATHLAB tools to solve LMIs (8) and (9), we may obtain* $g_2 = 7.8794$, *indicating that the neutral system under consideration is finite-time bounded. The activation function is described by* $f(\xi(t)) = 2|\cos(t)|$, *and we allow discrete time-varying delays to satisfy* $\mu(t) = 0.8 + 0.5|\sin(t)|$, $\rho(t) = 0.1|\sin(t)|$ *and* $\tau(t) = 0.2|\cos(t)|$.

Example 2. *Consider the following matrix parameters for the neutral-type neural networks matrix parameters:*

$$\dot{\xi}(t) - G_c\dot{\xi}(t-\tau(t)) = -A\xi(t) + G_b f(\xi(t)) + G_d f(\xi(t-\mu(t))) + H\kappa(t)$$
$$+ G_e\int_{t-\rho(t)}^{t} f(\xi(s))ds,$$
$$z(t) = G_1 f(\xi(t)) + G_2\kappa(t),$$

with the following.

$$A = \begin{bmatrix} 1.5 & 0 \\ 0 & 1.5 \end{bmatrix},\quad G_b = \begin{bmatrix} 1.1 & 0.2 \\ -0.1 & -1.1 \end{bmatrix},\quad G_d = \begin{bmatrix} 0.2 & 0 \\ 0.2 & -0.2 \end{bmatrix},$$

$$G_c = \begin{bmatrix} -0.5 & 0.3 \\ 0.2 & 0.1 \end{bmatrix},\quad H = \begin{bmatrix} 0.4 & -0.2 \\ 0.3 & -0.14 \end{bmatrix},\quad G_1 = \begin{bmatrix} 0.1 & 0.2 \\ -0.01 & 0.4 \end{bmatrix},$$

$$G_2 = \begin{bmatrix} 0.2 & -0.6 \\ 0.3 & 0.2 \end{bmatrix}.$$

Let the following be the case:

$$\begin{aligned} \mu_M &= 2.4, & \tau_M &= 1.2, & g_1 &= 0.5, & T &= 5, & g_2 &= 6, \\ \alpha &= 0.10, & \delta &= 1, & \mu_d &= 0.9, & \tau_d &= 0.2, & \rho_M &= 0.1, \end{aligned}$$

then $F_1 = \mathbf{diag}\{0,0\}, F_2 = \mathbf{diag}\{0.5,0.9\}$. Using the MATHLAB tools to solve LMIs (35) and (36), we may obtain $\gamma = 17.4493$, indicating that the neutral system under consideration is finite-time passive. The activation function is described by $f(\xi(t)) = [0.5|sin(t)|, 0.9|cos(t)|]$, and we allow discrete time-varying delays to satisfy $\mu(t) = 0.1 + 0.1|sin(t)|$, $\rho(t) = 1.1|sin(t)|$ and $\tau(t) = 1 + 0.2|cos(t)|$.

Example 3. *Consider the following matrix parameters for the neutral-type neural networks:*

$$\dot{\xi}(t) - G_c\dot{\xi}(t - \tau(t)) = -A\xi(t) + G_b f(\xi(t)) + G_d f(\xi(t - \mu(t))) + G_e \int_{t-\rho(t)}^{t} f(\xi(s))ds,$$

with the following.

$$A = \begin{bmatrix} 4 & 0 \\ 0 & 4 \end{bmatrix}, \qquad G_b = \begin{bmatrix} 1.3 & 0.4 \\ 0.9 & 0.2 \end{bmatrix}, \qquad G_d = \begin{bmatrix} 0.6 & 0.2 \\ 0.3 & -0.3 \end{bmatrix},$$

$$G_c = \begin{bmatrix} -0.5 & 0 \\ 0.2 & 0.5 \end{bmatrix}, \qquad H = \begin{bmatrix} 0.41 & 0.5 \\ 0.69 & -0.31 \end{bmatrix}, \qquad E = \begin{bmatrix} 0.4 & -0.2 \\ 0.3 & -0.3 \end{bmatrix}.$$

Let the following be the case:

$$\begin{aligned} \tau_M &= 0.2, & g_1 &= 3, & T &= 5, & \rho_M &= 1.1, \\ \alpha &= 0.001, & \delta &= 0.005, & \mu_d &= 0.1, & \tau_d &= 0.1, \end{aligned}$$

and $\mu_M = 0.1$, $F_1 = \mathbf{diag}\{0,0\}, F_2 = \mathbf{diag}\{2,2\}$. Using the MATHLAB tools to solve LMIs (8) and (9), we may obtain $g_2 = 0.5996$, indicating that the neutral system under consideration is finite-time stable. The activation function is described by $f(\xi(t)) = 4|cos(t)|$, and we allow discrete time-varying delays to satisfy $\mu(t) = 0.8 + 0.5|sin(t)|$, $\rho(t) = 0.1|sin(t)|$ and $\tau(t) = 0.1 + 0.1|cos(t)|$.

Example 4. *Consider the following matrix parameters for the neural networks matrix parameters:*

$$\dot{\xi}(t) = -A\xi(t) + G_b f(\xi(t)) + G_d f(\xi(t - \mu(t))),$$

with the following:

$$A = \begin{bmatrix} 2 & 0 \\ 0 & 2 \end{bmatrix}, \qquad G_b = \begin{bmatrix} 1 & 1 \\ -1 & -1 \end{bmatrix}, \qquad G_d = \begin{bmatrix} 0.88 & 1 \\ 1 & 1 \end{bmatrix}.$$

then $F_1 = \mathbf{diag}\{0,0\}, F_2 = \mathbf{diag}\{0.4,0.8\}$. Using the MATHLAB tools to solve LMIs (35) and (36), we indicate that the neutral system under consideration is finite-time passive. In addition, the acquired results are compared to previously published studies. The findings show that the stability conditions presented in this paper are more effective than those found in previous research. By solving Example 4 with LMI in Remark 3, we can obtain a maximum permissible upper bound μ_M for different μ_d, as shown in Table 1.

Figure 1 provides the state response of system (4) under zero input and the initial condition $[-3.5, 3.5]$. The interval time-varying delays are chosen as $\mu(t) = [3.6 + 0.9|sin(t)|]$, and the activation function is set as $f(\xi(t)) = [0.4tanh(x_1(t)), 0.8tanh(x_2(t))]^T$.

The permissible upper bound μ_M for various μ_d is shown in Table 1. Table 1 shows that the conclusions of Remark 3 in this study are less conservative than those in [45–48], demonstrating the effectiveness of our efforts. Table 1 shows the state variables' temporal responses. The allowable upper bounds of μ_M are listed in Table 1.

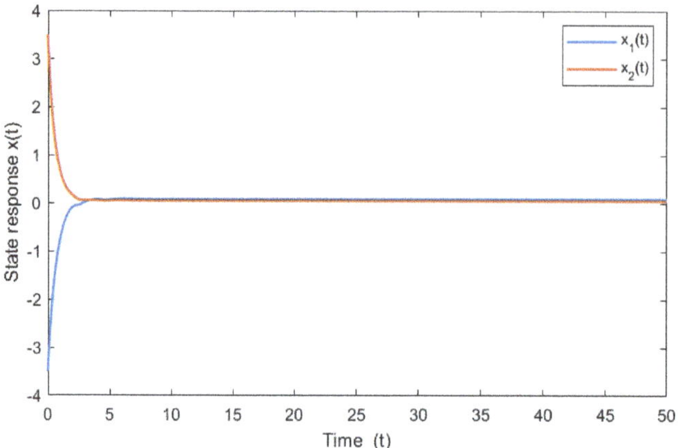

Figure 1. It provides the state response of system (4) under zero input and the initial condition $[-3.5, 3.5]$. The interval time-varying delays are chosen as $\mu(t) = [3.6 + 0.9|sin(t)|]$, and the activation function is set as $f(\xi(t)) = [0.4tanh(x_1(t)), 0.8tanh(x_2(t))]^T$.

Table 1. Allowable upper bound μ_M for various μ_d of Example 4.

Method	$\mu_d = 0.8$	$\mu_d = 0.9$	Number of Variables
[45]	4.5940	3.4671	$7.5n^2 + 8.5n$
[46]	4.8167	3.4245	$13.5n^2 + 13.5n$
[47]	5.4428	3.6482	-
[48]	5.6384	3.7718	$22n^2 + 14n$
Remark 3	6.5411	4.5074	$23n^2 + 23n$

Example 5. *Consider the following matrix parameters for the neural networks matrix parameters:*

$$\dot{\xi}(t) = -A\xi(t) + G_b f(\xi(t)) + G_d f(\xi(t - \mu(t))),$$

with the following:

$$A = \begin{bmatrix} 1.5 & 0 \\ 0 & 1.7 \end{bmatrix}, \quad G_b = \begin{bmatrix} 0.0503 & 0.0454 \\ 0.0987 & 0.2075 \end{bmatrix}, \quad G_d = \begin{bmatrix} 0.2381 & 0.9320 \\ 0.0388 & 0.5062 \end{bmatrix},$$

then $F_1 = diag\{0, 0\}, F_2 = diag\{0.3, 0.8\}$. The maximum delay bounds with μ calculated by Remark 3, as and the recommended criteria are presented in the Table 2.

Figure 2 provides the state response of system (4) under zero input and the initial condition $[-3.5, 3.5]$. The interval time-varying delays are chosen as $\mu(t) = [6.3190 + 0.55|sin(t)|]$, and the activation function is set as $f(\xi(t)) = [0.3tanh(x_1(t)), 0.8tanh(x_2(t))]^T$.

From Table 2, it follows that Remark 3 provides significantly better results than [49–52] in the case of $\mu_d = 0.4$ and $\mu_d = 0.45$. However, in cases where $\mu_d = 0.5$ and $\mu_d = 0.55$, the results are slightly worse than in [21]. Additionally, the acquired results are compared to previously published studies. The findings show that the stability conditions presented in this paper are more effective than those found in previous research.

Table 2. Allowable upper bound μ_M for various μ_d of Example 5.

Method	$\mu_d = 0.4$	$\mu_d = 0.45$	$\mu_d = 0.5$	$\mu_d = 0.55$	Number of Variables
[49]	4.6569	3.7268	3.4076	3.2841	$8n^2 + 12n$
[50]	4.5543	3.8364	3.5583	3.4110	$13.5n^2 + 21.5n$
[51]	7.6697	6.7287	6.4126	3.2569	$13.5n^2 + 13.5n$
[52]	8.3498	7.3817	7.0219	6.8156	$7n^2 + 11n$
Remark 3	9.7901	7.6470	6.7875	6.3190	$23n^2 + 23n$

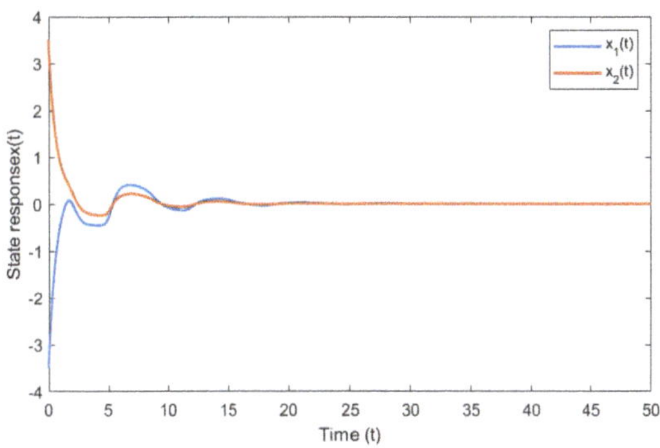

Figure 2. It provides the state response of system (4) under zero input and the initial condition $[-3.5, 3.5]$. The interval time-varying delays are chosen as $\mu(t) = [6.3190 + 0.55|sin(t)|]$, and the activation function is set as $f(\xi(t)) = [0.3tanh(x_1(t)), 0.8tanh(x_2(t))]^T$.

5. Conclusions

In this study, a novel result was presented. The new systems have been used to derive the analysis of finite-time passivity analysis of neutral-type neural networks with mixed time-varying delays. The time-varying delays are distributed, discrete and neutral, and the upper bounds for the delays are available. We are investigating the creation of sufficient conditions for finite boundness, finite-time stability and finite-time passivity, which has not been performed before. First, we create a new Lyapunov–Krasovskii functional, Peng–Park's integral inequality, descriptor model transformation and zero equation use, and then we used Wirtinger's integral inequality technique. New finite-time stability necessary conditions are constructed in terms of linear matrix inequalities to guarantee finite-time stability for the system. Finally, numerical examples are presented to demonstrate the result's effectiveness, and our proposed criteria are less conservative than prior studies in terms of larger time-delay bounds. By combining numerous integral components of the Lyapunov–Krasovskii function with inequality, our results offered wider bounds of time-delay than the previous literature (see Tables 1 and 2). Construction of an LMI variable number based on integral inequalities yields less conservative stability criteria for interval time-delay systems. We expect to be able to improve existing research and lead research into other areas of application.

Author Contributions: Conceptualization, I.K., N.S. and K.M.; methodology, I.K. and N.S.; software, N.S. and I.K.; validation, N.S., T.B. and K.M.; formal analysis, K.M.; investigation, N.S., I.K., T.B. and K.M.; writing—original draft preparation, N.S., I.K. and K.M.; writing—review and editing, N.S., I.K., T.B. and K.M.; visualization, I.K. and K.M.; supervision, N.S. and K.M.; project administration, I.K. and K.M.; funding acquisition, K.M. All authors have read and agreed to the published version of the manuscript.

Funding: This work is supported by the Faculty of Engineering, Rajamangala University of Technology Isan Khon Kaen Campus and Research and Graduate Studies, Khon Kaen University (Research Grant RP64-8-002).

Institutional Review Board Statement: Not applicable.

Informed Consent Statement: Not applicable.

Data Availability Statement: Not applicable.

Acknowledgments: The authors thank the reviewers for their valuable comments and suggestions, which resulted in the improvement of the content of the paper.

Conflicts of Interest: The authors declare no conflict of interest.

Appendix A

For $\Pi_{(i,j)} = \Pi_{(j,i)}$, $i, j = 1, 2, 3, \ldots, 23$ where the following is the case:

$$\begin{aligned}
\Pi_{(1,1)} &= -P_1 A - A^T P_1 - Q_1^T A - A^T Q_1 + Q_2^T + Q_2 + P_3 + R_1 + R_4 - F_1 H_1 + \mu_M^2 P_4 \\
&\quad + M1^T + M1^T + \mu_M M_3 - 9P_8 - 12P_9 + \frac{\mu_M^4}{4} P_{10} - 2\mu_M^2 P_{11} - P_2 A - AP_2 \\
&\quad + \mu_M^2 R_7 - R_9 - O_2^T A - A^T O_2 - O_6^T A - A^T O_6 - 4P_{12} - P_{13}, \\
\Pi_{(1,2)} &= -Q_1^T - A^T Q_{10} + Q_{11} + \mu_M^2 R_8 - A^T O_5 - O_6^T, \\
\Pi_{(1,3)} &= P_1 G_b + Q_1^T G_b + R_2 + R_5 + F_2 H_1 + P_2 G_b + O_2^T G_b - A^T O_3 + O_6^T G_b - A^T O_7, \\
\Pi_{(1,4)} &= P_1 G_c + Q_1^T G_d + P_2 G_d + O_2^T G_d - A^T O_4 + O_6^T G_d - A^T O_8, \\
\Pi_{(1,5)} &= -Q_2^T - Q_3^T - M_1^T + M_2 + \mu_M M_4 + R_9 + P_{13} - S, \\
\Pi_{(1,6)} &= Q_3^T + 3P_8 - 2P_{12} + S, \\
\Pi_{(1,7)} &= -A^T Q_4 - Q_2^T - Q_5, \\
\Pi_{(1,8)} &= -A^T Q_7 + Q_8 - Q_3^T, \\
\Pi_{(1,10)} &= -R_8^T, \\
\Pi_{(1,12)} &= -W_1 F_1 + W_2 F_2 - A^T N_2^T - O_1 - O_2^T, \\
\Pi_{(1,13)} &= 36P_8 + 12P_9 + 6P_{12}, \\
\Pi_{(1,14)} &= -60P_8 - 120P_9, \\
\Pi_{(1,15)} &= 360P_9, \\
\Pi_{(1,16)} &= 2\mu_M P_{11}, \\
\Pi_{(1,18)} &= Z_1, \\
\Pi_{(1,21)} &= O_6^T G_c + O_2^T G_c + P_1 G_c + P_2 G_c, \\
\Pi_{(1,22)} &= O_6^T G_e + O_2^T G_e + P_1 G_e + P_2 G_e, \\
\Pi_{(1,23)} &= O_6^T H + O_2^T H + P_1 H + P_2 H, \\
\Pi_{(2,2)} &= -Q_{10}^T - Q_1 0 + \mu_M P_5 + \mu_M^2 P_6 + \frac{\mu_M^2}{2} P_9 + \frac{\mu_M^4}{2} P_{11} + Q_{14}^T + Q_{14} + \mu_M^2 R_9 \\
&\quad - O_5^T - O_5 + \mu_M^2 P_8, \\
\Pi_{(2,3)} &= Q_{10}^T G_b + O_5^T G_b - O_7, \\
\Pi_{(2,4)} &= Q_{10}^T G_d + O_5^T G_d - O_8, \\
\Pi_{(2,5)} &= -Q_{11}^T - Q_{12}^T, \\
\Pi_{(2,6)} &= Q_{12}^T, \\
\Pi_{(2,7)} &= -Q_4 - Q_{11}^T, \\
\Pi_{(2,8)} &= -Q_7 - Q_{12}^T, \\
\Pi_{(2,12)} &= Q_{13}^T - Q_{14}, \\
\Pi_{(2,21)} &= O_5^T G_c, \\
\Pi_{(2,22)} &= O_5^T G_e, \\
\Pi_{(2,23)} &= O_5^T H, \\
\Pi_{(3,3)} &= R_3 + R_6 - H_1 + O_3^T G_b + G_b^T O_3 + O_7^T G_b + G_b^T O_7 + \mu_M^2 P_{12} + \mu_M^2 M P_{13} \\
&\quad + \rho^2 P_{16},
\end{aligned}$$

$$\Pi_{(3,4)} = O_3^T G_d + G_d^T O_4 + O_7^T G_d + G_b^T O_8,$$

$$\Pi_{(3,7)} = G_b^T Q_4,$$

$$\Pi_{(3,8)} = G_b^T Q_7,$$

$$\Pi_{(3,12)} = W_1 - W_2 + G_b^T N_2 + G_b^T O_1 - O_3^T,$$

$$\Pi_{(3,21)} = O_7^T G_c + O_3^T G_c,$$

$$\Pi_{(3,22)} = O_7^T G_e + O_3^T G_e,$$

$$\Pi_{(3,23)} = O_7^T H + O_3^T H,$$

$$\Pi_{(4,4)} = \mu_d G_d R_3 - R_3 - H_2 + O_4^T G_d + G_d^T O_4 + O_8^T G_d + G_d^T O_8,$$

$$\Pi_{(4,5)} = \mu_d G_d R_2^T - R_2^T + H2^T F2^T,$$

$$\Pi_{(4,7)} = G_d^T Q_4,$$

$$\Pi_{(4,8)} = G_d^T Q_7,$$

$$\Pi_{(4,12)} = G_d^T N_2 + G_d^T O_1 - O_4^T,$$

$$\Pi_{(4,21)} = O_8^T G_c + O_4^T G_c,$$

$$\Pi_{(4,22)} = O_8^T G_e + O_4^T G_e,$$

$$\Pi_{(4,23)} = O_8^T H + O_4^T H,$$

$$\Pi_{(5,5)} = \mu_d G_d R_1 - R_1 + M_1 + M_1^T - M_2 - M_2^T + \mu_M M_3 + \mu_M M_5 - F_1 H_2$$
$$- R_9 - R_9^T - 2P_{13} + S + S^T,$$

$$\Pi_{(5,6)} = M_2 - M_1^T + \mu_M M_4 + R_9 + P_{13} - S,$$

$$\Pi_{(5,7)} = -Q_5 - Q_6,$$

$$\Pi_{(5,8)} = -Q_8 - Q_9,$$

$$\Pi_{(5,10)} = R_8^T,$$

$$\Pi_{(5,11)} = -R_8^T,$$

$$\Pi_{(5,18)} = -Z_1,$$

$$\Pi_{(5,19)} = Z_2,$$

$$\Pi_{(6,6)} = -P_3 - R_4 - M_2 - M_2^T + \mu_M M_5 - 9P_8 - R_9 - 4P_{12} - P_{13},$$

$$\Pi_{(6,7)} = Q_6,$$

$$\Pi_{(6,8)} = Q_9,$$

$$\Pi_{(6,9)} = -R_5,$$

$$\Pi_{(6,11)} = R_8^T,$$

$$\Pi_{(6,13)} = -24P_8 + 6P_{12},$$

$$\Pi_{(6,14)} = 60P_8,$$

$$\Pi_{(6,19)} = -Z_2,$$

$$\Pi_{(7,7)} = -Q_5^T - Q_5 - P_6,$$

$$\Pi_{(7,8)} = -Q_8 - Q_6^T - P_6,$$

$$\Pi_{(8,8)} = -Q_9^T - Q_9 - P_6,$$

$$\Pi_{(9,9)} = -R_6,$$

$$\Pi_{(10,10)} = -P_4 - R_7,$$

$$\Pi_{(11,11)} = -P_4 - R_7,$$

$$\Pi_{(12,12)} = -N_2^T - N_2 + \mu_M^2 P_7 - Q_{13}^T - Q_{13} - O_1^T A - A^T O_1 + P_{14} + \tau_M P_{15},$$

$$\Pi_{(12,21)} = O_1^T G_c,$$

$$\Pi_{(12,22)} = O_1^T G_e,$$

$$\Pi_{(12,23)} = O_1^T H,$$

$$\Pi_{(13,13)} = -192P_8 - 72P_9 - 12P_{12},$$

$$\Pi_{(13,14)} = 360P_8 + 480P_9,$$

$$\Pi_{(13,15)} = -1080P_9,$$

$$\Pi_{(14,14)} = -720P_8 - 3600P_9,$$

$$\Pi_{(14,15)} = 8640P_9,$$

$$\Pi_{(15,15)} = -21600P_9,$$

$$\Pi_{(16,16)} = -2P_{11},$$

$$\Pi_{(17,17)} = -P_{10},$$

$$\Pi_{(18,18)} = -Z_1^T - Z_1 - P_7,$$

$$\Pi_{(18,19)} = -P_7,$$

$$\Pi_{(19,19)} = -Z_2^T - Z_2 - P_7,$$

$$\Pi_{(20,20)} = -\tau_M P_{15},$$

$$\Pi_{(21,21)} = -P_{14} + \tau_M P_{14},$$

$$\Pi_{(22,22)} = -P_{16},$$

$$\Pi_{(23,23)} = -\alpha I, \quad \text{and the other are equal zero.}$$

References

1. Wu, A.; Zeng, Z. Exponential passivity of memristive neural networks with time delays. *Neural Netw.* **2014**, *49*, 11–18. [CrossRef]
2. Gunasekaran, N.; Zhai, G.; Yu, Q. Sampled-data synchronization of delayed multi-agent networks and its application to coupled circuit. *Neurocomputing* **2020**, *413*, 499–511. [CrossRef]
3. Gunasekaran, N.; Thoiyab, N.M.; Muruganantham, P.; Rajchakit, G.; Unyong, B. Novel results on global robust stability analysis for dynamical delayed neural networks under parameter uncertainties. *IEEE Access* **2020**, *8*, 178108–178116. [CrossRef]
4. Wang, H.T.; Liu, Z.T.; He, Y. Exponential stability criterion of the switched neural networks with time-varying delay. *Neurocomputing* **2019**, *331*, 1–9. [CrossRef]
5. Syed Ali, M.; Saravanan, S. Finite-time stability for memristor base d switched neural networks with time-varying delays via average dwell time approach. *Neurocomputing* **2018**, *275*, 1637–1649.
6. Pratap, A.; Raja, R.; Cao, J.; Alzabut, J.; Huang, C. Finite-time synchronization criterion of graph theory perspective fractional-order coupled discontinuous neural networks. *Adv. Differ. Equ.* **2020**, *2020*, 97. [CrossRef]
7. Pratap, A.; Raja, R.; Alzabut, J.; Dianavinnarasi, J.; Cao, J.; Rajchakit, G. Finite-time Mittag-Leffler stability of fractional-order quaternion-valued memristive neural networks with impulses. *Neural Process. Lett.* **2020**, *51*, 1485–1526. [CrossRef]
8. Li, J.; Jiang, H.; Hu, C.; Yu, J. Analysis and discontinuous control for finite-time synchronization of delayed complex dynamical networks. *Chaos Solut. Fractals* **2018**, *114*, 219–305. [CrossRef]
9. Wang, H.; Liu, P.X.; Bao, J.; Xie, X.J.; Li, S. Adaptive neural output-feedback decentralized control for large-scale nonlinear systems with stochastic disturbances. *IEEE Trans. Neural Netw. Learn. Syst.* **2020**, *31*, 972–983. [CrossRef]
10. Zheng, M.; Li, L.; Peng, H.; Xiao, J.; Yang, Y.; Zhang, Y.; Zhao, H. Finite-time stability and synchronization of memristor-based fractional-order fuzzy cellular neural networks. *Commun. Nonlinear Sci. Numer. Simul.* **2018**, *59*, 272–297. [CrossRef]
11. Wu, A.; Zeng, Z. Lagrange stability of memristive neural networks with discrete and distributed delays. *IEEE Trans. Neural Netw. Learn. Syst.* **2014**, *25*, 690–703. [CrossRef]
12. Bai, C.Z. Global stability of almost periodic solutions of Hopfield neural networks with neutral time-varying delays. *Appl. Math. Comput.* **2008**, *203*, 72–79. [CrossRef]
13. Syed Ali, M.; Zhu, Q.; Pavithra, S.; Gunasekaran, N. A study on $\langle (Q,S,R),\gamma \rangle$-dissipative synchronisation of coupled reaction-diffusion neural networks with time-varying delays. *Int. J. Syst. Sci.* **2017**, *49*, 755–765.
14. Chen, Z.; Ruan, J. Global stability analysis of impulsive Cohen-Grossberg neural networks with delay. *Phys. Lett. A* **2005**, *345*, 101–111. [CrossRef]
15. Li, W.J.; Lee, T. Hopfield neural networks for affine invariant matching. *IEEE Trans. Neural Netw. Learn. Syst.* **2001**, *12*, 1400–1410.
16. Syed Ali, M.; Gunasekaran, N.; Zhu, Q. State estimation of T-S fuzzy delayed neural networks with Markovian jumping parameters using sampled-data control. *Fuzzy Sets Syst.* **2017**, *306*, 87–104. [CrossRef]

17. Syed Ali, M.; Narayanana, G.; Sevgenb, S.; Shekher, V.; Arik, S. Global stability analysis of fractional-order fuzzy BAM neural networks with time delay and impulsive effects. *Commun. Nonlinear Sci. Numer. Simul.* **2019**, *78*, 104853. [CrossRef]

18. Samorn, N.; Yotha, N.; Srisilp, P.; Mukdasai, K. LMI-based results on robust exponential passivity of uncertain neutral-type neural networks with mixed interval time-varying delays via the reciprocally convex combination technique. *Computation* **2021**, *9*, 70. [CrossRef]

19. Meesuptong, B.; Mukdasai, K.; Khonchaiyaphum, I. New exponential stability criterion for neutral system with interval time-varying mixed delays and nonlinear uncertainties. *Thai J. Math.* **2020**, *18*, 333–349.

20. Jian, J.; Wang, J. Stability analysis in lagrange sense for a class of BAM neural networks of neutral type with multiple time-varying delays. *Neurocomputing* **2015**, *149*, 930–939. [CrossRef]

21. Lakshmanan, S.; Park, J.H.; Jung, H.Y.; Kwon, O.M.; Rakkiyappan, R. A delay partitioning approach to delay-dependent stability analysis for neutral type neural networks with discrete and distributed delays. *Neurocomputing* **2013**, *111*, 81–89. [CrossRef]

22. Peng, W.; Wu, Q.; Zhang, Z. LMI-based global exponential stability of equilibrium point for neutral delayed BAM neural networks with delays in leakage terms via new inequality technique. *Neurocomputing* **2016**, *199*, 103–113. [CrossRef]

23. Syed Ali, M.; Saravanakumar, R.; Cao, J. New passivity criteria for memristor-based neutral-type stochastic BAM neural networks with mixed time-varying delays. *Neurocomputing* **2016**, *171*, 1533–1547. [CrossRef]

24. Klamnoi, A.; Yotha, N.; Weera, W.; Botmart, T. Improved results on passivity analysis of neutral-type neural networks with mixed time-varying delays. *J. Res. Appl. Mech. Eng.* **2018**, *6*, 71–81.

25. Ding, D.; Wang, Z.; Dong, H.; Shu, H. Distributed H_∞ state estimation with stochastic parameters and nonlinearities through sensor networks: The finite-horizon case. *Automatica* **2012**, *48*, 1575–1585. [CrossRef]

26. Carrasco, J.; Baños, A.; Schaft, A.V. A passivity-based approach to reset control systems stability January. *Syst. Control Lett.* **2010**, *59*, 18–24. [CrossRef]

27. Krasovskii, N.N.; Lidskii, E.A. Analytical design of controllers in systems with random attributes. *Autom. Remote Control* **1961**, *22*, 1021–1025.

28. Debeljković, D.; Stojanović, S.; Jovanović, A. Finite-time stability of continuous time delay systems: Lyapunov-like approach with Jensen's and Coppel's inequality. *Acta Polytech. Hung.* **2013**, *10*, 135–150.

29. Wang, L.; Shen, Y.; Ding, Z. Finite time stabilization of delayed neural networks. *Neural Netw.* **2015**, *70*, 74–80. [CrossRef]

30. Rajavela, S.; Samiduraia, R.; Caobc, J.; Alsaedid, A.; Ahmadd, B. Finite-time non-fragile passivity control for neural networks with time-varying delay. *Appl. Math. Comput.* **2017**, *297*, 145–158. [CrossRef]

31. Chen, X.; He, S. Finite-time passive filtering for a class of neutral time-delayed systems. *Trans. Inst. Meas. Control* **2016**, 1139–1145. [CrossRef]

32. Syed Ali, M.; Saravanan, S.; Zhu, Q. Finite-time stability of neutral-type neural networks with random time-varying delays. *Syst. Sci.* **2017**, *48*, 3279–3295.

33. Saravanan, S.; Syed Ali, M.; Alsaedib, A.; Ahmadb, B. Finite-time passivity for neutral-type neural networks with time-varying delays. *Nonlinear Anal. Model. Control* **2020**, *25*, 206–224.

34. Syed Ali, M.; Meenakshi, K.; Gunasekaran, N. Finite-time H_∞ boundedness of discrete-time neural networks normbounded disturbances with time-varying delay. *Int. J. Control Autom. Syst.* **2017**, *15*, 2681–2689. [CrossRef]

35. Phanlert, C.; Botmart, T.; Weera, W.; Junsawang, P. Finite-Time Mixed H_∞/passivity for neural networks with mixed interval time-varying delays using the multiple integral Lyapunov-Krasovskii functional. *IEEE Access* **2021**, *9*, 89461–89475. [CrossRef]

36. Amato, F.; Ariola, M.; Dorato, P. Finite-time control of linear systems subject to parametric uncertainties and disturbances. *Automatica* **2001**, *37*, 1459–1463. [CrossRef]

37. Song, J.; He, S. Finite-time robust passive control for a class of uncertain Lipschitz nonlinear systems with time-delays. *Neurocomputing* **2015**, *159*, 275–281. [CrossRef]

38. Kwon, O.M.; Park, M.J.; Park, J.H.; Lee, S.M.; Cha, E.J. New delay-partitioning approaches to stability criteria for uncertain neutral systems with time-varying delays. *SciVerse Sci. Direct* **2012**, *349*, 2799–2823. [CrossRef]

39. Seuret, A.; Gouaisbaut, F. Wirtinger-based integral inequality: Application to time-delay system. *Automatica* **2013**, *49*, 2860–2866. [CrossRef]

40. Peng, C.; Fei, M.R. An improved result on the stability of uncertain T-S fuzzy systems with interval time-varying delay. *Fuzzy Sets Syst.* **2013**, *212*, 97–109. [CrossRef]

41. Park, P.G.; Ko, J.W.; Jeong, C.K. Reciprocally convex approach to stability of systems with time-varying delays. *Automatica* **2011**, *47*, 235–238. [CrossRef]

42. Kwon, O.M.; Park, M.J.; Park, J.H.; Lee, S.M.; Cha, E.J. Analysis on robust H_∞ performance and stability for linear systems with interval time-varying state delays via some new augmented Lyapunov Krasovskii functional. *Appl. Math. Comput.* **2013**, *224*, 108–122.

43. Singkibud, P.; Niamsup, P.; Mukdasai, K. Improved results on delay-rage-dependent robust stability criteria of uncertain neutral systems with mixed interval time-varying delays. *IAENG Int. J. Appl. Math.* **2017**, *47*, 209–222.

44. Tian, J.; Ren, Z.; Zhong, S. A new integral inequality and application to stability of time-delay systems. *Appl. Math. Lett.* **2020**, *101*, 106058. [CrossRef]

45. Kwon, O.M.; Park, J.H.; Lee, S.M.; Cha, E.J. New augmented Lyapunov-Krasovskii functional approach to stability analysis of neural networks with time-varying delays. *Nonlinear Dyn.* **2014**, *76*, 221–236. [CrossRef]

46. Zeng, H.B.; He, Y.; Wu, M.; Xiao, S.P. Stability analysis of generalized neural networks with time-varying delays via a new integral inequality. *Neurocomputing* **2015**, *161*, 148–154. [CrossRef]
47. Yang, B.; Wang, J. Stability analysis of delayed neural networks via a new integral inequality. *Neural Netw.* **2017**, *88*, 49–57. [CrossRef] [PubMed]
48. Hua, C.; Wang, Y.; Wu, S. Stability analysis of neural networks with time-varying delay using a new augmented Lyapunov-Krasovskii functional. *Neurocomputing* **2019**, *332*, 1–9. [CrossRef]
49. Zhu, X.L.; Yue, D.; Wang, Y. Delay-dependent stability analysis for neural networks with additive time-varying delay components. *IET Control Theory Appl.* **2013**, *7*, 354–362. [CrossRef]
50. Li, T.; Wang, T.; Song, A.; Fei, S. Combined convex technique on delay-dependent stability for delayed neural networks. *IEEE Trans. Neural Netw. Learn. Syst.* **2013**, *24*, 1459–1466. [CrossRef]
51. Zhang, C.K.; He, Y.; Jiang, L.; Wu, M. Stability analysis for delayed neural networks considering both conservativeness and complexity. *IEEE Trans. Neural Netw. Learn. Syst.* **2015**, *27*, 1486–1501. [CrossRef] [PubMed]
52. Zhang, C.K.; He, Y.; Jiang, L.; Lin, W.J.; Wu, M. Delay-dependent stability analysis of neural networks with time-varying delay: A generalized free-weighting-matrix approach. *Appl. Math. Comput.* **2017**, *294*, 102–120. [CrossRef]

Article

Research on Intellectualized Location of Coal Gangue Logistics Nodes Based on Particle Swarm Optimization and Quasi-Newton Algorithm

Shengli Yang [1], Junjie Wang [2,*], Ming Li [1,*] and Hao Yue [1]

1   School of Energy and Mining Engineering, China University of Mining and Technology (Beijing), Beijing 100083, China; 108992@cumtb.edu.cn (S.Y.); bqt2100101020@student.cumtb.edu.cn (H.Y.)
2   China National Coal Group Corporation, Beijing 100120, China
*   Correspondence: wangjunjie5@chinacoal.com (J.W.); zqt1900101015g@student.cumtb.edu.cn (M.L.); Tel.: +86-13051178588 (J.W.); +86-17614131040 (M.L.)

Citation: Yang, S.; Wang, J.; Li, M.; Yue, H. Research on Intellectualized Location of Coal Gangue Logistics Nodes Based on Particle Swarm Optimization and Quasi-Newton Algorithm. *Mathematics* **2022**, *10*, 162. https://doi.org/10.3390/math10010162

Academic Editors: Quanxin Zhu and Simeon Reich

Received: 23 September 2021
Accepted: 23 December 2021
Published: 5 January 2022

Abstract: The optimization of an integrated coal gangue system of mining, dressing, and backfilling in deep underground mining is a multi-objective and complex decision-making process, and the factors such as spatial layout, node location, and transportation equipment need to be considered comprehensively. In order to realize the intellectualized location of the nodes for the logistics and transportation system of underground mining and dressing coal and gangue, this paper establishes the model of the logistics and transportation system of underground mining and dressing coal gangue, and analyzes the key factors of the intellectualized location for the logistics and transportation system of coal and gangue, and the objective function of the node transportation model is deduced. The PSO–QNMs algorithm is proposed for the solution of the objective function, which improves the accuracy and stability of the location selection and effectively avoids the shortcomings of the PSO algorithm with its poor local detailed search ability and the quasi-Newton algorithm with its sensitivity to the initial value. Comparison of the particle swarm and PSO–QNMs algorithm outputs for the specific conditions of the New Julong coal mine, as an example, shows that the PSO–QNMs algorithm reduces the complexity of the calculation, increases the calculation efficiency by eight times, saves 42.8% of the cost value, and improves the efficiency of the node selection of mining–dressing–backfilling systems in a complex underground mining environment. The results confirm that the method has high convergence speed and solution accuracy, and provides a fundamental basis for optimizing the underground coal mine logistics system. Based on the research results, a node siting system for an integrated underground mining, dressing, and backfilling system in coal mines (referred to as MSBPS) was developed.

Keywords: integration of mining–dressing–backfilling; coal gangue logistics system; node intelligent location; PSO–QNMs algorithm

1. Introduction

1.1. Study on the Integrated Technologies of Mining–Dressing–Backfilling Systems

In recent years, with the continuous increase in energy consumption and mining intensity, China's coal mining depth to an average of 10~25 m has sped to the deep extension. Furthermore, deep coal mines need to excavate a large number of rock alleys to meet the needs of the mine production system, surrounding rock stability control, and safe pressure relief mining, which produces a large amount of gangue, which will not only aggravate the contradiction of insufficient lifting capacity of deep shafts, but will also bring infill mining as a sustainable mining technology, which prevents or minimizes the adverse effects of mining coal resources on the environment and other resources from the perspective of mining, with the goal of achieving the best economic, environmental, and social benefits. In recent years, many achievements have been made in backfilling equipment, theories, and technologies, in order to further realize that gangue not ascend the shaft and reduce gangue lifting costs, etc. Relevant scholars [1] also put forward the integration technologies of the mining–dressing–backfilling system, which is to establish coal–gangue separation and selection centers, backfilling preparation centers, and gangue pockets in the underground; the coal and gangue products extracted from the underground working face are not lifted to the shaft, but sorted by the underground dressing system, and the sorted gangue is filled in place by the underground backfilling preparation system to realize the gangue backfilling in the extraction area. This technology can achieve safe and efficient recovery of coal resources that cannot be extracted by traditional coal mining methods and improves the resource recovery rate, while improving the effective lifting efficiency of the main shaft and relieving the load of the surface coal washing plant. Moreover, gangue technology can reduce the discharge of gangue on the ground and consequently make full use of gangue to effectively slow ground subsidence, and finally achieve the purpose of protecting the environment and land resources. This is why China strongly advocates these green mining and scientific mining methods.

Experts engaged in this area of research have conducted much research on process optimization and equipment improvement of integrated technologies of the mining–dressing–backfilling system, and on the theory of rock movement patterns caused by this green mining method, mainly from the perspective of traditional mine pressure and formation control and mining technology optimization. Wang Jiachen et al. analyzed the relationship between supports in backfilling mining and surrounding rocks and the movement characteristics of overlying rocks, established the roof load estimation method, used similar simulation and numerical calculation to simulate the process of workface retrieval and gangue backfilling, and verified by backfilling mining examples [2,3]. Zhang Jixiong et al. further proposed the sustainable mining system of "mining–dressing–backfilling + X" in coal mines, revealed the law of mineral pressure manifestation and rock movement control mechanism of solid filling mining, and performed much research on the theory and technology of mining–dressing–backfilling green mining of deep coal resources [4–8]. Tu Shihao developed a theoretical concept of the selective mining technology for the integration technologies of mining–dressing–backfilling systems, and analyzed the critical aspects of "mining–dressing–backfilling + controlling", "mining–dressing–backfilling + extraction", "mining–dressing–backfilling + prevention"," mining–dressing–backfilling + protection", and examined other key mining scientific issues from the perspectives of control backfilling rock movement, stress concentration, fracture field development, and stability of the entry [9]. The results of these studies are relatively mature and have been extensively disseminated in many mines in China.

However, since integrated technologies of the mining–dressing–backfilling system are proposed, it is destined to be a coordinated process of multiple systems in engineering application. On the basis of the existing research to clarify the system composition and structure function of the "mining–dressing–backfilling" system, it is of great significance for the future development of this technology to systematically analyze the operating characteristics of deep underground gangue logistics and the interfeeding linkage relationship. The study of this problem necessarily involves the efficient layout of the "mining–dressing–backfilling" system; first, we must choose the location of the crucial underground "logistics" node, which is the basis to ensure the efficient and coordinated transportation of coal and gangue logistics. The current research results on this issue are relatively few. Wang Jinfeng et al. combined the complex characteristics of a coal mine production logistics system and studied the safety resource allocation, safety evaluation, and production logistics efficiency; further systematic research was conducted on optimization methods to maximize the efficiency of coal mine production logistics systems and rationalize safety resources [10–13]. Based on exploring the key factors affecting the efficiency of the logistics system, Xia Dan et al. used a system dynamics approach to dynamically analyze and predict the efficiency of a complex production logistics system for the integrated technologies of mining–dressing–backfilling systems and calculated the true impact rate of different production steps on the production level [14]. Although these studies have dealt with the efficiency of the integrated production and logistics of "mining–dressing–backfilling" systems, they are all from the perspective of macromanagers and have not really achieved substantial research on the logistics node location selection and optimization of the system layout.

1.2. Study on the Logistics Node Location Selection

Although there is not much research in the field of coal mining, the location and positioning of key system nodes is very important for the supply chain and logistics transportation system, which need to consider the distance between nodes, cost, and the influence of multiple factors from the perspective of logistics systems, with the continuous development of applied mathematics, increasingly more factors are taken into account, and various siting methods are introduced into logistics node siting in order to select the best location.

The logistics location problem in the supply chain varies in the factors to be focused on in different fields and systems, but multiobjective decision-making oriented to consider multiple factors is an important research topic [15]. Zhang Guofang et al. proposed that the main influencing factors for evaluating the location of logistics nodes are infrastructure platform conditions, basic information platform conditions, and economic and sustainable development conditions of logistics nodes, and in this way subdivided into 28 specific indicators [16]. A multiobjective genetic algorithm (MOGA) was applied in supply chain decision-making for agricultural systems to find the best combination of agricultural inputs that minimize greenhouse gas emissions and maximizes output energy and benefit–cost ratio [17]. The importance of supply and demand on the location of distribution centers is argued [18]. Various factors such as politics, economy, environment, and the enterprise itself are also important for the location of logistics nodes [19]. Considering four criteria in supply chain planning—cost, quality, delivery, and supplier relationship management— a decision method considering quantity discounts and supplier capacity constraints is proposed, and TPSO, PSO, and GA are used for comparative numerical experiments [20]. As research expands, factors such as customer satisfaction, delivery time, service quality, and sustainability are also taken into account [21,22]. Under various operational constraints, cost minimization and profit maximization are the ultimate goals in most supply-chain planning [23,24].

The method of logistics node location selection is evolving as a result of the increasing number of factors to be considered, from the early center-of-gravity method [25–27], it has evolved to multiobjective site selection alternatives including fuzzy integrated analysis [16,28], analytic hierarchy process (AHP) [29–31], and data envelopment analysis (DEA) [32], which can consider more factors and are friendly to some hard-to-quantify factors. While it is computationally difficult to solve larger site planning problems, various heuristic and intelligent algorithms supported by big data and computers are applied to solve the logistics node site selection problem. The alternative location algorithm (ALA) and intelligent algorithms use parallel search techniques to solve the site selection problem, which overcome the difficulty of traditional solution methods and can select the global optimal solution efficiently and accurately. Commonly used algorithms include the ant colony algorithm (ACA) [33,34], genetic algorithm (GA) [35–37], tabu search algorithm (TS) [38], particle swarm optimization algorithm (PSO) [38–40], among others. The common methods and characteristics of logistics node site selection are shown in Table 1.

Recognizing the importance of spatial node layout planning for an integrated coal gangue system of "mining–dressing–backfilling" in the underground, it is necessary to reference logistics node location selection methods that have matured in the field of supply chain and apply them to the integrated technologies of mining–dressing–backfilling systems, and carry out intelligent site selection for the nodes of coal gangue logistics and transportation systems. The integrated production system model of mining, dressing, and backfilling was proposed in the literature [41], and the scientific siting of the nodes of the underground integration technologies of mining–dressing–backfilling systems was studied.

Due to the complex underground environment of the mine, a three-dimensional logistics space node siting model was developed for the logistics and transportation system under the premise of making appropriate scientific assumptions, and the objective function of multiobjective decision-making was established mainly from the perspective of production efficiency and economy. Furthermore, the PSO algorithm was proposed in solving the objective function. In this paper, the crucial factors of the complicated underground mining and coal gangue transportation system are elaborated from the perspective of logistics rationalization. The coal gangue logistics system location nodes model is further established, and the objective function is defined to identify the vital nodes based on "the highest efficiency and the lowest cost". The traditional evolutionary algorithms, such as the particle swarm algorithm, fully discuss solving this problem, but lack local area search capability, and the results are unstable because of stagnation at the later stage of calculation. The particle swarm and quasi-Newton algorithm (PSO–QNMs) is introduced to design a hybrid algorithm for solving the location of coal gangue logistics nodes by fully taking advantage of the global search capability of the PSO algorithm and the localized and detailed search capability of the quasi-Newton algorithm. The improved algorithm is applied to analyze the issue of coal gangue logistics node siting in a real case: New Julong coal mine.

Table 1. Common methods and characteristics of logistics node location selection.

	Logistics Node Location Selection Methods	Key Features
Classical solution methods	Center-of-gravity method	The distribution of the nodes of the logistics node system is placed on a plane, and the demand and resources of each node are seen as the weight of the point, and the best point for the location of logistics facilities is the center of gravity of the logistics system
	Integer programming method	Setting the objective function, parameters and variables, making assumptions and constraints simplify, establish a relatively idealized model, and solve it by an appropriate algorithm
Multiobjective solving methods	Analytic Hierarchy Process (AHP)	A discrete method for evaluating and analyzing alternatives to arrive at the optimal site by establishing an index evaluation system, usually used in conjunction with the fuzzy evaluation method
	Data Envelopment Analysis (DEA)	A system analysis method evolved on the basis of evaluating relative efficiency, adjusting the weight indicators of the evaluation model dynamically according to the inputs and outputs, evaluating the alternatives from the perspective of the decision unit, independent of the metric and subjective factors of the indicators, and applicable to the site selection decision of multiple input and output problems
	Fuzzy Integrated Evaluation	It can determine the weight of indicators and quantitative representation of indicators, combine qualitative and quantitative, make a comprehensive evaluation of a variety of factors, suitable for nondeterministic problem solving, cannot solve the problem of correlation between factors, and the transformation of indicators has a certain degree of subjectivity.

Table 1. *Cont.*

Logistics Node Location Selection Methods		Key Features
Heuristic and intelligent algorithms	Ant Colony Algorithm (ACA)	With fewer setup parameters and good convergence performance, it can generate solutions in a very short time, and is suitable for solving complex logistics node siting problems with great flexibility.
	Genetic Algorithm (GA)	Fast computation and easy combinations with other algorithms.
	Tabu Search Algorithm (TS)	Easy to understand and implement, strong generality, strong local development ability, fast convergence; based on single solution, and weak group development ability.
	Particle Swarm Optimization (PSO)	Simple operation, fast convergence, does not depend on the strict mathematical properties of the optimization problem itself, can achieve global optimality, and easy to combine with other algorithms.
	… …	… …

2. Coal Gangue Logistics and Transportation Systems in the Integration of Mining–Dressing–Backfilling

In order to realize the deep underground sorting and in situ filling technology model, an efficient, reliable, intelligent, and economical "coal mining–dressing–backfilling" integrated logistics production system for underground coal mines was established. To study the problem of optimal selection of nodes in the integrated coal gangue logistics production system of "mining–dressing–backfilling" in underground coal mining, the precondition is that the underground mining, sorting, filling, and transportation system links are analyzed separately from the perspective of logistics rationalization. The first and most important is the optimization and rationalization of the logistics system as a whole during the process of completing the underground cycle of mining, sorting, and backfilling from the working face. An underground mining and coal gangue transport system is relatively complex. The core of the two major systems for the coal gangue sorting system consists of the gangue and other waste filling system and the underground mining and coal gangue logistics production, which is to separate coal and gangue in the underground. Gangue is used as the main raw material for underground filling; due to the limited capacity of filling, the flow of gangue produced at the working face and the space relative position of mining and charging will determine the coordinated treatment capacity of mining and filling of the gangue transportation system. Insufficient gangue production will lead to obstruction of the underground filling work, and the surplus of gangue production will cause the excess gangue to be stacked randomly [42]. Therefore, it is necessary to reasonably design the spatial location relationship of key nodes for underground mining, selection, and filling coal gangue logistics systems to ensure efficient and coordinated transportation of mining, selection, and filling. In order to optimize the coal mine production logistics system, the coal production and operation process are transformed into a logistics and transportation process. The integrated coal gangue logistics system of mining, selection, and filling includes two parts: gangue production supply logistics and gangue production logistics. In the transportation part of coal gangue logistics, the normal output of coal gangue is the most fundamental and essential logistical component of the transportation link. In an underground coal mine, the coal and gangue produced from the working face are transported through the complex and extensive transport routes. As there are more logistics nodes in transportation, there will be a certain suspension in the sorting center, the filling preparatory center, and the underground gangue silo, based on the efficiency of the whole system, which can be improved by setting a reasonable key node.

3. Coal Gangue Logistics System Location Nodes Model

The intelligent selection of nodes in the underground integration of the mining–dressing–backfilling system is an optimum solution selected among many solutions to meet the actual engineering background. The establishment and solution of the nonlinear equation system occupies an important position in the optimization problem, especially in the field of industrial engineering, etc. For practical cases, a mathematical model needs to be built and transformed into a system of equations for the problem to be solved. Among many solution methods—particle swarm algorithm, genetic algorithm, ant colony algorithm, Newton's method—search better from the consideration of solution accuracy and convergence, but there are still some defects in the solution process for specific application cases [43–45]. The problem of intelligent output of key nodes in underground integration of mining–dressing–backfilling can actually be regarded mathematically as the problem of large flow, high efficiency, and minimum cost of gangue transportation, by modifying the relevant parameters and changing different constraints, the sum of costs such as construction and transportation is minimized, and the flow rate in the logistics system is maximized. The ultimate goal is to improve the operational efficiency of the coal gangue logistics system. The gangue that is used for filling the working face is partly from the gangue on the surface, and partly from the gangue produced during underground working-face mining and roadway excavation. This is especially important for the large number of rock roadways excavated in deep mining, which gangue is used to enhance the stability of the surrounding rock of the roadway. In the solution of the model, assuming that the surface gangue is transported to the underground gangue silo through the vertical feeding hole, the location of the key nodes of the system is sited from the perspective of maximum logistics and optimal cost, without considering the loss of coal gangue in the transportation process. As shown in Figure 1, the key transportation nodes include six positions, where *I* is the underground coal–gangue separation and selection center, *J* is the underground backfilling preparation center, *K* is the underground gangue pocket, *T* is the input port, *E* is the gangue mountain, and *D* is the shaft coal pocket.

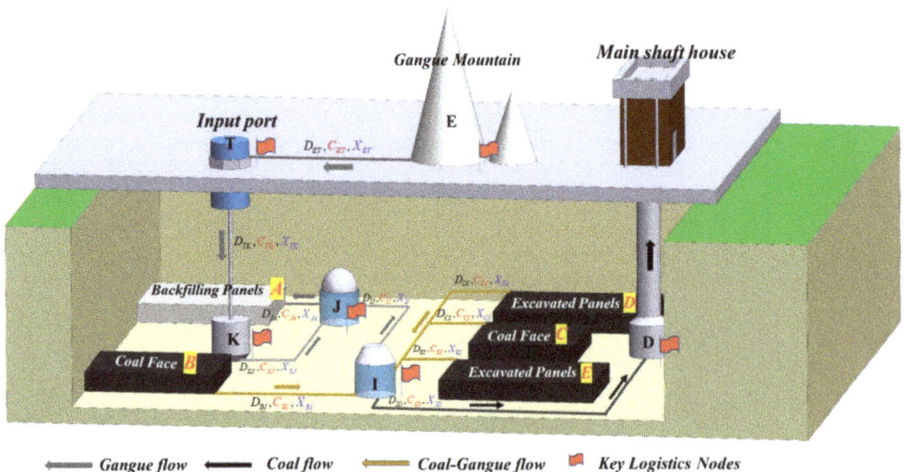

Figure 1. Node transportation model of the coal gangue backfilling system.

From the perspective of economy and logistics, the objective function is established based on the principle of "highest efficiency and lowest cost", i.e., the sum of transportation costs between the links is minimized. The objective function is established in Equation (1) and the constraints are given in Equations (2)–(4).

$$minZ = \sum_{i \in M} D_{M_i,I} C_{M_i,I} X_{M_i,I} + \sum_{i \in N} D_{N_i,J} C_{N_i,J} X_{N_i,J} + D_{E,T} C_{E,T} X_{E,T} + \\ D_{T,K} C_{T,K} X_{T,K} + D_{D,I} C_{D,I} X_{D,I} \tag{1}$$

$$\sum_{i \in M} X_{M_i,I} + \sum_{i \in N} X_{N_i,J} = X_{I,J} + X_{I,D} \tag{2}$$

$$X_{I,J} + X_{K,J} = \sum_{i \in M} X_{J,A} \tag{3}$$

$$X_{E,T} = X_{T,K} = X_{K,J} \tag{4}$$

where $D_{M_i,I}$ indicates position i and j; $C_{M_i,I}$ indicates transportation cost for each unit of coal between position i and j; $X_{M_i,I}$ is the transportation flow rate between position i and j; M_i is the i-th coal mining/excavation working face, where $(M_i = A, B, C, D, E \cdots)$; and N_i denote the key logistics nodes in integration of the mining–dressing–backfilling system.

4. Intelligent Algorithmic Optimization

4.1. Particle Swarm Algorithm

The particle swarm algorithm [22] is a swarm-based random optimization intelligence algorithm that originated from the study of bird feeding behavior, where the simplest and most finite strategy to find food is to search around the bird that is currently closest to the food. The algorithm is an abstraction of solving the objective decision function as the process of searching for the optimal in the decision space in a continuous iteration, which is one of the methods to solve the optimal solution of the multidimensional function. The mathematical description of the algorithm follows. Assuming that the position of particle I in N-dimensional space is represented as a vector: $X_i = (X_1, X_2, X_3, \ldots\ldots, X_N)$, and the velocity of the particle motion is represented as a vector: $V_i = (V_1, V_2, V_3, \ldots\ldots, V_N)$, then each particle has an adaptation value determined by the objective function and knows the best position experienced by the individual called the individual historical best position, defined as pbest and by its present position; each particle also knows the best position found by all particles in the whole population thus far, which is defined as gbest.

After initializing a group of random particles, the optimal solution is found by iteration. In each iteration, the particles update themselves mainly by tracking pbest and gbest, and the particles will iterate to update their velocity and position according to Equation (5):

$$V_{id}(k+1) = \omega V_{id}(k) + c_1 r_1 (P_{id}(k) - X_{id}(k)) + c_2 r_2 (P_{id}(k) - X_{id}(k)); \\ X_{id}(k+1) = V_{id}(k) + V_{id}(k) \tag{5}$$

where ω is the inertia weight factor; c_1 and c_2 are non-negative constants called the learning factor; and r_1 and r_2 are random numbers in $[0, 1]$ with independent uniform distribution.

The core code formulas for continuously updating the velocity New_v_{id} and position New_x_{id} of the particle for each particle motion are Equations (6) and (7), respectively:

$$New\_v_{id} = w * v_{id} + c_1 * rand() * (p_{id} - x_{id}) + c_2 * rand(p_{gd} - x_{id}) \tag{6}$$

$$New\_x_{id} = x_{id} + New\_v_{id} \tag{7}$$

where p_{id} is the individual known optimal solution; p_{gd} is the global known optimal solution; v_{id}, x_{id} denote the velocity and position of the particle updated by the last operation in the population, respectively; w is the inertia weight factor; $c_1 = c_2$ is one of the particle learning factors, which usually is valued between 0 and 2; and $rand()$ is a random number within $(0, 1)$. To reduce the possibility of particles leaving the search space during the search process, V is usually limited to a certain range.

The PSO is an efficient parallel search algorithm that retains a population-based global search strategy with a relatively simple operational model that preserves the individual historical extremes of each particle, which has been applied to the output of the key node location in the coal gangue logistics and transportation system of integrated underground mining, dressing, and backfilling The author references particle swarm algorithm in the literature to solve the node siting for the integrated logistics of the mining–dressing–backfilling system, and some of the core codes have also been reflected in the literature, the algorithm initially realized the automatic output of coal gangue logistics node siting under the role of complex factors. However, it was found in subsequent application that the algorithm lacks the ability of fine search in a local area, and the phenomenon of convergence stagnation that often occurs in the later stage of the search is not very sensitive to the population size and cannot obtain an unique and accurate solution.

4.2. Quasi-Newton Methods Algorithm

The quasi-Newton methods (QNMs) [46–50] were first described by the American physicist Davidson in the mid-1950s, and shortly thereafter it was proved by the operational scientists Fletcher and Powell to be both faster and more stable than the algorithms available at that time. In recent years, the QNMs have become an important research area for algorithms to solve both constrained and unconstrained optimization problems. The QNMs do not need to calculate the Hesse array of the objective function in the computational process as does the Newton method, yet it can have the same efficacy in some sense as when using the Hesse array, and has a second-order convergence speed. This not only simplifies the computational process, but also ensures algorithm convergence speed. Therefore, in recent decades, the QNMs are one of the most important methods for solving nonlinear systems of equations and optimization problems. The QNMs program code is shown in Equation (8):

$$function\ [k, x, val] = bfgs(fun, gfun, x_0, \text{varargin}) \tag{8}$$

where k is the number of iterations; x, val is the approximate optimal site and the optimal value, respectively; $fun, gfun$ is the objective function and its gradient, respectively; x_0 is the initial site; and $varargin$ is the input variable parameter.

The main characteristics of the QNMs are simple internal update rules, high accuracy, strong numerical stability, and fast convergence. However, the selection of the initial value of the method is challenging; if a random value is used for the solution, then in actual engineering background application it extremely difficult attain convergence. Therefore, the convergence depends on the selection of the initial value; thus, it is very important to provide an optimal initial value for QNMs.

4.3. Particle Swarm and Quasi-Newton Algorithm

In view of the poor local search ability of the PSO algorithm and the sensitivity of the QNMs algorithm to the initial value, in order to improve the localization accuracy and convergence, combining the characteristics of the two algorithms, a particle swarm optimization algorithm based on quasi-Newton algorithm (PSO–QNMs algorithm), is designed to precisely optimize the nodes of the coal and gangue system to achieve the effect of making full use of the advantages of the two algorithms. The process of the PSO–QNMs algorithm is shown in Figure 2. Firstly, the PSO algorithm is used to search the problem in a wide range within the feasible solution area to find the optimal algorithm to a certain extent, and to provide a good initial point for the QNMs algorithm, which is used as the initial value of the QNMs algorithm for continuous iteration, then the QNMs algorithm was used to search precisely until a more precise root of the equation is found. The program code is shown in Equation (9):

$$function\ y = funadd(n, x, fixed\text{point}, \cos t, traffic flow)$$
$$function\ y = funadd\ Gra(x) \tag{9}$$

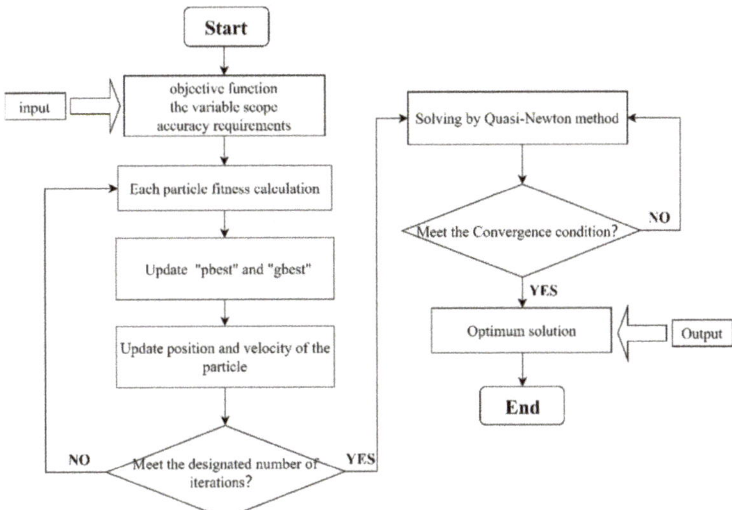

Figure 2. Flow chart of the PSO–QNMs hybrid algorithm.

The PSO–QNMs [25] runs the particle swarm optimization algorithm M generation at the initial stage (population size is m), reaches the termination condition, outputs the current global optimal individual (defined in the PSO algorithm as "zbest") and gives this value to the initial value of Newton method, as indicated in Equation (10):

$$x_0 = \text{zbest} \tag{10}$$

where $x_0 \in \mathrm{R}^n$, termination of the error $0 \leq \varepsilon \leq 1$. The initial positive definite matrix $H_0 \in R^{n \times n}, k := 0$; if $\|g_k\| \leq \varepsilon$, then the operation is stopped and the output x_k is taken as an approximate minimum point; the direction of calculation and search $d_k = -H_k g_k$, and α_k is solved by linear search along direction d_k;x_{k+1} : is expressed as $x_{k+1} := x_k + \alpha_k d_k$. Calibration H_k produces H_{k+1}, and $k = k + 1$ iterations are performed. Part of the core program code of the PSO–QNMs algorithm is shown in Equation (11):

$$
\begin{aligned}
\text{fminunc\_options} &= \text{optimoptions}(@\text{fminunc},'\text{Algorithm}','\text{quasi} - \text{newton}','\text{MaxFunEvals}', 100000, \\
&\quad '\text{PlotFcns}', @\text{optimplotfval}); \\
[\text{Gra\_best}, &\text{fval\_Gra}, \text{exitflag\_Gra}, \text{output\_Gra}] = \text{fminunc}(@\text{funadd\_Gra}, x0, \text{fminunc\_options})
\end{aligned} \tag{11}
$$

5. Case Analysis

5.1. Background

This algorithm comparison uses the Xinjulong coal mine of Shandong Energy Group as the basic background for engineering application to simulate the intellectualized location of coal gangue logistics nodes of integrated system for "mining–dressing–backfilling" in the underground. The working face of the Xinjulong coal mine, which is mainly a fully mechanized coal caving and backfilling face, and the production of the double mining district is mainly at the same gallery level. The coal seam thickness of the 1302N-2# backfilling face is 2.2–3.63 m, the average coal thickness is 2.73 m, the mining coefficient is 1, and the coal seam variation coefficient is 11.9%, which belongs to a medium thick coal seam with simple structure and stable thickness within the mining range of the working face. The 1302N-2# backfilling face is located north of the 1# direct track rise at the −810 level, east is the 1302N-1# gob, west is the unprepared 1303N-1# backfilling face, south is the village protective coal pillar, and north is the 1302N#gob protective coal pillar.

The gangue for backfilling in the working face comes from the fully mechanized caving face, the fully mechanized backfilling panel, and the excavated panel produced at the same time. In the simulation, three coal mining faces and two tunneling faces are set. The gangue produced from the 1302N-2# backfilling face goes through the sieving and smashing system; the excavated gangue is directly transported to the gangue pocket. The raw coal mixed with coal and gangue is transferred to the coal–gangue separation system, and the cleaned coal after separation is moved to the transportation roadway through the loading station of the transportation roadway, and finally is lifted to the ground through the main shaft. The gangue is transported to the gangue pocket of the first mining wing for storage along the gangue transport roadway, and transported to the backfilling face for gob through the 1302N-2#tailentry and the return-air rise. The coal gangue logistics and transportation system in the integrated mining–dressing–backfilling system at the Xinjulong coal mine is shown in Figure 3.

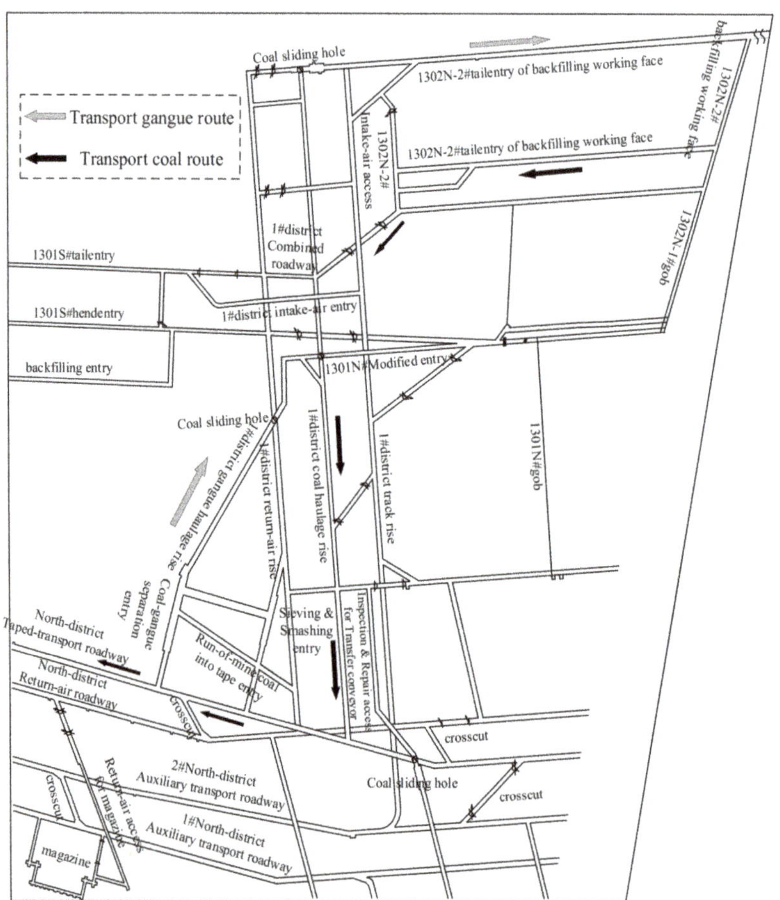

Figure 3. Schematic diagram of coal gangue logistics and transportation system in the Xinjulong coal mine.

The PSO algorithm and the PSO–QNMs algorithm are used to carry out the case study. In this calculation, the gangue selection rate is 95%; the fixed coordinates of transportation links for each logistics system and the flow parameters of working face are given in Table 2.

Furthermore, based on the project background, the input of ground gangue is also taken into account in this analysis.

Table 2. Input parameters.

Parameter	Coal Face 1#	Coal Face 2#	Coal Face 3#	Excavated Panel 1#	Excavated Panel 2#	Shaft Coal Pocket	Gangue Mountain
fixed coordinates	$(1000,3000,-800)$	$(3000,1000,-900)$	$(2000,2000,-850)$	$(1000,1500,-800)$	$(1500,1000,-950)$	$(1500,2000,-750)$	$(1000,2000,-0)$
flow/(t/h)	230	220	240	200	210	—	—
gangue percentage	0.32	0.3	0.3	0.35	0.33	—	—

5.2. Results

In this paper, three groups of experiments are designed. The PSO algorithm and the PSO–QNMs algorithm are set in each group, and the values of the output results are analyzed, based on the two solution algorithms. In the solution process, the same initial value is given to the two intelligent optimization algorithms. In the process of the PSO algorithm, after several iterative solutions, the fitness value of the node particles after each operation was compared with the population optimal solution in the particle swarm. The resulting optimal value was used to replace the population optimal value, and the particle with the most adaptive value was obtained; that is, the objective result of the optimal function was obtained and the calculation was then terminated. In the process of the PSO–QNMs algorithm, the preliminary calculation steps are the same as the PSO algorithm, except that the termination condition of the algorithm is when the calculated gradient value g_k meets the termination error ε of the algorithm. The two algorithms were programmed and solved by using MATLAB. The data of the three groups of experimental results are shown in Table 3. After the program operation, the iterative graphs of the two intelligent optimization algorithms are output, as shown in Figure 4. The spatial location relationship diagrams of logistics nodes of the three groups of intelligent optimization algorithms are output, as shown in Figures 5–7.

Table 3. Simulation experiment results.

Experiments	Algorithm		Separation and Selection Centers	Backfilling and Preparation Centers	Gangue Pockets	Optimal Values /$\times 10^8$	Maximum Number of Iterations	Operation Time /s
Group 1	PSO	XYZ	1675.449 1676.430 −676.581	865.714 1676.819 −331.045	−114.715 639.201 −232.421	3.293	2670	8.840
	PSO–QNMs	XYZ	1499.999 2000.000 −750.000	1460.048 2037.973 −717.835	1460.047 2037.973 −717.835	1.846	53	1.148
Group 2	PSO	XYZ	1133.104 1765.071 −816.158	678.695 697.2961 −866.207	888.8732 661.397 −768.231	3.359	2420	9.633
	PSO–QNMs	XYZ	1500.000 2000.000 −750.000	1454.982 2040.568 −715.002	1454.979 2040.568 −714.999	1.846	52	1.098
Group 3	PSO	XYZ	1592.962 1894.860 −753.395	298.096 1090.314 −225.351	270.873 1231.951 −376.964	3.019	2870	8.681
	PSO–QNMs	XYZ	1500.000 2000.000 −749.999	1456.274 2039.312 −715.927	1456.273 2039.312 −715.927	1.846	60	1.269

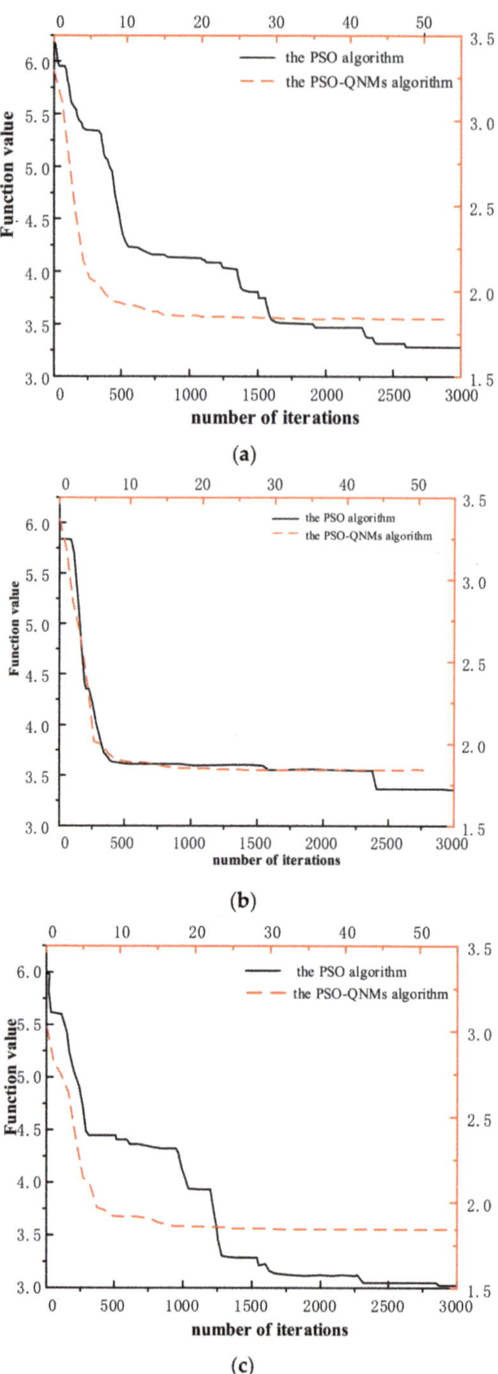

Figure 4. Intelligent optimization method iteration curve: (**a**) group 1, (**b**) group 2, (**c**) group 3.

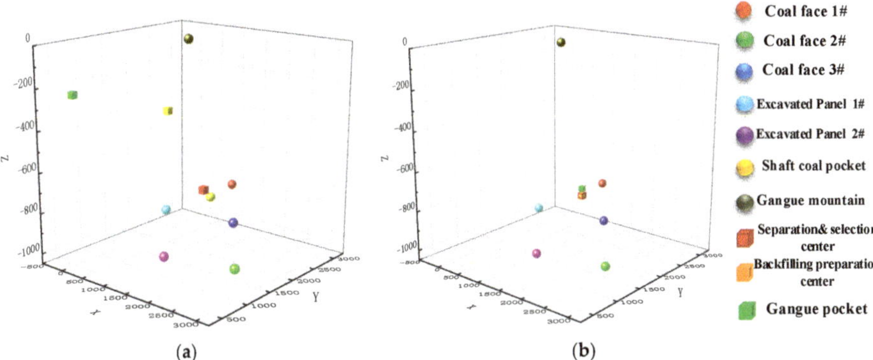

Figure 5. The spatial position relationship diagram of the logistics nodes (group 1): (**a**) PSO algorithm, (**b**) PSO–QNMs algorithm.

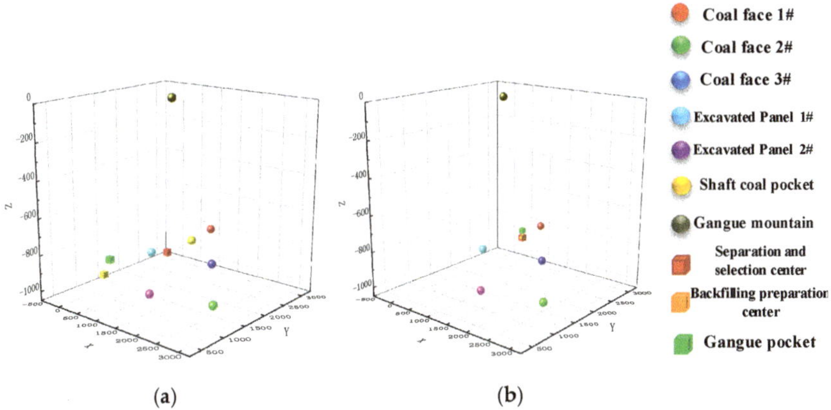

Figure 6. The spatial position relationship diagram of the logistics nodes (group 2): (**a**) PSO algorithm, (**b**) PSO–QNMs algorithm.

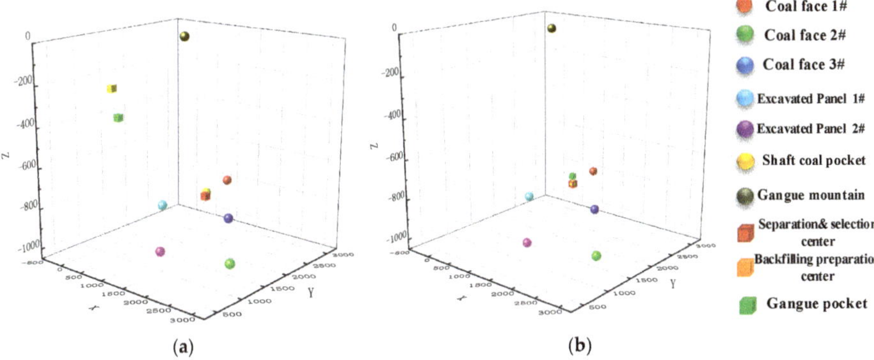

Figure 7. The spatial position relationship diagram of the logistics nodes (group 3): (**a**) PSO algorithm, (**b**) PSO–QNMs algorithm.

5.3. Analysis

(1) Combined with the simulation results (Table 2) and the intelligent optimization algorithm iterative graph (Figure 4), it can be concluded that the PSO–QNMs algorithm, compared with the PSO algorithm, has stronger ability to converge to the global optimal solution, and its convergence speed is significantly higher than that of PSO. The solution time of the whole system operation is about one-eighth of that of the latter. Therefore, the PSO–QNMs algorithm reduces the computational complexity and ensures the global convergence of the algorithm when solving the coal gangue logistics nodes in the underground integration technologies of the mining–dressing–backfilling system. At the same time, when the number of iterations of PSO algorithm is larger, the accuracy of the optimal solution is relatively higher, and then the initial value assigned to the QNMs is better, and the PSO–QNMs algorithm can give full play to its global fine search performance, and the search effect of the optimal solution is better.

(2) According to the optimal function value output from the simulation experiment results in Table 2, it can be concluded that the PSO–QNMs reduces the cost value by about 42.8% compared with the PSO algorithm, indicating that the former has a good approximation effect on the extreme value of the objective function of the coal gangue logistics nodes model, and its accuracy can be improved by 100%.

(3) Comparing the three groups of the intelligent optimization algorithm logistics node location diagram (Figures 5–7), it can be concluded that the PSO selects nonunique coal gangue logistics system location nodes, and the output is not stable. However, the PSO–QNMs algorithm is able to select basically similar but extremely stable results. Therefore, the PSO–QNMs algorithm is more accurate and stable than the PSO, which also demonstrates the superiority of the PSO in solving the nodes in the coal and gangue transportation system of mining and separation within a complex environment.

6. Node Location Decision System for Integration Technologies of the Mining–Dressing–Backfilling System

Based on the research results, a node siting system for an integrated underground mining–dressing–backfilling system in coal mines, referred to as MSBPS, was developed as shown in Figure 8. This is a logistics system node siting output system that is closely related to the research content of integrated deep underground mining, dressing, and backfilling. It only needs to input the spatial location coordinates of specific known nodes and the unit cost between each link, and calculates the solution independently to output the requested optimal node location information, i.e., it can output the spatial coordinates of node locations with one key. At the same time, it can obtain the three-dimensional spatial location relationship between the requested node location and specific known nodes, with a data export function and Chinese and English interface operation, which greatly simplifies the complexity of outputting node locations. It provides a fast, convenient, and accurate image and data export platform for locating nodes deep underground in an integrated mining, dressing, and backfilling logistics system, and is an effective three-dimensional dynamic system platform design tool for conducting further research on deep mining, dressing, and backfilling.

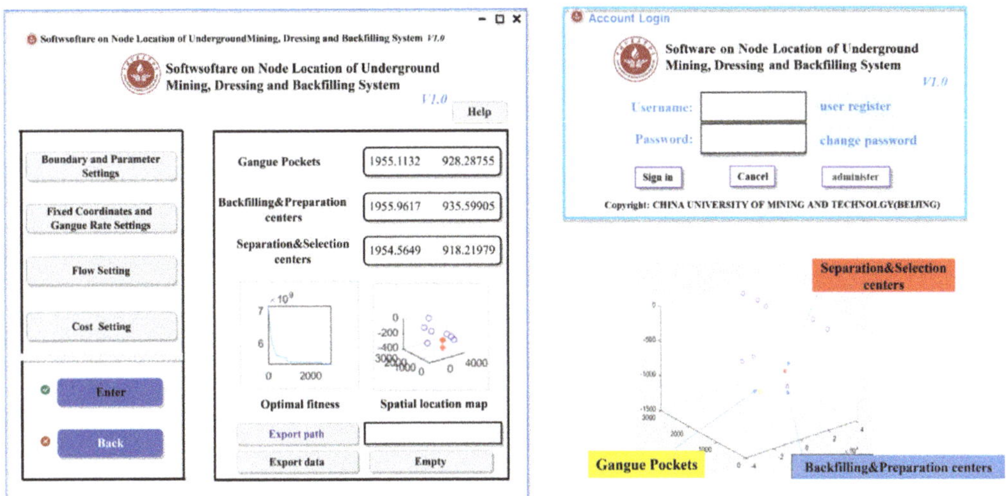

Figure 8. Main Interfaces of the node location decision system.

7. Conclusions

Based on the site selection of coal and gangue logistics nodes in the integration of mining, separation, and filling, this paper provides an in-depth study and proposes an improvement in the precise intelligent optimization algorithm of nodes in underground integration technologies of a mining–dressing–backfilling system. Based on the poor local search ability of the PSO algorithm and the instability of node output results, the PSO–QNMs algorithm was proposed. This algorithm realized a fast and fine search of nodes in the integrated logistics system of coal mining–dressing–backfilling under the action of complex factors. The following main conclusions were obtained:

(1) Based on the complex coal mine environment in China, the relative positions of coal–gangue separation and selection center, backfilling preparation center, and gangue pocket are provided by combining the following components: practical engineering background; mining–dressing–backfilling underground transportation system analysis; construction of coal gangue system node location model; and "high efficiency, lowest cost" as the principal function of a variety of intelligent optimization algorithms. It has practical guiding value in future use of the optimization algorithm of node intelligence for complex underground systems.

(2) Based on the poor global convergence of the PSO algorithm, the PSO-QNMs algorithm is proposed. The results of several groups of simulation experiments have shown that the PSO-QNMs algorithm has stronger convergence than the PSO algorithm in solving the node position of the mining-dressing-backfilling system in complex environment, and the whole operation time is only 1/8 of that of the PSO.

(3) Based on the poor global convergence of the PSO algorithm, the PSO–QNMs algorithm is proposed. The results of several groups of simulation experiments have shown that the PSO–QNMs algorithm has stronger convergence than the PSO algorithm in solving the node position of the mining–dressing–backfilling system in a complex environment, and the whole operation time is only one-eighth that of the PSO algorithm.

(4) In terms of objective function value, the PSO–QNMs algorithm reduces the cost value by about 42.8% compared with the PSO, which optimizes the objective function value, and improves the node optimization efficiency of the mining–dressing–backfilling system within a complex underground environment.

(5) By comparing the performance of the PSO and PSO–QNMs algorithms in the spatial position coordinates and optimal solution of objective function of logistics nodes, it is further confirmed that the PSO–QNMs algorithm is of high precision and provides stable experimental output results. The superiority of the PSO–QNMs algorithm to solve the intellectualized location of coal gangue logistics nodes under complex environment is proven.

(6) Based on the research results, a node siting system for integrated underground mining, processing, and charging systems in coal mines (referred to as MSBPS) was developed, which is an effective tool for further research on three-dimensional dynamic platform design for an integrated deep mining, processing, and charging system.

Author Contributions: S.Y.: conceptualization, data curation, funding acquisition, methodology, project administration, investigation, software, writing—original draft, and writing—review and editing; J.W.: data curation, formal analysis, investigation, software, and writing—original draft; M.L.: data curation, formal analysis, investigation, and software; H.Y.: formal analysis and investigation. All authors have read and agreed to the published version of the manuscript.

Funding: This research was funded by the National Key Research and Development Program of China (no. 2018YFC0604701).

Institutional Review Board Statement: Not applicable.

Informed Consent Statement: Not applicable.

Data Availability Statement: All data used during the study appear in the submitted article.

Conflicts of Interest: The authors declare no conflict of interest.

References

1. He, Q. Research on Key Technology of Integration of Mining, Dressing, Backfilling and Mining in Underground Coal Mine. Ph.D. Thesis, China University of Mining and Technology, Xu Zhou, China, 2014.
2. Wang, J.; Yang, S. Research on support-rock system in solid backfill mining methods. *J. China Coal Soc.* **2010**, *35*, 1821–1826.
3. Wang, J.; Yang, S.; Yang, B. Simulation experiment of overlying strata movement features of longwall with gangue backfill mining. *J. China Coal Soc.* **2012**, *37*, 1256–1262.
4. Zhang, J. Sthdy on Strata Movement Controlling by Raw Waste Backfilling with Fully-Mechanized Coal Winning Technology and Its Applications. Ph.D. Thesis, China University of Mining & Technology, Xu Zhou, China, 2008.
5. Zhang, J.; Zhou, Y.; Huang, Y. Integrated technology of fully mechanized solid backfill mining. *Coal Sci. Technol.* **2012**, *40*, 10–13.
6. Zhang, J.; Zhang, Q.; Ju, F.; Zhou, N.; Li, M.; Sun, Q. Theory and technique of greening mining integrating mining, separating and backfilling in deep coal resources. *J. China Coal Soc.* **2018**, *43*, 377–389.
7. Zhang, J.; Zhang, Q.; Ju, F.; Zhou, N.; Li, M.; Zhang, W. Practice and technique of green mining with integration of mining, dressing, backfilling and X in coal resources. *J. China Coal Soc.* **2019**, *44*, 64–73.
8. Miu, X.; Zhang, J. Key technology of integration of coal mining gangue washing backfilling and coal mining. *J. China Coal Soc.* **2014**, *39*, 1424–1433.
9. Tu, S.; Hao, D.; Li, W.; Liu, X.; Miao, K.; Yang, Z.; Ban, J. Construction of the theory and technology system of selective mining in "mining, dressing, backfilling and X" integrated mine. *J. Min. Saf. Eng.* **2020**, *37*, 81–92.
10. Wang, J.; Zhu, Y.; Feng, L. Research on Resource Allocation Efficiency of Coal Mine Production Logistics System. *Coal Eng.* **2014**, *46*, 114–117.
11. Wang, J.; Du, X.; Feng, L.; Zhai, X. Safety resource allocation model of coal mine production logistics system using stochastic programming. *China Saf. Sci. J.* **2015**, *25*, 16–22.
12. Wang, J.; Zhai, X.; Feng, L. Efficiency Optimization of Coal Mine Production Logistics Under Safety Hard Constraint. *Chin. J. Manag. Sci.* **2014**, *22*, 59–66.
13. Jia, Y.; Du, X.; Wang, J.; Feng, L. Research of security resource allocation model of coal mine production logistics system based on maximum information entropy. *Ind. Mine Autom.* **2016**, *42*, 33–37.
14. Xia, D.; Wang, X. Efficiency simulation of integrated production logistics system for underground coal mining, dressing and filling. *China Min. Mag.* **2020**, *29*, 75–81.
15. Dou, Z.; Shao, Y.; Yuan, Z.; Ji, M. Review on Logistics Node Location Selection. *Logist. Eng. Manag.* **2020**, *42*, 1–4.
16. Zhang, G.; Bao, F. Research on Logistics Centre with Fuzzy Synthesis Evaluation of Entropy Weights Model. *J. Wuhan Univ. Technol.* **2005**, *7*, 91–93.

17. Shamshirband, S.; Khoshnevisan, B.; Yousefi, M.; Bolandnazar, E.; Anuar, N.B.; Wahab, A.W.A.; Khan, S.U.R. A multi-objective evolutionary algorithm for energy management of agricultural systems-A case study in Iran. *Renew. Sustain. Energy Rev.* **2015**, *44*, 457–465. [CrossRef]

18. Chu, T.C.; Lai, M.T. Selecting distribution centre location using an improved fuzzy MCDM approach. *Int. J. Adv. Manuf. Technol.* **2005**, *26*, 293–299. [CrossRef]

19. Zak, J.; Weglinski, S. The selection of the logistics center location based on MCDM/A methodology. In Proceedings of the 17th Meeting of the EURO-Working-Group on Transportation (EWGT), Seville, Spain, 2–4 July 2014; pp. 555–564.

20. Che, Z.H.; Chiang, T.-A.; Che, Z.-G. Using analytic network process and turbo particle swarm optimization algorithm for non-balanced supply chain planning considering supplier relationship management. *Trans. Inst. Meas. Control.* **2012**, *34*, 720–735. [CrossRef]

21. Zhang, R.; Wang, Z. Bi-level Programming Model and Solution Algorithm for Layout of Terminals of City Distribution. *J. Tongji Univ. (Nat. Sci.)* **2012**, *40*, 1035–1040.

22. Guo, X.; Wang, Y. Multi-objective Model for Logistics Distribution Programming Considering Logistics Service Level. *J. Southwest Jiaotong Univ.* **2012**, *47*, 874–880.

23. Cui, L.X. Joint optimization of production planning and supplier selection incorporating customer flexibility: An improved genetic approach. *J. Intell. Manuf.* **2016**, *27*, 1017–1035. [CrossRef]

24. Griffis, S.E.; Bell, J.E.; Closs, D.J. Metaheuristics in Logistics and Supply Chain Management. *J. Bus. Logist.* **2012**, *33*, 90–106. [CrossRef]

25. Niu, D.; Wu, J. A Case Study of Multiple Gravity Method Site Selection. *Mark. Manag. Rev.* **2021**, 152–153. [CrossRef]

26. Velazquez-Marti, B.; Fernandez-Gonzalez, E. Mathematical algorithms to locate factories to transform biomass in bioenergy focused on logistic network construction. *Renew. Energy* **2010**, *35*, 2136–2142. [CrossRef]

27. Almetova, Z.; Shepelev, V.; Shepelev, S. Cargo Transit Terminal Locations According to the Existing Transport Network Configuration. In Proceedings of the 2nd International Conference on Industrial Engineering (ICIE), Chelyabinsk, Russia, 19–20 May 2016; pp. 1396–1402.

28. Huang, M.; Cui, Y.; Yang, S.; Wang, X. Fourth party logistics routing problem with fuzzy duration time. *Int. J. Prod. Econ.* **2013**, *145*, 107–116. [CrossRef]

29. Fang, L.; He, J. Combining the Analytic Hierarchy Process and Goal Programming for Location M odel of Emergency Systems. *Syst. Eng.-Theory Pract.* **2003**, *62*, 116–120.

30. Bi, K.; Yang, M.; Zahid, L.; Zhou, X. A New Solution for City Distribution to Achieve Environmental Benefits within the Trend of Green Logistics: A Case Study in China. *Sustainability* **2020**, *12*, 8312. [CrossRef]

31. Xi, S.-E. Evaluation of Nodes Importance in of Highway Transportation Hubs Planning Based on Fuzzy-AHP. In Proceedings of the 2nd International Conference on Civil Engineering and Transportation (ICCET 2012), Guilin, China, 27–28 October 2013; pp. 1181–1187.

32. Zhang, S.; Cui, R. Logistics Efficiency Network Spatial Structure Based on Coastal City Shandong. *J. Coast. Res.* **2020**, 322–327. [CrossRef]

33. Piao, C.; Hu, H.; Zhang, Y. Logistics distribution vehicle path planning research. In Proceedings of the IEEE International Conference on Artificial Intelligence and Information Systems (ICAIIS), Dalian, China, 20–22 March 2020; pp. 396–399.

34. Zou, Q. Research on Hybrid Hub-Spoke Express Network Based on Multiple Transportation Modes. In Proceedings of the IEEE 7th International Conference on Industrial Engineering and Applications (ICIEA), Bangkok, Thailand, 16–21 April 2020; pp. 905–911.

35. Jing, C.; Destech Publications, I. Research on Location of Sales Logistics Network Nodes Based on Chaotic Optimization Algorithm. In Proceedings of the 3rd International Conference on Vehicle, Mechanical and Electrical Engineering (ICVMEE), Wuhan, China, 30–31 July 2016; pp. 71–75.

36. Sun, Y.; Geng, N.; Gong, S.; Yang, Y. Research on improved genetic algorithm in path optimization of aviation logistics distribution center. *J. Intell. Fuzzy Syst.* **2020**, *38*, 29–37. [CrossRef]

37. Hu, Y.; Zhang, K.; Yang, J.; Wu, Y. Application of Hierarchical Facility Location-Routing Problem with Optimization of an Underground Logistic System: A Case Study in China. *Math. Probl. Eng.* **2018**, *2018*. [CrossRef]

38. Deng, Y.; Zheng, Y.; Li, J. Route optimization model in collaborative logistics network for mixed transportation problem considered cost discount based on GATS. *J. Ambient. Intell. Humaniz. Comput.* **2019**, *10*, 409–416. [CrossRef]

39. Pinho, T.M.; Coelho, J.P.; Veiga, G.; Moreira, A.P.; Boaventura-Cunha, J. Soft computing optimization for the biomass supply chain operational planning. In Proceedings of the 13th APCA International Conference on Control and Soft Computing (CONTROLO), Ponta Delgada, Portugal, 4–6 June 2018; pp. 259–264.

40. Guo, J.; Fang, J.; Gen, M. Dynamic Joint Construction And Optimal Strategy Of Multi-Objective Multi Period Multi-Stage Reverse Logistics Network: A Case Study Of Lead Battery In Shanghai. In Proceedings of the 28th International Conference on Flexible Automation and Intelligent Manufacturing (FAIM)-Global Integration of Intelligent Manufacturing and Smart Industry for Good of Humanity, Columbus, OH, USA, 11–14 June 2018; pp. 1171–1178.

41. Yang, S.; Wang, J.; Deng, X. Research on node location of underground mining, dressing and backfilling system based on particle swarm optimization. *J. Min. Saf. Eng.* **2020**, *37*, 359–365.

42. Cao, Y.; Liu, M.; Xing, Y.; Li, G.; Luo, J.; Gui, X. Current situation and prospect of underground coal preparation technology. *J. Min. Saf. Eng.* **2020**, *37*, 192–201.

43. Alexandridis, A.; Famelis, I.T.; Tsitouras, C. Particle Swarm Optimization for Complex Nonlinear Optimization Problems. In Proceedings of the International Conference On Numerical Analysis And Applied Mathematics 2015 (ICNAAM-2015), Rhodes, Greece, 23–29 September 2015.

44. Xu, M.; Zhang, W.; Qu, R.; Wang, J. An Improved Partial Swarm Optimization Algorithm for Solving Nonlinear Equation Problems. In Proceedings of the 2014 11th World Congress On Intelligent Control And Automation (WCICA), Shenyang, China, 29 June–4 July 2014; pp. 3600–3604.

45. Mo, Y.; Liu, H.; Wang, Q. Conjugate direction particle swarm optimization solving systems of nonlinear equations. *Comput. Math. Appl.* **2009**, *57*, 1877–1882. [CrossRef]

46. Sahu, D.R.; Agarwal, R.P.; Singh, V.K. A Third Order Newton-Like Method and Its Applications. *Mathematics* **2019**, *7*, 31. [CrossRef]

47. Lai, K.K.; Mishra, S.K.; Ram, B. On q-Quasi-Newton's Method for Unconstrained Multiobjective Optimization Problems. *Mathematics* **2020**, *8*, 616. [CrossRef]

48. Ezquerro, J.A.; Hernandez-Veron, M.A. The Newtonian Operator and Global Convergence Balls for Newton's Method. *Mathematics* **2020**, *8*, 1074. [CrossRef]

49. Wang, X.; Tao, Y. A New Newton Method with Memory for Solving Nonlinear Equations. *Mathematics* **2020**, *8*, 108. [CrossRef]

50. Antonio Ezquerro, J.; Angel Hernandez-Veron, M. How to Obtain Global Convergence Domains via Newton's Method for Nonlinear Integral Equations. *Mathematics* **2019**, *7*, 553. [CrossRef]

 *mathematics*

Article

Symbolic Regulator Sets for a Weakly Nonlinear Discrete Control System with a Small Step

Yulia Danik * and Mikhail Dmitriev

Federal Research Center "Computer Science and Control" of the Russian Academy of Sciences, 119333 Mocow, Russia; mdmitriev@mail.ru
* Correspondence: yuliadanik@gmail.com

Abstract: For a class of discrete weakly nonlinear state-dependent coefficient (SDC) control systems, a suboptimal synthesis is constructed over a finite interval with a large number of steps. A one-point matrix Padé approximation (*PA*) of the solution of the initial problem for the discrete matrix Riccati equation is constructed based on the state-dependent Riccati equation (SDRE) approach and the asymptotics by the small-step of the boundary layer functions method. The symmetric gain coefficients matrix for Padé control synthesis is constructed based on the one-point *PA*. As a result, the parametric closed-loop control is obtained. The results of numerical experiments illustrate, in particular, the improved extrapolation properties of the constructed regulator, which makes the algorithm applicable in control systems for a wider range of parameter variation.

Keywords: discrete control systems; weakly nonlinear systems; small step; the SDRE approach; matrix discrete Riccati equation; the boundary layer functions method; Padé approximation; finite time interval

Citation: Danik, Y.; Dmitriev, M. Symbolic Regulator Sets for a Weakly Nonlinear Discrete Control System with a Small Step. *Mathematics* **2022**, *10*, 487. https://doi.org/10.3390/math10030487

Academic Editors: Ioannis G. Stratis and Quanxin Zhu

Received: 9 November 2021
Accepted: 30 January 2022
Published: 2 February 2022

Publisher's Note: MDPI stays neutral with regard to jurisdictional claims in published maps and institutional affiliations.

1. Introduction

In the literature, much attention is paid to the construction of optimal control laws for nonlinear systems and the corresponding approximate methods for their calculation. This can be explained by several factors; on the one hand, the greater accuracy of the description of dynamic systems in applications leads to the increase of their mathematical models' dimension, and on the other hand, the calculations often need to be carried out in real time. This is especially true for finding feedback laws in nonlinear control systems, where the consideration of even weak nonlinearity in constructing synthesis laws based on linear control laws can lead to a significant improvement in the value of the quality criterion. For linear control systems, the Kalman algorithm is often used, which allows stabilizing feedbacks to be built that, in addition to the asymptotic stability of closed systems, provide the optimality by the quadratic quality criterion.

The application field of the Kalman algorithm has been expanded to nonlinear control systems by the so-called state-dependent Riccati equation (SDRE) approach for continuous (see [1–3]) and discrete (see [4–9]) cases, where the systems are formally represented as linear systems in terms of state and control, the coefficients of the matrices are the functions of the state vector (state-dependent coefficients (SDC) systems), and the quality criterion is quadratic, but the quadratic forms matrices in the criterion can also be state-dependent. In [1], the design procedure for the SDRE approach is described and its capabilities are illustrated on a benchmark problem. Reference [2] contains a detailed review on SDRE: the theory developed to date, characteristics, advantages, and open issues. In [4–6], the discrete state-dependent Riccati equation (D-SDRE) approach is described, and the corresponding control algorithms are presented and tested. In [3,7–9], the development of the SDRE approach is proposed based on the asymptotic theory.

Often, the mathematical models in applied control problems contain parameters, the variation of which generates a family of admissible controls and the corresponding trajectories, which leads to new problems of their approximate analytical study and description. An approximation is a procedure for choosing the optimal approximating function from a certain class of functions, that describes the behavior of the given function. Approximation allows the numerical characteristics and qualitative properties of an object to be studied, reducing the problem into studying simpler or more convenient objects. Moreover, the solution of these problems significantly reduces the time for rational control synthesis and selection. Various methods of approximation theory including splines and fractional-rational Padé approximations [10,11], where the latter are constructed using asymptotic expansions by the corresponding small parameters, can be used for such problems. In particular, in the works [10,11], the possibility of the construction of symbolic families of stabilizing regulators for some classes of regularly and singularly perturbed nonlinear continuous control systems are demonstrated by constructing Padé approximations (*PA*) of the gain coefficients matrices [10,11], associated with asymptotic expansions of the matrix algebraic Riccati equations solutions into regular series in integer powers of the corresponding small parameter, both for the case of one-point and two-point *PA*, where the latter uses not only the asymptotic expansions by integer powers of the parameter but also the asymptotic expansions over the inverse powers of a positive small parameter. A regulator is a closed-loop system, used to maintain the desired system output, usually, zero.

In this paper, discrete nonlinear control systems with a small step are considered within the framework of the SDRE approach for continuous systems on a finite time interval [12]. It is shown that an approximate symbolic description of the family of gain matrices in the feedback control loop using *PA* can be obtained if the step in the discrete system is sufficiently small. Note that the optimal control problem for a discrete-time system with a small step is singularly perturbed and that is why the zones of exponentially decreasing boundary layers can arise near the boundary points in the trajectories and in the solution of problems for finding gain coefficients in the feedback circuit. This happens because the degenerate solution, for zero value of the parameter, does not account for the boundary conditions and there is a large discrepancy from the exact solution near the boundary points. The boundary functions (members of the boundary series) are significant in the neighborhood of the boundary points and together with the members of the regular series, describe the behavior of the solution in the boundary layer, and outside the boundary layer, they rapidly decrease. The solution of the perturbed problem outside these boundary layer zones is in a small neighborhood of the solution of degenerate (limiting) problems, one of which, in this case, is the matrix algebraic Riccati equation of the corresponding discrete linear-quadratic optimal control problem [13], where the coefficients of all the matrices may be the functions of the state vector.

For the first time, the discrete problems with a small step were considered as singularly perturbed by A.B. Vasil'eva with the help of the boundary layer functions method (BLFM) (see [14,15]). The asymptotic methods and singular perturbations theory can also be successfully applied to control problems, for example, see reviews [16,17]. The discrete optimal control problem was first considered as singularly perturbed in [18] on the example of a linear-quadratic control problem with a small step, where its asymptotics by BLFM is constructed. The asymptotics of the solution of singularly perturbed discrete nonlinear optimal control problems with a small step was constructed by BLFM in [19], using the so-called direct scheme [20].

In this paper, the asymptotic solution of the corresponding initial singularly perturbed problem for the discrete matrix Riccati equation with coefficients weakly dependent on the state and the corresponding one-point *PA* regulator is constructed using the SDRE approach. Only one asymptotic expansion of the matrix Riccati equation solution by a small parameter, which equals the time step size and the multiplier in front of the non-linearities, is used to construct the *PA* for the gain matrix in the feedback loop. *PAs* are constructed based on asymptotic expansions and are actively used in applications since they often extend the range of the parameter variation, where *PAs* approach the exact solutions with given accuracy and reproduce the qualitative picture of the solutions. The use of a qualitative picture allows the approximate solutions with higher accuracy to be obtained by using the results of the asymptotic analysis as an initial approximation for traditional nonlinear programming algorithms to minimize the residuals of the equations.

The results of numerical experiments are presented, which demonstrate the possibility of using this approach for the construction of stabilizing regulators for nonlinear discrete systems, and also show that the proposed algorithm is applicable for discrete control systems for a wider interval of the step value and the perturbation parameter variation.

The outline of the paper is as follows. In Section 2, we will describe the discrete state-dependent Riccati equation approach (SDRE) for small step discrete pseudo-linear control systems and construct the uniform second-order asymptotic approximation of the solution of the singularly perturbed initial problem for the difference Riccati equation using the boundary layer functions method (BLFM). In Section 3, we formulate the algorithm for the discrete one-point Padé regulator design. In Section 4, we will review the numerical experiments that demonstrate the work of the proposed control algorithms.

We start by listing definitions and notations: $R := (-\infty, \infty)$, $R^{n \times n}$—vector space of n-by-n matrices, ^{T}—the transpose operation, $\otimes$—the Kronecker product, E_n—the identity matrix of size $n \times n$, $\overline{P}$—the regular series of P, ΠP—the boundary series of P, $\overline{P}_k$—kth term of the $\overline{P}$ series, $\| \ \|$ the L_2 matrix norm.

2. An SDRE Approach for Small Step Discrete Control Systems

2.1. Asymptotic Expansion of the Discrete Riccati Equation Solution

Let's consider an affine (linear in control) discrete system,

$$x(t + \varepsilon) = \widehat{A}(x) + B(x)u, \ x(0) = x_0, \tag{1}$$

where $x(t)$ is an n-dimensional state, $u(t)$—is an r-dimensional control, $t \in T = \{t : t = k\varepsilon, k = 0, 1, \ldots, N - 1\} \subset \{t : 0 \leq t \leq 1\}$, $N = \frac{T}{\varepsilon}$, $\varepsilon > 0$, is a small time step, used as a small parameter. Further, to transform the system into a formally linear form, we use the method of "extended linearization" (see [1,21,22]), which gives a non-unique representation in the vector case, by presenting $\widehat{A}(x)$ in the form $\widehat{A}(x) = A(x)x$. By analogy, let us call the well-known heuristic technique for the introduction of a small parameter in the system right-hand side the "extended perturbation" method. In the last case, a small parameter is introduced in matrices $A(x)$ and $B(x)$ instead of a selected coefficient that approximately equals 1.

Let's demonstrate the last approach by transforming an arbitrary nonlinear function $\Theta(x)$, for example, as follows: $\Theta(x) = \Theta_0 + 1 \cdot (\Theta(x) - \Theta_0) = \Theta_0 + \frac{\varepsilon}{\delta} \cdot (\Theta(x) - \Theta_0) = \Theta_0 + \varepsilon\Theta_1(x)$, where $\Theta_1(x) = \frac{1}{\delta} \cdot (\Theta(x) - \Theta_0)$ and $1 \equiv \frac{\varepsilon}{\delta}$, but, at the same time, $\delta = \varepsilon$, and now δ becomes a constant independent of ε. So, this technique by analogy with the previous one will be called the "extended" perturbation technique.

After application of these two approaches—"extended" linearization and "extended" perturbation—system (1) may be presented in the form

$$x(t + \varepsilon) = A(x, \varepsilon)x + B(x, \varepsilon)u, \ x(0) = x_0,$$
$$x(t) \in X \in R^n, \ u \in R^r, \ t = k\varepsilon, k = 0, 1, \ldots, N - 1, \ 0 < \varepsilon \leq \varepsilon_0, \tag{2}$$

where $A(x, \varepsilon) = A_0 + \varepsilon A_1(x)$, $B(x, \varepsilon) = B_0 + \varepsilon B_1(x)$, ε_0 is a small enough positive number, A_0, B_0 are constant matrices, $A_0, A_1(x) \in R^{n \times n}$, $B_0, B_1(x) \in R^{n \times r}$, $\forall x \in X \subset R^n$ and X is a certain fixed bounded state space subset. For sufficiently small values of ε, systems (2) are called singularly perturbed and pseudo-linear in state and control with regularly perturbed coefficients so that near the interval endpoints the solutions of (2) may have the boundary layers behavior.

Let's consider a cost function that measures the system performance for different controls in order to compare them and select the most favorable one;

$$J(u) = \frac{1}{2}x^T(N)Fx(N) + \frac{1}{2}\sum_{k=0}^{N-1}\left(x^T(k\varepsilon)Q(x,\varepsilon)x(k\varepsilon) + u^T(k\varepsilon)Ru(k\varepsilon)\right) \to \min_u, \qquad (3)$$

where $Q(x,\varepsilon) = Q_0 + \varepsilon Q_1(x) + \varepsilon^2 Q_2(x) > 0$, $R > 0$, Q_0, R are constant matrices, $Q_0 > 0$, $Q_1(x) > 0$, $Q_2(x) > 0 \ \forall x \in X$, $\varepsilon \in (0,\varepsilon_0)$ and $F > 0$.

To construct a feedback control for discrete systems on a finite time interval, we will use the necessary optimality conditions here, in contrast to [12], where a dynamic programming scheme was used in the continuous case.

For problems (1) and (2), we introduce the Hamiltonian

$$H(x,u,\psi,t) = \psi^T(t)[A(x,\varepsilon)x(t) + B(x,\varepsilon)u(t)] - \frac{1}{2}[x^T(t)Q(x,\varepsilon)x(t) + u^T(t)R_0u(t)]. \quad (4)$$

Using the necessary optimality conditions, we have

$$x(t+\varepsilon) = A(x,\varepsilon)x + B(x,\varepsilon)u, \qquad (5)$$

$$
\begin{aligned}
\psi(t) &= \frac{\partial H(x(t),u(t),\psi(t+\varepsilon),t)}{\partial x} = \left\{\frac{\partial[A(x,\varepsilon)x(t)]}{\partial x} + \frac{\partial[B(x,\varepsilon)u(t)]}{\partial x}\right\}^T \psi(t+\varepsilon) - \\
&- \frac{1}{2}\frac{\partial[x^T(t)Q(x,\varepsilon)x(t)]}{\partial x(t)} = \left\{(x^T\otimes E_n)_{n\times n^2}\left[\frac{\partial A(x,\varepsilon)}{\partial x}\right]_{n^2\times n} + A(x,\varepsilon)_{n\times n} + (u^T\otimes E_n)_{n\times nr}\left[\frac{\partial B(x,\varepsilon)}{\partial x}\right]_{nr\times n}\right\}^T \psi(t+\varepsilon) - \\
&- \frac{1}{2}Q(x,\varepsilon)_{n\times n}x - \frac{1}{2}\left[\frac{\partial Q(x,\varepsilon)}{\partial x}\right]^T_{n\times n^2}(x\otimes E_n)_{n^2\times n}x - \frac{1}{2}Q(x,\varepsilon)^T x \\
0 &= \frac{\partial H(x(t),u,\psi(t+\varepsilon),t)}{\partial u} = B^T(x,\varepsilon)\psi(t+\varepsilon) - Ru,
\end{aligned}
\qquad (6)
$$

where $\otimes$ stands for Kronecker matrix product and E_n is a $n \times n$ unity matrix.

From (6) we obtain the next expression for the control

$$u(t) = R^{-1}B^T(x,\varepsilon)\psi(t+\varepsilon). \qquad (7)$$

Using the representation $\psi(t) = -P_\varepsilon(x,t)x(t)$, we obtain the following expression instead of (7):

$$u(x,t,\varepsilon) = -\left\{R + B^T(x,\varepsilon)P(x,t+\varepsilon)B(x,\varepsilon)\right\}^{-1}B^T(x,\varepsilon)P(x,t+\varepsilon)A(x,\varepsilon)x(t) = K(x,\varepsilon,t+\varepsilon)x(t), \qquad (8)$$

where $P(x,t,\varepsilon)$ must satisfy the discrete matrix Riccati-type equation with the zero-control matrix and as a result the missing quadratic nonlinearity part

$$
\begin{aligned}
&-P(x,t,\varepsilon) + \left\{(x^T\otimes E_n)_{n\times n^2}\left[\frac{\partial A(x,\varepsilon)}{\partial x}\right]_{n^2\times n} + A(x,\varepsilon)_{n\times n} + (x^T K(x,\varepsilon)^T\otimes E_n)_{n\times nr}\left[\frac{\partial B(x,\varepsilon)}{\partial x}\right]_{nr\times n}\right\}^T \times \\
&\times P(x,t+\varepsilon,\varepsilon) \times \\
&\times \left\{E - B(x,\varepsilon)[R + B^T(x,\varepsilon)P(x,t+\varepsilon,\varepsilon)B(x,\varepsilon)]^{-1}B^T(x,\varepsilon)P(x,t+\varepsilon,\varepsilon)\right\}A(x,\varepsilon) + \\
&+ Q(x,\varepsilon)_{n\times n} + \frac{1}{2}\left[\frac{\partial Q(x,\varepsilon)}{\partial x}\right]^T_{n\times n^2}(x\otimes E_n)_{n^2\times n} = \Phi(P,t,\varepsilon) = 0,
\end{aligned}
$$

with the initial condition at the end of the interval $P(x,T,\varepsilon) = F$. The latter can be represented as

$$
\begin{aligned}
\Phi(P,t,\varepsilon) &= -P(x,t,\varepsilon) + A^T(x,\varepsilon)_{n\times n}P(x,t+\varepsilon,\varepsilon)A(x,\varepsilon) - \\
&- A^T(x,\varepsilon)_{n\times n}P(x,t+\varepsilon,\varepsilon)B(x,\varepsilon)[R + B^T(x,\varepsilon)P(x,t+\varepsilon,\varepsilon)B(x,\varepsilon)]^{-1} \times \\
&\times B^T(x,\varepsilon)P(x,t+\varepsilon,\varepsilon)A(x,\varepsilon) + Q(x,\varepsilon)_{n\times n} + \varepsilon\Omega(P(x,t+\varepsilon),x,t+\varepsilon) = \\
&= \Psi(P,t,\varepsilon) + \varepsilon\Omega(P(x,t+\varepsilon),x,t+\varepsilon) = 0, \quad P(x,T,\varepsilon) = F,
\end{aligned}
\qquad (9)
$$

where $\Psi(P,t,\varepsilon)$ is the usual discrete matrix Riccati operator and
$K(x,\varepsilon) = -\{R + B^T(x,\varepsilon)P(x,t+\varepsilon,\varepsilon)B(x,\varepsilon)\}^{-1}B^T(x,\varepsilon)P(x,t+\varepsilon,\varepsilon)A(x,\varepsilon).$

The difference from the regular discrete difference Riccati equation on a finite interval lies in an additional, but cumbersome, term on the right-hand side with the function

$$\Omega(P(x,t+\varepsilon),x,t+\varepsilon) =$$

$$= \left\{ \begin{array}{l} (x^T \otimes E_n)_{n \times n^2}\left[\dfrac{\partial A_1(x)}{\partial x}\right]_{n^2 \times n} - \\[2mm] -\left(x^T\left(\{R + B^T(x,\varepsilon)P(x,t+\varepsilon,\varepsilon)B(x,\varepsilon)\}^{-1}B^T(x,\varepsilon)P(x,t+\varepsilon,\varepsilon)A(x,\varepsilon)\right)^T \otimes E_n\right)_{n \times nr} \times \\[2mm] \times \left[\dfrac{\partial B_1(x)}{\partial x}\right]_{nr \times n} \end{array} \right\}^T \times$$

$$\times P(x,t+\varepsilon,\varepsilon)\left\{E - B(x,\varepsilon)[R + B^T(x,\varepsilon)P(x,t+\varepsilon,\varepsilon)B(x,\varepsilon)]^{-1}B^T(x,\varepsilon)P(x,t+\varepsilon,\varepsilon)\right\}A(x,\varepsilon) +$$

$$+\tfrac{1}{2}\left[\dfrac{\partial(Q_1(x)+\varepsilon Q_2(x))}{\partial x}\right]^T_{n \times n^2}(x \otimes E_n)_{n^2 \times n}.$$

Let's denote the matrix of the closed-loop (cl) system for (1) as
$A_{cl}(x,\varepsilon) = (A(x,\varepsilon) - B(x,\varepsilon)R^{-1}B^T(x,\varepsilon)K(x,\varepsilon))x.$

Taking into account the dependency of matrices $Q(x,\varepsilon)$, $A(x,\varepsilon)$, $B(x,\varepsilon)$ on the small parameter a uniform asymptotic approximation of the second order for the solution of the singularly perturbed initial problem (9) is constructed using the boundary layer functions method (BLFM) [14,15]. The solution is found in the reverse time in the following form:

$$P(x,t,\varepsilon) = \overline{P}_2(x,t,\varepsilon) + \Pi_2 P(x,\tau,\varepsilon), \ \tau = (t-T)/\varepsilon = -1,-2,\ldots,-N, \ N = T/\varepsilon,$$

where $\overline{P}_2(x,t,\varepsilon) = \overline{P}_0 + \varepsilon\overline{P}_1(x,t) + \varepsilon^2\overline{P}_2(x,t)$,
$\Pi_2 P(x,\tau,\varepsilon) = \Pi_0 P(x,\tau) + \varepsilon\Pi_1 P(x,\tau)) + \varepsilon^2\Pi_2 P(x,\tau))$, are the second-order partial sums of the regular and boundary power series by parameter ε, respectively.

Now the operator Φ is transformed in the following way:

$$\Phi(P,t,\varepsilon) = \Phi\big(\overline{P}(x,t,\varepsilon),t,\varepsilon\big) + \Phi\big(\overline{P}(x,\tau\varepsilon + T,\varepsilon) + \Pi P(x,\tau,\varepsilon),\tau,\varepsilon\big) - \Phi\big(\overline{P}(x,\tau\varepsilon + T,\varepsilon),\tau\varepsilon + T,\varepsilon\big) =$$
$$= \overline{\Phi} + \Pi\Phi,$$

where $\overline{\Phi}(x,t,\varepsilon) = \Phi(\overline{P}(x,t,\varepsilon),t,\varepsilon) = \Phi_0(\overline{P}_0,t) + \varepsilon\Phi_1(\overline{P}_0,t) + \ldots$, $\Pi\Phi(x,\tau\varepsilon + T,\varepsilon) = \Phi\big(\overline{P}(x,\tau\varepsilon + T,\varepsilon) + \Pi P(x,\tau,\varepsilon),\tau\varepsilon + T,\varepsilon\big) - \Phi\big(\overline{P}(x,\tau\varepsilon + T,\varepsilon),\tau\varepsilon + T,\varepsilon\big)$. Substituting the representations for $\overline{P}_2(x,t,\varepsilon)$, $\Pi_2 P(x,\tau,\varepsilon)$ into the equation and the initial condition in (9) and equating the terms with the same powers of the small parameter, we obtain a discrete algebraic Riccati equation (DARE) for the zero term of the regular series $\overline{P}_0$

$$A_0^T\overline{P}_0 A_0 - \overline{P}_0 - A_0^T\overline{P}_0 B_0(R_0 + B_0^T\overline{P}_0 B_0)^{-1}B_0^T\overline{P}_0 A_0 + Q_0 = 0, \tag{10}$$

and for the first term of the regular series $\overline{P}_1(x)$, we obtain the matrix discrete algebraic Lyapunov equation

$$A_{cl}^{0\,T}\overline{P}_1(x)A_{cl}^0 - \overline{P}_1(x) = -G_1(x), \tag{11}$$

where $A_{cl}^0 = A_0 + B_0 K_0 = A_0 - B_0[R_0 + B_0^T\overline{P}_0 B_0]^{-1}B_0^T\overline{P}_0 A_0 -$ is the matrix of the linear closed-loop system (cl) and $G_1(x) = D_0(x) + D_1(x) - D_2(x) - D_3(x) + D_4(x)$,

$$D_0(x) = \left\{\left[\dfrac{\partial A_1(x)}{\partial x}\right]^T(x \otimes E_n) - \left[\dfrac{\partial B_1(x)}{\partial x}\right]^T\big([(R + B_0^T\overline{P}_0 B_0)^{-1}B_0^T\overline{P}_0 A_0]x \otimes E_n\big)\right\}\overline{P}_0[A_0 - B_0[R + B_0^T\overline{P}_0 B_0]^{-1}B_0^T\overline{P}_0 A_0] +$$

$$+Q_1(x) + \tfrac{1}{2}\left[\dfrac{\partial Q_1(x)}{\partial x}\right]^T_{n \times n^2}(x \otimes E_n)_{n^2 \times n},$$

$$D_1(x) = A_0^T\overline{P}_0 A_1 + A_1^T(x)\overline{P}_0 A_0,$$

$$D_2(x) = A_0^T\overline{P}_0 B_1[R + B_0^T\overline{P}_0 B_0]^{-1}B_0^T\overline{P}_0 A_0 + A_0^T\overline{P}_0 B_0[R + B_0^T\overline{P}_0 B_0]^{-1}B_1^T\overline{P}_0 A_0,$$

$$D_3(x) = A_0^T\overline{P}_0 B_0[R + B_0^T\overline{P}_0 B_0]^{-1}B_0^T\overline{P}_0 A_1 + A_1^T(x)\overline{P}_0 B_0[R + B_0^T\overline{P}_0 B_0]^{-1}B_0^T\overline{P}_0 A_0,$$

$$D_4(x) = A_0^T\overline{P}_0 B_0[R + B_0^T\overline{P}_0 B_0]^{-1}B_0^T\overline{P}_0 B_1[R + B_0^T\overline{P}_0 B_0]^{-1}B_0^T\overline{P}_0 A_0 +$$

$$+A_0^T\overline{P}_0 B_0[R + B_0^T\overline{P}_0 B_0]^{-1}B_1^T\overline{P}_0 B_0[R + B_0^T\overline{P}_0 B_0]^{-1}B_0^T\overline{P}_0 A_0.$$

For the second term of the regular series $P_2(x)$, the matrix discrete algebraic Lyapunov equation has the same form as Equation (11) for $P_1(x)$,

$$A_{cl}^{0}{}^{T}\overline{P}_2(x)A_{cl}^{0} - \overline{P}_2(x) = -G_2(x),$$

where $G_2(x)$ has a similar structure as $G_1(x)$ but it is a more complex function of the found expansion terms and matrices of the system, and we omit its representation here.

For the zero term of the boundary series $\Pi_0 P(\tau)$, we obtain the difference initial problem

$$\begin{aligned}
\Pi_0 P(\tau) &= -\overline{P}_0 + A_0{}^T(\overline{P}_0 + \Pi_0 P(\tau+1))A_0 - A_0{}^T(\overline{P}_0 + \Pi_0 P(\tau+1))B_0 \times \\
&\quad \times [R + B_0{}^T(\overline{P}_0 + \Pi_0 P(\tau+1))B_0]^{-1} B_0{}^T(\overline{P}_0 + \Pi_0 P(\tau+1))A_0 + Q_0, \\
\Pi_0 P(0) &= F - \overline{P}_0,
\end{aligned} \tag{12}$$

and for the first term of the boundary series $\Pi_1 P(x,\tau)$, we have the following discrete problem:

$$\begin{aligned}
\Pi_1 P(x,\tau) &= \zeta^T(\tau)\Pi_1 P(\tau+1)\zeta(\tau) + \Pi_1 G(x,\tau), \\
\Pi_1 P(0) &= -\overline{P}_1(x), \quad \tau = -1, -2, \ldots,
\end{aligned} \tag{13}$$

where $\zeta(\tau) = A_0 - B_0[R + B_0{}^T(\overline{P}_0 + \Pi_0 P(\tau+1))B_0]^{-1} B_0{}^T(\overline{P}_0 + \Pi_0 P(\tau+1))A_0$, and $\Pi_1 G(x,\tau)$ has a complex structure

$$\begin{aligned}
\Pi_1 G(x,\tau) &= A_0{}^T(\overline{P}_0 + \Pi_0 P(\tau+1))B_0(R_0 + B_0{}^T(\overline{P}_0 + \Pi_0 P(\tau+1))B_0)^{-1} B_0{}^T \overline{P}_1(x)B_0 \times \\
&\quad \times (R_0 + B_0{}^T(\overline{P}_0 + \Pi_0 P(\tau+1))B_0)^{-1} B_0{}^T(\overline{P}_0 + \Pi_0 P(\tau+1))A_0 - \\
&\quad - A_0{}^T(\overline{P}_0 + \Pi_0 P(\tau+1))B_0(R_0 + B_0{}^T(\overline{P}_0 + \Pi_0 P(\tau+1))B_0)^{-1} B_0{}^T \overline{P}_1(x)A_0 \\
&\quad + A_0{}^T \overline{P}_1(x)A_0 - B_0(R_0 + B_0{}^T(\overline{P}_0 + \Pi_0 P(\tau+1))B_0)^{-1} B_0{}^T(\overline{P}_0 + \Pi_0 P(\tau+1))A_0) + \\
&\quad - \overline{P}_1(x) + A_0{}^T(\overline{P}_0 + \Pi_0 P(\tau+1))A_1(x) + \\
&\quad + A_0{}^T(\overline{P}_0 + \Pi_0 P(\tau+1))(-B_0(R_0 + B_0{}^T(\overline{P}_0 + \Pi_0 P(\tau+1))B_0)^{-1} \times \\
&\quad \times (B_1{}^T(x)(\overline{P}_0 + \Pi_0 P(\tau+1))A_0 + B_0{}^T(\overline{P}_0 + \Pi_0 P(\tau+1))A_1(x)) + \\
&\quad + B_0(R_0 + B_0{}^T(\overline{P}_0 + \Pi_0 P(\tau+1))B_0)^{-1} \times \\
&\quad \times (B_1{}^T(x)(\overline{P}_0 + \Pi_0 P(\tau+1))B_0 + B_0{}^T(\overline{P}_0 + \Pi_0 P(\tau+1))B_1(x)) \times \\
&\quad \times (R_0 + B_0{}^T(\overline{P}_0 + \Pi_0 P(\tau+1))B_0)^{-1} B_0{}^T(\overline{P}_0 + \Pi_0 P(\tau+1))A_0 - \\
&\quad - B_1(x)(R_0 + B_0{}^T(\overline{P}_0 + \Pi_0 P(\tau+1))B_0)^{-1} B_0{}^T(\overline{P}_0 + \Pi_0 P(\tau+1))A_0) + \\
&\quad + \left(\left[\frac{\partial A_1(x)}{\partial x}\right]^T (x \otimes E_n) + A_1{}^T(x) - \right. \\
&\quad - \left[\frac{\partial B_1(x)}{\partial x}\right]^T \left(\left((R_0 + B_0{}^T(\overline{P}_0 + \Pi_0 P(\tau+1))B_0)^{-1} B_0{}^T(\overline{P}_0 + \Pi_0 P(\tau+1))A_0\right) x \otimes E_n\right) \times \\
&\quad \times (\overline{P}_0 + \Pi_0 P(\tau+1))\left(A_0 - B_0(R_0 + B_0{}^T(\overline{P}_0 + \Pi_0 P(\tau+1))B_0)^{-1} B_0{}^T(\overline{P}_0 + \Pi_0 P(\tau+1))A_0\right).
\end{aligned}$$

For $\Pi_2 P(x,\tau)$, we obtain a discrete initial problem similar to (12), but $\Pi_2 G(x,\tau)$ has an even more cumbersome structure than $\Pi_1 G(x,\tau)$ and we omit this representation here.

Let us introduce the conditions:

I. *Coefficients of matrices $A_1(x)$, $B_1(x)$, $Q_1(x)$, $Q_2(x)$ are continuous and bounded functions on* X, $g(0,\varepsilon) \equiv 0$ ($g(x,\varepsilon) = A_{cl}(x(t),\varepsilon)x(t))$ *and the parameter ε belongs to a bounded interval $(0, \varepsilon_0]$, the solution to problems (1) and (2) exists and is bounded for all admissible x_0, ε, t;*

II. *The triple of matrices $(A_0, B_0, Q_0^{\frac{1}{2}})$ is controllable and observable.*

By definition, a system is said to be controllable, if it is possible to transfer the system state from any initial state to any desired state within a finite interval of time. A system is said to be observable if every state can be completely identified by measurements of the outputs at a finite time interval.

The following is true.

Theorem 1. *Under conditions I–II, there is a sufficiently small value of $\varepsilon_0 > 0$, such that for all $x \in X$, $t \in [0, T]$, $\varepsilon \in (0, \varepsilon_0]$, $\tau = -1, -2, \ldots, -N, \ldots$ the following statements hold:*

1. *All eigenvalues of the matrix $A_{cl}^0 = A_0 - B_0[R_0 + B_0{}^T \overline{P}_0 B_0]^{-1} B_0{}^T \overline{P}_0 A_0$ are inside the unit circle, where $\overline{P}_0$ is a positive definite solution to Equation (10).*
2. *Solution of (9) exists, is unique and the following estimate for the remainder of the second-order asymptotics is valid (the L_2 norm is used):*

$$\|P(x, t, \varepsilon) - (\overline{P}_2(x, t, \varepsilon) + \Pi_2 P(x, \tau, \varepsilon))\| = O(\varepsilon^3)$$
$$x \in X, \ t \in [0, T], \ \varepsilon \in (0, \varepsilon_0], \ \tau = -1, -2, \ldots, -N, \ldots \tag{14}$$

Proof of Theorem 1. The statements of the theorem for each $x \in X$ follow from the corresponding proof schemes of the statements in paragraph 2 of paper [18], where the similar linear-quadratic problem is considered, but with time-dependent matrix coefficients. The statements here generalize the similar results presented in [18] since in comparison with [18] the initial problem (9) is additionally regularly perturbed with continuously differentiable perturbation $\varepsilon\Omega$ on the right-hand side of the matrix discrete Riccati equation. Therefore, here we present only the new components of the proof.

As the associated system for the singularly perturbed problem (9) for $\varepsilon = 0$, $\Omega \equiv 0$ coincides with the analogous one in [18], where it is shown that the positive definite root $\overline{P}_0$ of the limiting algebraic discrete matrix Riccati equation is an asymptotically stable equilibrium point of system (9) for $\tau \to -\infty$ and matrix F belongs to the domain of influence of this root. Further, in [18], the form of the main functional matrix Γ of the associated system to problem (9) is established, which is calculated by setting $\varepsilon = 0$ in the right-hand side and using the Kronecker product of matrices can be represented in the form $\Gamma = A_{cl}^0{}^T \otimes A_{cl}^0$. From the properties of the spectrum of the constant matrix A_{cl}^0 (the matrix of the corresponding closed-loop system) under condition *II*, it follows that its eigenvalues λ_i, $i = 1, 2, \ldots, n$ are inside the unit circle. Taking into account the properties of the spectrum of the Kronecker product of matrices, we find that the eigenvalues of matrix Γ are representable in the form $\lambda_i \lambda_j$, $i, j = 1, 2, \ldots, n$. Thus, it is established that the spectrum of the main functional matrix of the associated system has eigenvalues inside the unit circle.

Because the presence of a disturbance $\varepsilon\Omega$ will not fundamentally change the form of inhomogeneities in the corresponding Riccati and Lyapunov equations, then, as in Theorem 2.1 from [18], it can be proved that there exist such $\alpha > 0$, $\beta > 0$, that the corresponding estimates for the boundary functions hold

$$\|\Pi_i P(\tau)\| \leq \alpha \exp(\beta\tau), \ \tau \leq 0, \ i = 0, 1, 2. \tag{15}$$

Despite the presence of regular perturbations $\varepsilon\Omega$ in Equation (9) the next estimate for the residual term of the asymptotics $\eta(x, t, \tau, \varepsilon) = P(x, t, \varepsilon) - \sum\limits_{i=0}^{2} \varepsilon^i (\overline{P}_i(x, t) + \Pi_i P(\tau))$, $\tau = \frac{t-T}{\varepsilon}$ can be obtained using the scheme from [18],

$$\|\eta(x, t, \tau, \varepsilon)\| = O(\varepsilon^3), \tag{16}$$

for all $x \in X$, $\varepsilon \in (0, \varepsilon_0]$, $t \in [0, T]$, $\tau = -1, -2, \ldots, -N$.

Moreover, using the smoothness of all functions included in Φ, Ω and assuming that the norm $\|P(x, t, \varepsilon)\|$ is uniformly bounded for all $x \in X$, $t \in [0, T]$, $\varepsilon \in (0, \varepsilon_0]$, $\tau = -1, -2, \ldots, -N, \ldots$, here we can follow the proof as in Lemma 6.1 and Theorem 6.1 given in [14].

This, in turn, will lead to the fact that when choosing a sufficiently small $\varepsilon_0 > 0$, one can obtain the existence, as well as the uniqueness of the solution in the problem (9). The last statements follow from the application of the principle of contractive mappings to the equation for the residual term of asymptotics $\eta(x, t, \tau, \varepsilon)$. $\square$

2.2. Symmetrization

Since the matrix $G_1(x)$ may not be symmetric, we introduce the transposed equation, and by adding the equation $A_{cl}^{0\,T}\overline{P}_1(x)A_{cl}^0 - \overline{P}_1(x) = G_1(x)A_{cl}^{0\,T}\overline{P}_1(x)A_{cl}^0 - \overline{P}_1(x) = G_1(x)$ to $A_{cl}^{0\,T}\overline{P}_1{}^T(x)A_{cl}^0 - \overline{P}_1{}^T(x) = G_1{}^T(x)$, we get

$$A_{cl}^{0\,T}(\overline{P}_1(x) + \overline{P}_1{}^T(x))A_{cl}^0 - (\overline{P}_1(x) + \overline{P}_1{}^T(x)) = G_1(x) + G_1{}^T(x).$$

Further, in this paper the next equations with "averaged" right-hand sides will be used for the regular and the boundary layer terms of the first and second order (the terms of the power series ε and ε^2). For example, instead of Equation (11) the following equation is introduced for a symmetric matrix $\widetilde{\overline{P}}_1(x)$

$$A_{cl}^{0\,T}\widetilde{\overline{P}}_1(x)A_{cl}^0 - \widetilde{\overline{P}}_1(x) = \widetilde{G}_1(x), \tag{17}$$

where $\widetilde{G}_1(x) = \frac{1}{2}(G_1(x) + G_1{}^T(x))$, $\widetilde{\overline{P}}_1(x,t) = \frac{1}{2}(\overline{P}_1(x) + \overline{P}_1{}^T(x))$. A similar operation is performed for the second approximation term.

In addition, $\widetilde{\Pi}_1 P(x,\tau)$ is found in the form

$$\widetilde{\Pi}_1 P(x,\tau) = \zeta^T(x,\tau)\widetilde{\Pi}_1 P(\tau+1)\zeta(x,\tau) + \widetilde{\Pi}_1 G(x,\tau),$$
$$\widetilde{\Pi}_1 P(0) = -\overline{P}_1(x),\ \tau = -1, -2, \ldots,$$

where $\widetilde{\Pi}_1 G(x) = \frac{1}{2}(\Pi_1 G(x) + \Pi_1 G^T(x))$, $\widetilde{\Pi}_1 P(x,\tau) = \frac{1}{2}(\Pi_1 P(x,\tau) + \Pi_1 P^T(x,\tau))$.

A similar equation is obtained for $\widetilde{\Pi}_2 P(x,\tau)$, where the inhomogeneity in the right-hand side equals to $\widetilde{\Pi}_2 G(x) = \frac{1}{2}(\Pi_2 G(x) + \Pi_2 G^T(x))$.

3. Discrete One-Point Padé Regulator

The asymptotic analysis for small parameter values can lead to an acceptable quality approximation of the exact solution, but with an increase of the small parameter value the asymptotic representations can strongly deviate from the exact solutions and their use in numerical analysis for larger values of the parameter is limited and at best they can serve only to restore the qualitative nature of the solution behavior. *PA* often increases the interval of parameter variation for which it can provide the approximation of the exact solution and restore its qualitative picture in comparison with the asymptotics. Thus, *PA* demonstrates better extrapolation properties [23].

In general, a particular system of algebraic equations, which, generally speaking, is in a certain way selected from some redefined system, is solved to find the *PA*.

Here a one-point Padé regulator of an order [1/2] is constructed, which contains two asymptotic approximations: the first order uniform asymptotic approximation in the "numerator" of the *PA*—for reproducing the boundary layer in the general construction, and the second-order approximation of some regular series in the "denominator", i.e., the proposed construction has the following form:

$$PA_{[1/2]}(x,t,\tau,\varepsilon) = (M_0(x) + \varepsilon M_1(x) + \Pi M_0(x,\tau) + \varepsilon\Pi M_1(x,\tau)) \times$$
$$\times (E + \varepsilon N_1(x) + \varepsilon^2 N_2(x))^{-1}. \tag{18}$$

Note that the form of the "denominator" in (18) is less complex than the "numerator" which makes it easier to overcome the "denominator" zeros problem, which is the main reason for the quality decline of the approximations of the exact solution using *PA*.

So, we do not introduce the boundary functions in the "denominator" of *PA*. Decomposing the matrix in (18) in a series of integer powers of parameter ε and equating the terms with the same powers of the parameter ε of the resulting decomposition and the corresponding terms of the expansions $\bar{\phi}$ and $\Pi\Phi$, separately for the terms dependent on t and τ, we get an inhomogeneous linear system of six equations with matrix coefficients depending on $x \in X$, $t \in [0, T]$, $\varepsilon \in (0, \varepsilon_0]$, $\tau = -1, -2, \ldots, -N$ for the six unknown matrices, the coefficients of PA.

$$
\begin{aligned}
&M_0(x) = \overline{P}_0 \\
&\Pi M_0(x, \tau) = \Pi_0 P(\tau) \\
&M_1(x) - \overline{P}_0 N_1(x) = \overline{P}_1(x) \\
&\Pi M_1(x, \tau) - \Pi_0 P(\tau) N_1(x) = 0 \\
&\overline{P}_1(x) N_1(x) + \overline{P}_0 N_2(x) = -\overline{P}_2(x) \\
&\Pi_1 P(x, \tau) N_1(x) + \Pi_0 P(\tau) N_2(x) = -\Pi_2 P(x, \tau).
\end{aligned}
\tag{19}
$$

The first matrices $M_0(x)$, $\Pi M_0(x, \tau)$ are immediately determined from the first two equations, and for the remaining four matrices the following linear system is obtained:

$$
\begin{pmatrix}
E_n & 0 & -\overline{P}_0 & 0 \\
0 & E_n & -\Pi_0 P(\tau) & 0 \\
0 & 0 & \overline{P}_1(x) & \overline{P}_0 \\
0 & 0 & \Pi_1 P(x, \tau) & \Pi_0 P(\tau)
\end{pmatrix}
\begin{pmatrix}
M_1(x) \\
\Pi M_1(x, \tau) \\
N_1(x) \\
N_2(x)
\end{pmatrix}
=
\begin{pmatrix}
\overline{P}_1(x) \\
0 \\
-\overline{P}_2(x) \\
-\Pi_2 P(x, \tau)
\end{pmatrix},
\tag{20}
$$

where E_n—is an identity matrix of the size $n \times n$. Next by denoting the matrix of a linear system (20) by $Y = \begin{pmatrix} Y_{11} & Y_{12} \\ Y_{21} & Y_{22} \end{pmatrix}$, where the corresponding blocks are $Y_{11} = E_{2n}$, $Y_{21} = 0$, $Y_{12} = \begin{pmatrix} -\overline{P}_0 & 0 \\ -\Pi_0 P(\tau) & 0 \end{pmatrix}$, $Y_{22} = \begin{pmatrix} \overline{P}_1(x) & \overline{P}_0 \\ \Pi_1 P(x, \tau) & \Pi_0 P(\tau) \end{pmatrix}$ and by taking into account that block $Y_{11} = E_{2n}$ is a nondegenerate $(2n \times 2n)$ identity matrix, we find that for the existence of the matrix inverse $Y^{-1} = \begin{pmatrix} E & -Y_{12} Y_{22}^{-1} \\ 0 & \Delta^{-1} \end{pmatrix}$ it is necessary and sufficient [24] that matrix $\Delta = Y_{22} - Y_{21} Y_{11}^{-1} Y_{12}$ is nondegenerate and $\Delta = Y_{22}$ since $Y_{21} \equiv 0$. Now the solution of the last two equations in (20) is explicitly defined by

$$
\begin{pmatrix} N_1(x) \\ N_2(x) \end{pmatrix} = \Delta^{-1} \begin{pmatrix} -\overline{P}_2(x) \\ -\Pi_2 P(x, \tau) \end{pmatrix}.
\tag{21}
$$

As $M_1(x)$ and $\Pi M_1(x, \tau)$ are found from the first two equations in (20) and are the functions of $N_1(x)$, where $N_1(x)$, $N_2(x)$ are determined from the last two equations in (20) and take the form (21).

As some of the matrices in (18)–(20) that are the regular series terms in the asymptotic representation of $P(x, t, \varepsilon)$ are positive definite, it is possible to make the other matrices sign-definite to guarantee the solvability of (20) and positive definiteness of PA on the entire interval, for example, by a special choice of criteria matrices F, $Q_0 > 0$, $Q_1(x) > 0$, $Q_2(x) > 0$, $x \in X$. In the latter case, we are dealing not with a problem of the criterion (2) minimization along the trajectories of (1), but with a synthesis construction problem using the SDRE algorithm.

Let us introduce a modified PA, defined as $\widetilde{P}A_{[1/2]}(x, t, \tau, \varepsilon)$, for which the system of equations is obtained from system (20) by replacing the elements of the first and second-order terms of the approximation with their symmetric, "averaged" values.

$$
\begin{pmatrix}
E_n & 0 & -\overline{P}_0 & 0 \\
0 & E_n & -\Pi_0 P(\tau) & 0 \\
0 & 0 & \widetilde{\overline{P}}_1(x) & \overline{P}_0 \\
0 & 0 & \widetilde{\Pi}_1 P(x, \tau) & \Pi_0 P(\tau)
\end{pmatrix}
\begin{pmatrix}
M_1(x) \\
\Pi M_1(x, \tau) \\
N_1(x) \\
N_2(x)
\end{pmatrix}
=
\begin{pmatrix}
\widetilde{\overline{P}}_1(x) \\
0 \\
-\overline{P}_2(x) \\
-\widetilde{\Pi}_2 P(x, \tau)
\end{pmatrix}.
\tag{22}
$$

By analogy with the study of system (20), the last two equations are firstly solved and the following matrices are formally introduced, $\widetilde{\Delta} = \widetilde{Y}_{22} = \begin{pmatrix} \widetilde{\overline{P}}_1(x) & \overline{P}_0 \\ \widetilde{\Pi}_1 P(x, \tau) & \Pi_0 P(\tau) \end{pmatrix}$

and $Z = \Pi_0 P - \tilde{\Pi}_1 P(x,\tau)\widetilde{\overline{P}}_1(x)^{-1}\overline{P}_0$, $L = \widetilde{\overline{P}}_1(x) - \overline{P}_0\Pi_0 P(\tau)^{-1}\tilde{\Pi}_1 P(x,\tau)$, which, in the case of their non-degeneracy allow us to present the solution for the "denominator" matrices in an explicit form

$$\begin{pmatrix} N_1(x) \\ N_2(x) \end{pmatrix} = \begin{pmatrix} -L^{-1}\widetilde{\overline{P}}_2(x) + \widetilde{\overline{P}}_1(x)^{-1}\overline{P}_0 Z^{-1}\tilde{\Pi}_2 P(x,\tau) \\ Z^{-1}[\tilde{\Pi}_1 P(x,\tau)\widetilde{\overline{P}}_1(x)^{-1}\widetilde{\overline{P}}_2(x) - \tilde{\Pi}_2 P(x,\tau)] \end{pmatrix},$$

$$\tilde{\Delta}^{-1} = \begin{pmatrix} K^{-1} & -\widetilde{\overline{P}}_1(x)^{-1}\overline{P}_0 H^{-1} \\ -H^{-1}\tilde{\Pi}_1 P(x,\tau)\widetilde{\overline{P}}_1(x)^{-1} & H^{-1} \end{pmatrix}.$$

For consistency with the Kalman regulator (the linear-quadratic regulator named after R.E. Kalman who posed and solved the corresponding control problem for nonstationary linear systems in 1960), which leads to stabilization and also to the optimal trajectory according to the quadratic criterion in a closed-loop linear system on the semi-axis, the following symmetric gain matrix is introduced to ensure the symmetry of the gain matrix of the regulator based on the *PA* for $P(x,t,\varepsilon)$:

$$K_{[1/2]}(x,\tau,\varepsilon) = \frac{1}{2}(\tilde{P}A_{[1/2]}(x,\tau,\varepsilon) + \tilde{P}A_{[1/2]}^{T}(x,\tau,\varepsilon)), \tag{23}$$

which leads to a modified Padé regulator for the *PA* obtained from the system (22)

$$u(x,t,\varepsilon) = -\left\{R + B^T(x,\varepsilon)K_{[1/2]}(x,\tau,\varepsilon)B(x,\varepsilon)\right\}^{-1}B^T(x,\varepsilon)K_{[1/2]}(x,\tau,\varepsilon)A(x,\varepsilon)x(t). \tag{24}$$

Remark 1. *In numerical experiments, it becomes possible to modify the proposed PA structure (18) by the introduction of multipliers in front of the matrix coefficients in systems (19)–(20) and search for the values of these multipliers.*

The following conditions are additionally introduced

III. Matrices F, $Q_0 > 0$, $Q_1(x) > 0$, $Q_2(x) > 0$ can be selected such that $F - \overline{P}_0 > 0$, $\tilde{G}_1(x) > 0$, $\tilde{G}_2(x) > 0$, matrix $\tilde{\Delta}^{-1}$ exists and is uniformly bounded and $\Pi_0 P(\tau) > 0$ for all $x \in X$, $t \in [0,T]$, $\varepsilon \in (0,\varepsilon_0]$, $\tau = -1,-2,\ldots,1-T$.

IV. Matrix $\left(E + \varepsilon N_1(x) + \varepsilon^2 N_2(x)\right)^{-1}$ exists, where

$$N_1(x) = -L^{-1}\widetilde{\overline{P}}_2(x) + \widetilde{\overline{P}}_1(x)^{-1}\overline{P}_0 Z^{-1}\tilde{\Pi}_2 P(x,\tau),$$
$$N_2(x) = Z^{-1}[\tilde{\Pi}_1 P(x,\tau)\widetilde{\overline{P}}_1(x)^{-1}\widetilde{\overline{P}}_2(x) - \tilde{\Pi}_2 P(x,\tau)],$$
$$Z = \Pi_0 P(\tau) - \tilde{\Pi}_1 P(x,\tau)\widetilde{\overline{P}}_1(x)^{-1}\overline{P}_0, \quad L = \widetilde{\overline{P}}_1(x) - \overline{P}_0\Pi_0 P(\tau)^{-1}\tilde{\Pi}_1 P(x,\tau)$$

for all $x \in X$, $t \in [0,T]$, $\varepsilon \in (0,\varepsilon_0]$, $\tau = -1,-2,\ldots,1-T$.

For the construction of *PA* [1/2] and the solvability of the corresponding system for the coefficients of *PA* an asymptotic expansion of the second order is required, where some of the terms, in particular of the second order, can be found approximately.

The next statement takes place.

Theorem 2. *If conditions I–IV are satisfied, then there is a sufficiently small $\varepsilon_0 > 0$, such that for all $x \in X$, $t \in [0,T]$, $\varepsilon \in (0,\varepsilon_0]$, $\tau = -1,-2,\ldots,1-T$ there is a unique solution of the matrix system of Equations (20) and the corresponding one-point matrix PA [1/2] of form (18) with a symmetric matrix $K_{[1/2]}(x,\tau,\varepsilon)$ (23) exists.*

Proof of Theorem 2. From condition II $\overline{P}_0 > 0$ we get $M_0(x) > 0$, then by condition III $\Pi_0 P(\tau) > 0$ $\forall \tau$ and from here $\Pi M_0(x,\tau) > 0$. Let us consider the discrete linear Lyapunov Equation (17) for $\widetilde{\overline{P}}_1(x)$. It is known [24] that if $\tilde{G}_1(x) > 0$ (condition III), the solution $\widetilde{\overline{P}}_1(x)$ of this equation is positive definite. From condition III, $\tilde{G}_2(x) > 0$ and there exist $\widetilde{\overline{P}}_2(x) > 0$. It is easy to show that $\zeta(x,\tau)$, $\tilde{\Pi}_1 G(x)$ are found from (13) and the corresponding matrices $\tilde{\Pi}_1 P(x,\tau)$ are obtained as the solutions of difference Lyapunov equations with the initial condition $\tilde{\Pi}_1 P(0) = -\overline{P}_1(x)$, $\tau = -1,-2,\ldots.$ By analogy, the

$\tilde{\Pi}_2 P(x,\tau)$ term is found. By condition *III*, matrix $\tilde{\Delta} = \begin{pmatrix} \tilde{\bar{P}}_1(x) & \bar{P}_0 \\ \tilde{\Pi}_1 P(x,\tau) & \Pi_0 P(\tau) \end{pmatrix}$ has an inverse for all $x \in X$, $t \in [0,T]$, $\varepsilon \in (0,\varepsilon_0]$, $\tau = -1,-2,\dots,1-T$ and it follows that $\tilde{Y}^{-1} = \begin{pmatrix} E & -Y_{12}Y_{22}^{-1} \\ 0 & \tilde{\Delta}^{-1} \end{pmatrix}$ exists and system (19), (20) is uniquely solvable with a solution

$$\begin{pmatrix} M_1(x) \\ \Pi M_1(x,\tau) \\ N_1(x) \\ N_2(x) \end{pmatrix} = \tilde{Y}^{-1} \begin{pmatrix} \bar{P}_1(x) \\ 0 \\ -\bar{P}_2(x) \\ -\tilde{\Pi}_2 P(x,\tau) \end{pmatrix}.$$

Thus, the remaining coefficients of the Padé approximation $M_1(x)$, $\Pi M_1(x,\tau)$, $N_1(x)$, $N_2(x)$ are found in the form

$$\begin{pmatrix} N_1 \\ N_2 \end{pmatrix} = \tilde{\Delta}^{-1} \begin{pmatrix} -\tilde{\bar{P}}_2(x) \\ -\tilde{\Pi}_2 P(x,\tau) \end{pmatrix},$$

$$\begin{pmatrix} M_1(x) \\ \Pi M_1(x,\tau) \end{pmatrix} = \begin{pmatrix} \tilde{\bar{P}}_1(x) \\ 0 \end{pmatrix} - \begin{pmatrix} -\bar{P}_0 & 0 \\ -\Pi_0 P(\tau) & 0 \end{pmatrix} \tilde{\Delta}^{-1} \begin{pmatrix} -\tilde{\bar{P}}_2(x) \\ -\tilde{\Pi}_2 P(x,\tau) \end{pmatrix}. \tag{25}$$

Under condition *IV*, there exists a corresponding one-point matrix $PA\ [1/2]$ of form (18), and the symmetric matrix of the gain coefficients of the regulator is found from (24).
$\square$

4. Computational Experiments

One of the ways to concretize the coefficients of the system of linear equations for the Padé approximation can be associated with the analysis of the coefficients of the system, and the assumption that the Padé structures form a certain framework, within which the coefficients can be improved from the point of view of the optimality criterion of the control problem. This approach is illustrated below on an example of a simple pendulum [25].

$t \in [0,1]$, $\varepsilon = 0.05$, $N = 20$,

$$A(x) = \begin{pmatrix} 1 & 0.01 \\ \frac{-10\sin(x_1)}{x_1} & 1 \end{pmatrix}, \ B(x) = \begin{pmatrix} 0.1 \\ 0.1 \end{pmatrix}, \ A_0 = \begin{pmatrix} 1 & 0.01 \\ -10 & 1 \end{pmatrix}, A_1(x) = 1/\varepsilon \begin{pmatrix} 0 & 0 \\ \frac{-10\sin(x_1)}{x_1} + 10 & 0 \end{pmatrix}$$

$x_0 = [0.1; -0.1]^T$

$$Q_0 = \begin{pmatrix} 10 & 0 \\ 0 & 0.05 \end{pmatrix}, \ Q_1 = \begin{pmatrix} 300 + x_1^2 & 0 \\ 0 & 300 + x_2^2 \end{pmatrix}, \ Q_2 = \begin{pmatrix} 0 & 0 \\ 0 & 0 \end{pmatrix}, R = 1, \ F = \begin{bmatrix} 300 & 0 \\ 0 & 300 \end{bmatrix}.$$

Here, x_1 denotes the pendulum angle, and x_2 is the angular velocity.

Using this example, taking into account the fulfillment of the conditions for the existence of matrices in the representations (18), we will demonstrate the Algorithm 1 for discrete modified Padé regulator construction and the results of its work.

In Table 1, the criterion values for the regulators based on the uniform first-order asymptotic approximation $(P_0, \tilde{\bar{P}}_1(x), M_0 P(\tau), \tilde{\Pi}_1 P(x,\tau))$ and the modified Padé $[1/2]$ from (22) are presented. The comparison is made with the D-SDRE regulator which uses the solution of Equation (9) by the algorithm from [5].

Table 1. Regulators comparison by criterion values for $\varepsilon = 0.05$.

D-SDRE	Uniform First-Order Asymptotic Approximation	Modified Padé [1/2] Approximation
19.2226	16.3533	16.0882089921279

Algorithm 1: Discrete modified Padé regulator construction.

1. The terms $P_0, \widetilde{\overline{P}}_1(x), \widetilde{\overline{P}}_2(x), \Pi_0 P(\tau)$ are found based on the uniform asymptotic approximation of the second order using formulas (10), (12), and (17).
2. The members of the boundary layer terms of the first and second order, $\hat{\Pi}_1 P(x, \tau)$ and $\hat{\Pi}_2 P(x, \tau)$, are found approximately, for example, as matrix exponentials with the unknown decay rates constants.
3. The terms of the modified Padé approximation of the solution of the Riccati equation are found from the following system of equations with unknown parameters $\lambda_1, \lambda_2, \lambda_3, \lambda_4, \lambda_5, \lambda_6$ which are the scalar multipliers of the asymptotic expansion terms and are selected using the quality criterion optimization.

$$
\begin{pmatrix}
E & 0 & -\lambda_5 \overline{P}_0 & 0 \\
0 & E & -\lambda_6 \Pi_0 P & 0 \\
0 & 0 & \lambda_1 \widetilde{\overline{P}}_1(x) & \lambda_2 \overline{P}_0 \\
0 & 0 & \lambda_3 \hat{\Pi}_1 P(x, \tau) & \lambda_4 \Pi_0 P(\tau)
\end{pmatrix}
\begin{pmatrix}
M_1(x) \\
\Pi M_1(x, \tau) \\
N_1(x) \\
N_2(x)
\end{pmatrix}
=
\begin{pmatrix}
\widetilde{\overline{P}}_1(x) \\
0 \\
-\widetilde{\overline{P}}_2(x) \\
-\hat{\Pi}_2 P(x, \tau)
\end{pmatrix}.
$$

The introduction of multipliers allows us to correct the *PA* system matrix coefficients within the obtained structure. Such a technique can be used as a basis for the correction of the resulting Padé regulator if the result by the optimality criterion is better than the corresponding results along the regulators using only the asymptotics and regulators built based on the SDRE technique, which is demonstrated in the calculations given below. A regulator built based on this approach will be called a modified Padé regulator.

4. The resultant modified Padé regulator gain is found from (23).

Table 2 shows that the modified Padé [1/2] regulator is closer to the D-SDRE solution by optimality criterion values on a larger interval of parameter variation in comparison with the regulator based on the uniform first-order asymptotic expansion, i.e., demonstrates good quality approximation for larger values of the parameter and has better extrapolation properties. In this example, the uniform first-order asymptotic approximation works only for small parameter values and fails to stabilize the system when the value of the parameter increases. Moreover, the Padé regulator is significantly better by criterion value than the D-SDRE regulator. Thus, here the modified Padé regulator is sufficiently better by optimality criterion than the two other control algorithms (D-SDRE, asymptotic approximation) in the selected parameter variation interval from 0.05 to 0.25 and the asymptotic approximation has a restricted area of application and provides worse quality of the approximation.

Table 2. Demonstration of extrapolation properties of the modified Padé regulator.

Parameter ε	Modified Padé [1/2] Approximation	Uniform First-Order Asymptotic Approximation	D-SDRE
0.05	16.0882	16.3533	19.2226
0.1	31.7171	29.2683	38.2234
0.125	36.1462	38.6610	47.7203
0.2	64.6920	1630.8255	76.2066
0.25	76.3028	2576.7654	95.1958

The corresponding closed-loop trajectories are presented in Figure 1. It can be seen that the Modified Padé [1/2] approximation brings the system to the neighborhood of the zero-equilibrium point and the trajectories are similar to the D-SDRE solution.

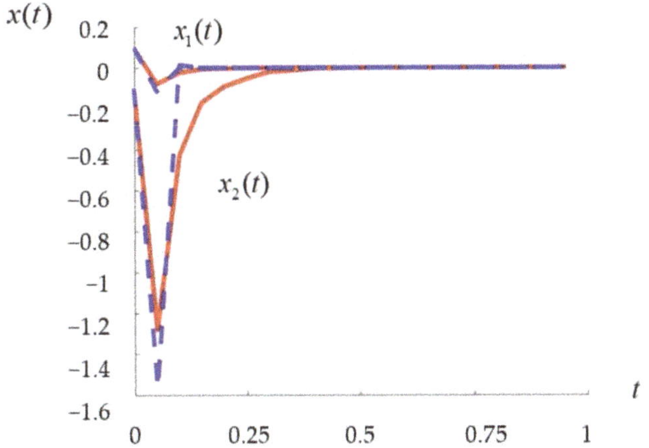

Figure 1. Closed-loop system trajectories: Modified Padé [1/2] approximation (red lines) and D-SDRE (blue lines).

5. Conclusions

Using the SDRE approach, the asymptotics of the solution of the corresponding initial singularly perturbed control problem for the matrix discrete Riccati equation with coefficients weakly dependent on the state is constructed and the corresponding one-point *PA* regulator is proposed, i.e., only one asymptotic approximation of the Riccati equation solution is used to construct the *PA* for the feedback gain matrix of the regulator. The results of numerical experiments illustrate, in particular, the improved extrapolation properties of the constructed regulator, which makes the algorithm applicable in control systems for a wider range of parameter variation. An approach for modified *PA* construction is also demonstrated, which consists of the correction of the system of equations for finding the *PA* coefficients taking into account the structure of the matrix of the original system and the properties of the terms of the asymptotic approximation.

Author Contributions: Conceptualization, M.D.; methodology, M.D. and Y.D.; software, Y.D.; formal analysis, M.D.; writing—original draft preparation, M.D. and Y.D.; writing—review and editing, M.D. and Y.D. All authors have read and agreed to the published version of the manuscript.

Funding: Research is supported by the Russian Science Foundation (Project No. 21-11-00202).

Institutional Review Board Statement: Not applicable.

Informed Consent Statement: Not applicable.

Data Availability Statement: Not applicable.

Conflicts of Interest: The authors declare no conflict of interest.

References

1. Mracek, C.P.; Cloutier, J.R. Control designs for the non-linear benchmark problem via the state-dependent Riccati equation method. *Int. J. Rob. Nonlin. Contr.* **1998**, *8*, 401–433. [CrossRef]
2. Cimen, T. State dependent Riccati Equation (SDRE) control: A Survey. In Proceedings of the 17th the International Federation of Automatic Control World Congress, Seoul, Korea, 6–11 July 2008; Volume 41, pp. 3761–3775.
3. Dmitriev, M.G.; Makarov, D.A. Smooth nonlinear controller in a weakly nonlinear control system with state depended coefficients. *Trud. ISA RAN* **2014**, *64*, 53–58. (In Russian)
4. Chang, I.; Bentsman, J. Constrained discrete-time state-dependent Riccati equation technique: A model predictive control approach. In Proceedings of the 52nd IEEE Conference on Decision and Control, Florence, Italy, 10–13 December 2013; pp. 5125–5130.
5. Dutka, A.S.; Ordys, A.W.; Grimble, M.J. Optimized discrete-time state dependent Riccati equation regulator. In Proceedings of the American Control Conference, Portland, OR, USA, 8–10 June 2005; pp. 2293–2298.

6. György, K.; Dávid, L.; Kelemen, A. Theoretical Study of the Nonlinear Control Algorithms with Continuous and Discrete-Time State Dependent Riccati Equation. *Procedia Technol.* **2016**, *22*, 582–591. [CrossRef]
7. Danik, Y.E.; Dmitriev, M.G. The robustness of the stabilizing regulator for quasilinear discrete systems with state dependent coefficients. In Proceedings of the International Siberian Conference on Control and Communications, Moscow, Russia, 12–14 May 2016.
8. Emel'yanov, S.V.; Danik, Y.E.; Dmitriev, M.G.; Makarov, D.A. Stabilization of nonlinear discrete-time dynamic control systems with a parameter and state dependent coefficients. *Dokl. Mathem.* **2016**, *93*, 121–123. [CrossRef]
9. Danik, Y.E.; Dmitriev, M.G. The comparison of numerical algorithms for discrete-time state dependent coefficients control systems. In Proceedings of the 21st International Conference on System Theory, Control and Computing, Sinaia, Romania, 19–21 October 2017; pp. 401–406.
10. Danik, Y.E.; Dmitriev, M.G. Construction of Parametric Regulators for Nonlinear Control Systems Based on the Padé Approximations of the Matrix Riccati Equation Solution. In Proceedings of the 17th the International Federation of Automatic Control Workshop on Control Applications of Optimization, Yekaterinburg, Russia, 15–19 October 2018; Volume 51, pp. 815–820.
11. Danik, Y.E.; Dmitriev, M.G. The construction of stabilizing regulators sets for nonlinear control systems with the help of Padé approximations. In *Nonlinear Dynamics of Discrete and Continuous Systems*; Abramian, A., Andrianov, I., Gaiko, V., Eds.; Springer: Cham, Switzerland, 2021; pp. 45–62.
12. Heydari, A.; Balakrishnan, S.N. Approximate closed-form solutions to finite-horizon optimal control of nonlinear systems. In Proceedings of the American Control Conference, Montreal, QC, Canada, 27–29 June 2012; pp. 2657–2662.
13. Kvakernaak, H.; Sivan, R. *Linear Optimum Control Systems*; Wiley-Interscience: New York, NY, USA, 1972.
14. Vasil'eva, A.B.; Butuzov, V.F. *Asymptotic Expansions of a Solutions of Singularly Perturbed Equations*; Nauka: Moscow, Russia, 1973. (In Russian)
15. Vasil'eva, A.B.; Butuzov, V.F.; Kalachev, L.V. *The Boundary Function Method for Singular Perturbation Problems*; Society for Industrial and Applied Mathematics: University City, MO, USA; Philadelphia, PA, USA, 1995.
16. Vasil'eva, A.B.; Dmitriev, M.G. Singular perturbations in optimal control problems. *J. Sov. Mathem.* **1986**, *34*, 1579–1629. [CrossRef]
17. Kurina, G.A.; Dmitriev, M.G.; Naidu, D.S. Discrete singularly perturbed control problems (A survey). *Dyn. Contin. Discr. Impuls. Syst. Ser. B Appl. Algor.* **2017**, *24*, 335–370.
18. Glizer, V.Y.; Dmitriev, M.G. Asymptotics of a solution of some discrete optimal control problems with small step. *Different. Equat.* **1979**, *15*, 1681–1691. (In Russian)
19. Gaipov, M.A. Asymptotics of the solution of a nonlinear discrete optimal control problem with small step without constraints on the control (formalism). *I. Izvest. Akad. Nauk TurkmenSSR* **1990**, *1*, 9–16. (In Russian)
20. Belokopytov, S.V.; Dmitriev, M.G. Direct scheme in optimal control problems with fast and slow motions. *Syst. Contr. Lett.* **1986**, *8*, 129–135. [CrossRef]
21. Afanas'ev, V.N. Control of nonlinear plants with state-dependent coefficients. *Autom. Remote Control* **2011**, *72*, 713–726. [CrossRef]
22. Afanas'ev, V.N.; Presnova, A.P. Parametric Optimization of Nonlinear Systems Represented by Models Using the Extended Linearization Method. *Autom. Remote Control* **2021**, *82*, 245–263. [CrossRef]
23. Baker, G.A.; Baker, G.A., Jr.; Baker, G.; Graves-Morris, P.; Baker, S.S. *Padé Approximants: Encyclopedia of Mathematics and It's Applications*, 2nd ed.; Cambridge University Press: Cambridge, UK, 1996.
24. Balandin, D.; Kogan, M. *Synthesis of Control Laws Based on Linear Matrix Inequalities*; Fizmatlit: Moscow, Russia, 2007. (In Russian)
25. ElBsat, M.N. Finite-Time Control and Estimation of Nonlinear Systems with Disturbance Attenuation. Ph.D. Thesis, Marquette University, Milwaukee, WI, USA, August 2012.

 *mathematics*

Article

Operator Methods of the Maximum Principle in Problems of Optimization of Quantum Systems

Alexander Buldaev * and Ivan Kazmin

Department of Applied Mathematics, Buryat State University, 670000 Ulan-Ude, Russia; kazminvanya@mail.ru
* Correspondence: buldaev@mail.ru; Tel.: +7-924-658-0183

Abstract: In the class of optimal control problems for quantum systems, operator optimality conditions for control are constructed in the form of fixed-point problems in the control space. The equivalence of the obtained operator optimality conditions to the well-known Pontryagin maximum principle is shown. Based on the obtained operator forms of optimality conditions, new iterative methods for finding extreme equations satisfying the maximum principle are developed. A comparative analysis of the effectiveness of the proposed operator methods of the maximum principle with known methods is carried out on model examples of optimization of quantum systems.

Keywords: controlled quantum systems; control optimality conditions; fixed-point problem; optimization method

Citation: Buldaev, A.; Kazmin, I. Operator Methods of the Maximum Principle in Problems of Optimization of Quantum Systems. *Mathematics* **2022**, *10*, 507. https://doi.org/10.3390/math10030507

Academic Editor: Jan Sładkowski

Received: 30 November 2021
Accepted: 3 February 2022
Published: 5 February 2022

Publisher's Note: MDPI stays neutral with regard to jurisdictional claims in published maps and institutional affiliations.

1. Introduction

Mathematical formulations of topical problems related to the optimal control of quantum systems have been considered in the works of many researchers [1–6]. In the works of V.F. Krotov, V.I. Gurman, and of their followers [7–9], there are studied classes of controlled quantum systems described by ordinary differential controls linear in state and control with nonlinear optimality criteria. In [7], the main features of the selected class of problems are indicated. The first feature is the high dimension of the system state vector ($n \approx 10^4 - 10^6$). The second feature is the absence of restrictions on the state, including terminal restrictions. The third feature is the use of a scalar control function characterizing the electric field. In this class, the search for an optimal solution based on standard necessary optimality conditions in the form of a boundary value problem of the maximum principle [10,11] causes significant difficulties due to the large dimension. In [7,8], the global Krotov method [12] was used as a tool for finding solutions to problems, which was compared in efficiency with the known gradient method.

In this paper, we consider and modify a new approach to finding a solution in the considered class of problems. This approach is based on the representation of optimality conditions for control in the form of operator problems about a fixed point in the space of admissible controls. This representation makes it possible to apply and modify the known methods of fixed points to find solutions to the considered problems related to the optimal control of quantum systems.

The new fixed-point approach has been used and developed for more than ten years for various classes of continuous, discrete, and discrete-continuous optimal control problems, including those involving terminal and phase constraints, mixed control functions and parameters, unfixed control process termination time, and other features.

In [13], the fixed-point approach is used to represent conditions for nonlocal improvement of control in a general class of nonlinear optimal control problems with control functions and parameters. In [14], the fixed-point approach for representing conditions for the nonlocal improvement of control is modified to the class of problems considered in [7–9]. The modification of the approach consists of taking into account the characteristic property of the singularity of the solutions of the problems under consideration, due to the above features.

The paper [15] describes the fixed-point approach for representing optimality conditions for control in a general class of nonlinear optimal control problems with control functions. In this paper, this approach for representing optimality conditions for control is modified and studied taking into account the characteristic property of the singularity of solutions in the considered class of optimization problems for quantum control systems.

2. Conditions for Optimality of Control

To illustrate the proposed fixed-point approach, we consider a model class of optimal control problems for quantum systems with a quadratic optimality criterion similar to the paper [14], in which new operator forms of optimality conditions have a relatively simple description:

$$\dot{x}(t) = (A + u(t)B)x(t), \quad x(t_0) = x^0, \quad u(t) \in U \subset R, \quad t \in T = [t_0, t_1], \tag{1}$$

$$\Phi(u) = \langle x(t_1), Lx(t_1) \rangle \to \inf_{u \in V}. \tag{2}$$

The vector $x(t) = (x_1(t), \ldots, x_n(t))$ describes the state of the system. The control $u(t)$, $t \in T$ is modeled by a piecewise continuous scalar function with values in a compact and convex set $U \subset R$. The set V denotes the corresponding set of admissible controls. The matrices A, B and L have real coefficients. The matrix L is symmetric. The initial state x^0 and the time interval T have fixed values.

The Pontryagin function with a conjugate variable ψ in problem (1) and (2) has the form:

$$H(\psi, x, u, t) = \langle \psi, (A + uB)x \rangle, \quad \psi \in R^n.$$

The standard conjugate system is represented as:

$$\dot{\psi}(t) = -(A^T + u(t)B^T)\psi(t), \quad t \in T, \quad \psi(t_1) = -2Lx(t_1). \tag{3}$$

Let $v \in V$. Let us introduce the following notation:

- $x(t, v)$, $t \in T$, the solution of the system (1) for $u(t) = v(t)$;
- $\psi(t, v)$, $t \in T$, the solution of the standard conjugate system (3) for $x(t) = x(t, v)$, $u(t) = v(t)$.

Additionally, we will use the notation P_Y for the projection operator on to a set $Y \subset R^k$ in the Euclidean norm:

$$P_Y(z) = \arg\min_{y \in Y}(\|y - z\|), \quad z \in R^k.$$

The projection operation is characterized by an important property that can be represented as an inequality:

$$\langle y - P_Y(z), z - P_Y(z) \rangle \leq 0, \quad y \in Y.$$

The known [10,11] necessary optimality conditions for an admissible control (maximum principle and differential maximum principle) in problems (1) and (2) are equivalent.

The condition of the maximum principle for control $u \in V$ can be represented in the form:

$$u(t) = \arg\max_{w \in U}\langle \psi(t, u), Bx(t, u) \rangle w, \quad t \in T. \tag{4}$$

The condition of the differential maximum principle using the projection operation can be written as the following relation with the parameter $\alpha > 0$:

$$u(t) = P_U(u(t) + \alpha\langle \psi(t, u), Bx(t, u) \rangle), \quad t \in T. \tag{5}$$

To fulfill the maximum principle condition (4), it suffices to check condition (5) for at least one $\alpha > 0$. Conversely, condition (4) implies the fulfillment of condition (5) for all $\alpha > 0$.

We define the mapping u^* as follows:

$$u^*(\psi, x) = \arg\max_{w \in U} \langle \psi, Bx \rangle w, \quad \psi \in R^n, \quad x \in R^n.$$

We introduce a mapping u^α with a parameter $\alpha > 0$ using the relation:

$$u^\alpha(\psi, x, w) = P_U(w + \alpha \langle \psi, Bx \rangle), \quad x \in R^n, \quad \psi \in R^n, \quad w \in U.$$

Using the introduced mappings, the maximum principle condition (4) can be written as:

$$u(t) = u^*(\psi(t, u), x(t, u)), \quad t \in T.$$

The condition of the differential maximum principle (5) takes the following form:

$$u(t) = u^\alpha(\psi(t, u), x(t, u), u(t)), \quad t \in T. \tag{6}$$

The well-known [10,11] approach to the search for extremal controls, i.e. satisfying the necessary optimality conditions, is the solution of the boundary value problem of the maximum principle. This problem in the considered classes, classes (1) and (2), takes the following form:

$$\dot{x}(t) = (A + u^*(\psi(t), x(t))B)x(t), \quad x(t_0) = x^0, \tag{7}$$

$$\dot{\psi}(t) = (-A^T - u^*(\psi(t), x(t))B^T)\psi(t), \quad \psi(t_1) = -2Lx(t_1). \tag{8}$$

Let the pair $(x(t), \psi(t))$, $t \in T$, be a solution to the boundary value problems (7) and (8). Let us construct the output control $v(t) = u^*(\psi(t), x(t))$, $t \in T$. Then, by construction, we obtain the relations:

$$x(t) = x(t, v), \quad \psi(t) = \psi(t, v), \quad t \in T.$$

Consequently, the control $v(t)$ satisfies condition (4).

Conversely, let the control $v \in V$ be a solution to Equation (4). Let us form a pair of functions $(x(t, v), \psi(t, v))$, $t \in T$. Then, by definition, these functions satisfy the boundary value problem (7) and (8).

Thus, the boundary value problems (7) and (8) are equivalent to the maximum principle condition (4).

Difficulties in solving the boundary value problems of the maximum principle, (7) and (8), in the general case are associated with the possible discontinuity and many meanings of the right-hand sides of the problem for the variables x, ψ. Even in the case of smoothness and uniqueness of the right-hand sides of the boundary value problem (7) and (8), its numerical solution by known methods (shooting method, linearization method, and finite difference method) [11] is computationally unstable due to the presence of positive real values of the eigenvalues of the corresponding Jacobian matrices.

In this paper, we consider a new approach to the search for extremal controls based on the transition from the boundary value problem of the maximum principle in the state space to equivalent operator problems on the fixed point of the maximum principle in the space of controls.

3. Operator Forms of the Maximum Principle

We define three mappings, X, Ψ and V^*, using the following relations:

$$X(v) = x, \quad v \in V, \quad x(t) = x(t, v), \quad t \in T,$$

$$\Psi(v) = \psi, \quad v \in V, \quad \psi(t) = \psi(t, v), \quad t \in T,$$

$$V^*(\psi, x) = v^*, \quad \psi \in C(T), \quad x \in C(T), \quad v^*(t) = u^*(\psi(t), x(t)), \quad t \in T,$$

where $C(T)$ is the space of functions continuous on T.

Using the above mappings, the maximum principle condition (4) can be represented as an operator equation in the form of a fixed-point problem in the control space:

$$v = V^*(\Psi(v), X(v)) = G_1^*(v), \quad v \in V. \tag{9}$$

Construct new operator equations in the form of fixed-point problems equivalent to condition (4). Introduce the mapping X^* as follows:

$$X^*(\psi) = x, \quad \psi \in C(T), \quad x \in C(T).$$

Here $x(t), t \in T$, is the solution of the special Cauchy problem:

$$\dot{x}(t) = (A + u^*(\psi(t), x(t))B)x(t), \quad x(t_0) = x^0.$$

Based on the introduced mapping X^*, consider the following operator equation:

$$v = V^*(\Psi(v), X^*(\Psi(v))) = G_2^*(v), \quad v \in V. \tag{10}$$

We define the following mapping:

$$\Psi^*(x) = \psi, \quad x \in C(T), \quad \psi \in C(T).$$

Here $\psi(t), t \in T$ is the solution to the special Cauchy problem:

$$\dot{\psi}(t) = (-A^T - u^*(\psi(t), x(t))B^T)\psi(t), \quad \psi(t_1) = -2Lx(t_1).$$

Consider the operator equation:

$$v = V^*(\Psi^*(X(v)), X(v)) = G_3^*(v), \quad v \in V. \tag{11}$$

Following the work [15], the operator equations, Equations (9)–(11), are equivalent to the set of admissible controls. Thus, we obtain the following statement:

Theorem 1. *Operator fixed-point problems (9)–(11) are equivalent to the boundary value problems of the maximum principle, (7) and (8).*

The condition of the maximum principle in projection form (5) can also be represented in the form of equivalent operator equations on the set of admissible controls.

Introduce an additional operator V^α, $\alpha > 0$, by the relation:

$$V^\alpha(\psi, x, v) = v^\alpha, \quad \psi \in C(T), \quad x \in C(T), \quad v \in V,$$

$$v^\alpha(t) = u^\alpha(\psi(t), x(t), v), \quad t \in T.$$

Define the operator X^α, $\alpha > 0$:

$$X^\alpha(\psi, v) = x^\alpha, \quad \psi \in C(T), \quad v \in V, \quad x^\alpha(t) = x^\alpha(t, \psi, v), \quad t \in T,$$

where $x^\alpha(t, \psi, v), t \in T$, is the solution of the Cauchy problem:

$$\dot{x}(t) = (A + u^\alpha(\psi(t), x(t), v(t))B)x(t), \quad x(t_0) = x^0.$$

Construct the operator Ψ^α, $\alpha > 0$:

$$\Psi^\alpha(x, v) = \psi^\alpha, \quad x \in C(T), \quad v \in V, \quad \psi^\alpha(t) = \psi^\alpha(t, x, v),$$

where $\psi^\alpha(t, x, v)$, $t \in T$, is the solution of the conjugate Cauchy problem:

$$\dot{\psi}(t) = (-A^T - u^\alpha(\psi(t), x(t), v(t))B^T)\psi(t), \quad \psi(t_1) = -2Lx(t_1).$$

Consider three operator equations in the form of fixed-point problems:

$$v = V^\alpha(\Psi(v), X(v), v) = G_1^\alpha(v), \quad v \in V, \quad \alpha > 0, \tag{12}$$

$$v = V^\alpha(\Psi(v), X^\alpha(\Psi(v), v), v) = G_2^\alpha(v), \quad v \in V, \quad \alpha > 0, \tag{13}$$

$$v = V^\alpha(\Psi^\alpha(X(v), v), X(v), v) = G_3^\alpha(v), \quad v \in V, \quad \alpha > 0. \tag{14}$$

Similarly, following [15], these equations are equivalent to the set of admissible controls. Thus, the following statement holds:

Theorem 2. *Projection operator fixed-point problems (12)–(14) are equivalent to the boundary value problem of the maximum principle (7) and (8).*

Let us note the following important features of the constructed projection problems on a fixed point.

1. Projection control operators, due to the properties of the projection operation, are continuous and satisfy the Lipschitz condition, in contrast to discontinuous and generally multivalued control operators based on the maximum operation in problems (9)–(11).

2. The search for extremal controls, which are solutions to the projection problems on a fixed point, (12)–(14), can be carried out for any given values of the projection parameter $\alpha > 0$, including sufficiently small values.

These features of projection problems on a fixed point are essential factors for increasing the efficiency of the numerical search for extremal controls.

4. Operator Methods of the Maximum Principle

We consider the general fixed-point problem for an operator $G_E : V_E \to V_E$, acting on a set V_E in a complete normalized space E with a norm $\| \cdot \|_E$,

$$v = G_E(v), \quad v \in V_E.$$

To solve it numerically, one can apply the well-known simple iteration method with an index $k \geq 0$, which has the form:

$$v^{k+1} = G_E(v^k), \quad v^0 \in V_E.$$

The convergence of the iterative process can be analyzed using the well-known principle of compressive mappings [16].

Each operator equation from relations (9)–(14) can be considered as a fixed-point problem on the set of admissible controls in the following general form:

$$v = G(v), \quad v \in V. \tag{15}$$

To solve the problem (15), an iterative process with the index $k \geq 0$ is proposed:

$$v^{k+1} = G(v^k), \quad v \in V. \tag{16}$$

As an illustration of processes of the form (16), consider iterative processes for searching for extremal controls based on projection problems about a fixed point of the maximum principle (12)–(14), which, respectively, take the form with index $k \geq 0$:

$$v^{k+1} = V^\alpha(\Psi(v^k), X(v^k), v^k) = G_1^\alpha(v^k), \quad v^0 \in V, \quad \alpha > 0, \tag{17}$$

$$v^{k+1} = V^\alpha(\Psi(v^k), X^\alpha(\Psi(v^k), v^k), v^k) = G_2^\alpha(v^k), \quad v^0 \in V, \quad \alpha > 0, \tag{18}$$

$$v^{k+1} = V^\alpha(\Psi^\alpha(X(v^k), v^k), X(v^k), v^k) = G_3^\alpha(v^k), \quad v^0 \in V, \quad \alpha > 0. \tag{19}$$

In the considered projection methods of the maximum principle, the projection parameter $\alpha > 0$ is fixed in the iterative process of successive approximations of the control.

The complexity of implementing one iteration of each of the processes (17)–(19) is two Cauchy problems for phase and conjugate variables.

Indeed, this is obvious for the process (17). In this case, process (17) is written in the pointwise form as:

$$v^{k+1}(t) = u^\alpha(\psi(t, v^k), x(t, v^k), v^k(t), t), \quad t \in T.$$

For process (18), at each k-th iteration with $k \geq 0$, we obtain the following.

After calculating the solution of the Cauchy problem $\psi(t, v^k)$, $t \in T$, we find the solution $x(t)$, $t \in T$ of the special Cauchy problem for the phase system:

$$\dot{x}(t) = (A + u^\alpha(\psi(t, v^k), x(t), v^k(t))B)x(t), \quad x(t_0) = x^0.$$

Simultaneously, together with the solution of the Cauchy problem, we determine the output control according to the rule:

$$v^{k+1}(t) = u^\alpha(\psi(t, v^k), x(t), v^k(t)), \quad t \in T.$$

Then, by construction, we obtain the relation:

$$x(t) = x(t, v^{k+1}), \quad t \in T.$$

By of this equality, the iterative process (18) in pointwise form can be written in the following implicit form:

$$v^{k+1}(t) = u^\alpha(\psi(t, v^k), x(t, v^{k+1}), v^k(t)), \quad t \in T. \tag{20}$$

Similarly, at the k-th iteration of the process (19), after calculating $x(t, v^k)$, $t \in T$, we find the solution $\psi(t)$, $t \in T$, of the special Cauchy problem for the conjugate system:

$$\dot{\psi}(t) = (-A^T - u^\alpha(\psi(t), x(t, v^k), v^k(t))B^T)\psi(t), \quad \psi(t_1) = -2Lx(t_1, v^k).$$

Simultaneously the output control is constructed according to the rule:

$$v^{k+1}(t) = u^\alpha(\psi(t), x(t, v^k), v^k(t)), \quad t \in T.$$

The theoretical conditions for the convergence of iterative processes (17)–(19) for sufficiently small projection parameters $\alpha > 0$ can be substantiated similarly to [17] based on the formulation of requirements in problems (1) and (2), ensuring the application of the indicated principle of compressive mappings in the complete space of continuous controls or the extended complete space of measurable controls:

$$V \subset V_L = \{v \in L_\infty(T): \quad v(t) \in U, \quad t \in T\}$$

with the norm $\|v\|_\infty = \operatorname*{ess\,sup}_{t \in T} \|v(t)\|, \quad v \in V_L.$

Unlike the standard gradient projection method, at each iteration of the proposed projection methods of the maximum principle, relaxation for the objective functional is not guaranteed.

In contrast to the global Krotov method, at each iteration of the proposed projection methods, Cauchy problems with a continuous and uniquely defined right-hand side are solved.

Let us single out other features of the proposed projection methods that are important for increasing their computational efficiency:

- The non-locality of successive approximations of control, due to the fixed choice of the projection parameter $\alpha > 0$;
- The absence of the operation of varying control in the vicinity of the current approximation to provide improved control, which is characteristic of gradient methods;
- The possibility of obtaining extreme controls for sufficiently small values of the projection parameter $\alpha > 0$, which ensure the fundamental convergence of iterative processes.

Simple iteration methods for solving fixed-point problems (9)–(11) based on the maximization operation have a similar structure. In particular, the iterative process with the index $k \geq 0$ for searching for extremal controls based on the fixed-point problem (9) takes the form:

$$v^{k+1} = V^*(\Psi(v^k), X(v^k)) = G_1^*(v^k), \quad v^0 \in V, \quad \alpha > 0.$$

In pointwise form, this process is written as:

$$v^{k+1}(t) = u^*(\psi(t, v^k), x(t, v^k)), \quad t \in T. \tag{21}$$

Note that method (21) is essentially equivalent to the well-known method of successive approximations of phase and conjugate variables [18] for solving the boundary value problems of the maximum principle, (7) and (8).

In contrast to the global Krotov method, at each iteration of the considered method (21), two simple Cauchy problems with a precomputed control are solved. In the Krotov method, at each iteration, in the general case, a special Cauchy problem with a discontinuous and multivalued right-hand side is solved.

5. Examples

The main feature of the considered class of optimal control problems for quantum systems is the property of singularity of extremal controls. This property is expressed in the existence of singular time intervals of non-zero measure for extremal controls, on which the derivative of the Pontryagin function becomes equal to zero:

$$H_u(\psi(t, u), x(t, u), u(t), t) = \langle \psi(t, u), Bx(t, u) \rangle = 0.$$

On such singular time intervals, it becomes impossible to determine the values of the extremal control from the condition of the maximum principle (4).

The proposed operator methods of the maximum principle, taking into account the indicated property of singularity, are modified in specific examples under consideration and compared in terms of computational efficiency with known methods.

The computational implementation of the proposed methods of the maximum principle is characterized by the following features.

The numerical solution of phase and conjugate Cauchy problems was performed by the Runge–Kutta–Werner method of variable (5–6) order of accuracy using the DIVPRK program of the IMSL Fortran PowerStation 4.0 library [19]. The values of the controlled, phase, and conjugate variables were stored in the nodes of a fixed uniform grid T_h with a sampling step $h > 0$ on the interval T. In the intervals between neighboring grid nodes T_h, the control value was assumed to be constant and equal to the value in the left node. The numerical calculation of the fixed-point problem was carried out before the condition was fulfilled:

$$max\{|v^{k+1}(t) - v^k(t)|, t \in T_h\} \leq \varepsilon_m,$$

in which $\varepsilon_m > 0$ is the given accuracy of calculating the fixed-point problem.

Example 1. *(projection methods (17) and (18)).*

The well-known model problem of control of the system of spins of quantum particles is considered [20], which can be represented in the following form:

$$\Phi(u) = 1 - \langle x(t_1), Lx(t_1)\rangle \to inf,$$

$$L = \begin{pmatrix} a_1^2 + b_1^2 & a_1a_2 + b_1b_2 & 0 & a_1b_2 - b_1a_2 \\ a_1a_2 + b_1b_2 & a_2^2 + b_2^2 & a_2b_1 - b_2a_1 & 0 \\ 0 & a_2b_1 - b_2a_1 & b_1^2 + a_1^2 & b_1b_2 + a_1a_2 \\ a_1b_2 - b_1a_2 & 0 & b_1b_2 + a_1a_2 & b_2^2 + a_2^2 \end{pmatrix},$$

$$\dot{x}_1(t) = u(t)x_3(t) + x_4(t), \quad \dot{x}_2(t) = x_3(t) - u(t)x_4(t),$$

$$\dot{x}_3(t) = -u(t)x_1(t) - x_2(t), \quad \dot{x}_4(t) = -x_1(t) + u(t)x_2(t),$$

$$x_1(0) = \frac{1}{\sqrt{2}}, \quad x_2(0) = \frac{1}{\sqrt{2}}, \quad x_3(0) = 0, \quad x_4(0) = 0, \quad t \in T = [0, t_1], \quad t_1 = 1.5,$$

$$a_1 = 0.6, \quad b_1 = -0.3, \quad a_2 = 0.1, \quad b_2 = \sqrt{0.54}.$$

The vector $x(t)$ describes the state of the quantum system, the function $u(t)$ characterizes the effect of an external field, $u(t) \in U = [-30, 30], t \in T$.

In [20], to calculate the optimal control problem under consideration, the global Krotov method was used, the efficiency of which was compared with the well-known gradient method. The control determined from physical considerations was chosen as the initial control approximation for the specified iterative methods:

$$u(t) = tg(2\gamma(2t - 1.5)), \quad t \in T, \quad \gamma = -\frac{1}{3}arctg(-30).$$

The Pontryagin function in the problem has the form:

$$H(\psi, x, u, t) = \psi_1(ux_3 + x4) + \psi_2(x_3 - ux_4) + \psi_3(-ux_1 - x_2) + \psi_4(-x_1 + ux_2).$$

The standard conjugate system is written as:

$$\dot{\psi}_1(t) = u(t)\psi_3(t) + \psi_4(t), \quad \dot{\psi}_2(t) = \psi_3(t) - u(t)\psi_4(t), \quad t \in T,$$

$$\dot{\psi}_3(t) = -u(t)\psi_1(t) - \psi_2(t), \quad \dot{\psi}_4(t) = u(t)\psi_2(t) - \psi_1(t), \quad t \in T,$$

$$\psi_1(t_1) = 2(a_1^2 + b_1^2)x_1(t_1) + 2(a_1a_2 + b_1b_2)x_2(t_1) + 2(a_1b_2 - b_1a_2)x_4(t_1),$$

$$\psi_2(t_1) = 2(a_1a_2 + b_1b_2)x_1(t_1) + 2(a_2^2 + b_2^2)x_2(t_1) + 2(a_2b_1 - b_2a_1)x_3(t_1),$$

$$\psi_3(t_1) = 2(a_2b_1 - b_2a_1)x_2(t_1) + 2(b_1^2 + a_1^2)x_3(t_1) + 2(b_1b_2 + a_1a_2)x_4(t_1),$$

$$\psi_4(t_1) = 2(a_1b_2 - b_1a_2)x_1(t_1) + 2(b_1b_2 + a_1a_2)x_3(t_1) + 2(b_2^2 + a_2^2)x_4(t_1).$$

The fixed-point projection problems (12) and (13) have the same pointwise form:

$$v(t) = P_U(v(t) + \alpha(\psi_1(t, v)x_3(t, v) - \psi_2(t, v)x_4(t, v) - \psi_3(t, v)x_1(t, v) + \psi_4(t, v)x_2(t, v))).$$

The explicit iterative method of the maximum principle (17) for solving this fixed-point problem at index $k \geq 0$ has a pointwise form:

$$v^{k+1}(t) = P_U(v^k(t) + \alpha(\psi_1(t, v^k)x_3(t, v^k) - \psi_2(t, v^k)x_4(t, v^k)$$
$$- \psi_3(t, v^k)x_1(t, v^k) + \psi_4(t, v^k)x_2(t, v^k))).$$

Accordingly, the implicit iterative method (20) at index $k \geq 0$ takes the form:

$$v^{k+1}(t) = P_U(v^k(t) + \alpha(\psi_1(t, v^k)x_3(t, v^{k+1}) - \psi_2(t, v^k)x_4(t, v^{k+1})$$
$$- \psi_3(t, v^k)x_1(t, v^{k+1}) + \psi_4(t, v^k)x_2(t, v^{k+1}))).$$

The calculation was carried out on a sampling grid with a step $h = 10^{-5}$ and the criterion for stopping the calculation $\varepsilon_m = 10^{-3}$.

Table 1 shows the comparative results of the first four iterations of improving the objective functional with an index $s \geq 0$, starting from the starting control $v^0 = u$ specified in [20]. We compare the results of the explicit projection method (PPM1) and the implicit projection method (PPM2) for the parameter $\alpha = 10^{-2}$ with known [20] calculation data by the global method (GlM) and the gradient method (GRM).

Table 1. The comparative results of the first four iterations.

Number s	$\Phi(u^s)$ GlM	$\Phi(u^s)$ GrM	$\Phi(v^s)$ PPM1	$\Phi(v^s)$ PPM2
0	0.7681	0.7681	0.7680	0.7680
1	0.1401	0.6911	0.6705	0.6610
2	0.0040	0.6107	0.4049	0.3881
3	0.0021	0.5421	0.2404	0.2175
4	0.0015	0.4913	0.1718	0.1191

Figure 1 shows the final computational control $v1(t)$, $t \in T$ obtained by the PPM1 method with the number of control improvement iterations equal to 14, and the value of the functional $\Phi^* \approx 0.001421$.

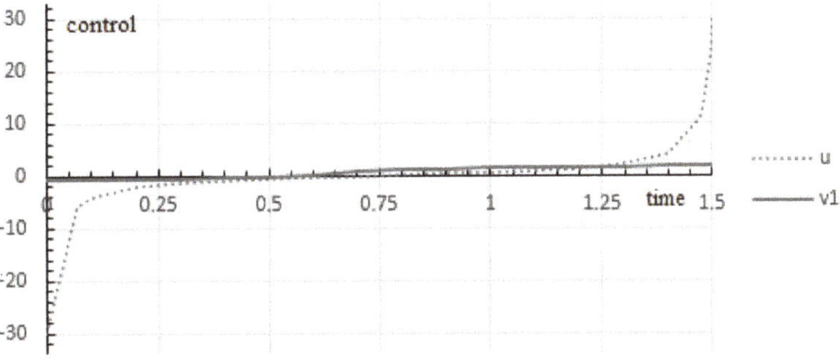

Figure 1. u—starting control; $v1$—computational control obtained by the PPM1 method.

Figure 2 shows the final computational control $v2(t)$, $t \in T$, obtained by the PPM2 method with the number of control improvement iterations equal to 26 and the value of the functional $\Phi^* \approx 0.000704$.

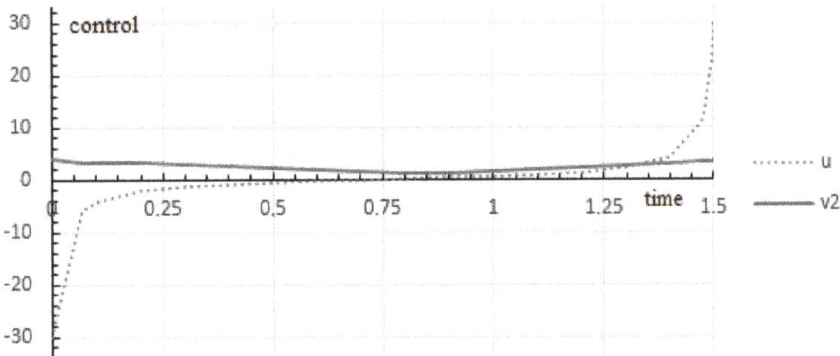

Figure 2. u—starting control; $v2$—computational control obtained by the PPM2 method.

In [20], the final calculated value of the functional $\Phi^* \approx 0.000952$ obtained by the global Krotov method at the ninth iteration of control improvement is indicated.

Thus, within the framework of the considered example, the proposed projection methods of the maximum principle allow one to achieve similar results in terms of the value of the functional with the global Krotov method. The methods differ in the number of calculated control improvement iterations. However, at each iteration of the global method, it is necessary to solve the complex Cauchy problem for phase variables with a special right-hand side based on a multivalued and discontinuous operation to the maximum for calculating control values. In the proposed fixed-point method, at each iteration, much simpler Cauchy problems with a uniquely defined and continuous right-hand side are solved, which makes the considered projection methods of the maximum principle much easier to implement than the global Krotov method. Due to the simplicity of implementation and easy adjustment of the convergence, controlled by the choice of the projection parameter $\alpha > 0$, these methods can be successfully used to obtain practical initial approximations for subsequent refinement by other iterative methods for solving optimal control problems of the class under consideration.

Example 2. *(method (21)).*

To illustrate the work of the maximum principle method based on the maximization operation (21), the problem from the previous example is considered.

The corresponding fixed-point problem of the maximum principle (9) has the following form:

$$v(t) = u^*(\psi(t, v), x(t, v)), \quad t \in T,$$

$$u^*(\psi, x) = \begin{cases} +30, g(\psi, x) > 0, \\ -30, g(\psi, x) < 0, \\ w \in U, g(\psi, x) = 0, \end{cases}$$

where $g(\psi, x) = \psi_1 x_3 - \psi_2 x_4 - \psi_3 x_1 + \psi_4 x_2$. The iterative process (21) for solving this fixed-point problem at $k \geq 0$, respectively, takes the form:

$$v^{k+1}(t) = u^*(\psi(t, v^k), x(t, v^k)), \quad t \in T,$$

with the above switching function $g(\psi, x)$.

In the case of the existence of a time interval $[\Theta_1, \Theta_2] \subset T$ of a non-zero measure, where $g(\psi(t), x(t)) = 0, t \in [\Theta_1, \Theta_2]$, the control u^* is called singular on this interval. Singular controls are determined by the sequential differentiation by an argument $t \in T$ of the identity $g(\psi(t), x(t)) = 0$ taking into account the phase and conjugate systems. In practical calculations, similarly to the work [20], the equality of the switching function $g(\psi, x)$, to zero, which determines a singular mode, is understood in the sense of belonging to some small ε, neighborhood of zero, where $\varepsilon > 0$. Thus, we obtain the following practical calculation formula for the simple iteration method:

$$v^{k+1}(t) = \begin{cases} +30, g(\psi(t, v^k), x(t, v^k)) > \varepsilon, \\ -30, g(\psi(t, v^k), x(t, v^k)) < -\varepsilon, \\ w \in U, |g(\psi(t, v^k), x(t, v^k))| \leq \varepsilon. \end{cases}$$

If at the time t the condition is satisfied:

$$|g(\psi(t, v^k), x(t, v^k))| \leq \varepsilon,$$

the value $w \in U$ is determined by the following rule.
The value is calculated:

$$g_k = g(\psi(t + \delta, v^k), x(t + \delta, v^k)),$$

for a given $\delta > 0$.
If $g_k > \varepsilon$, then $w = 30$.
If $g_k < -\varepsilon$, then $w = -30$.

If $|g_k| \leq \varepsilon$, then the value $w \in U$ is determined by the special control calculation rule as follows.

The value is calculated as:

$$a_k = -\psi_1(t, v^k)x_4(t, v^k) - \psi_2(t, v^k)x_3(t, v^k) + \psi_3(t, v^k)x_2(t, v^k) + \psi_4(t, v^k)x_1(t, v^k).$$

If $|a_k| > \varepsilon$, then

1. *The value is calculated as $c_k = \frac{b_k}{a_k}$, where:*

$$b_k = -\psi_1(t, v^k)x_3(t, v^k) + \psi_2(t, v^k)x_4(t, v^k) + \psi_3(t, v^k)x_1(t, v^k) - \psi_4(t, v^k)x_2(t, v^k).$$

If $|c_k| \leq 30$, then the value $w = c_k$.
If $|c_k| > 30$, then go to step 2.
If $a_k \leq \varepsilon$, then:

2. *The value is calculated as*

$$d_k = \psi_1(t, v^k)x_2(t, v^k) - \psi_2(t, v^k)x_1(t, v^k) + \psi_3(t, v^k)x_4(t, v^k) - \psi_4(t, v^k)x_3(t, v^k).$$

If $|d_k| > \varepsilon$, then the value $w = 0$.
If $|d_k| \leq \varepsilon$, then the value $w \in U$ is chosen randomly from the interval U.
In the numerical implementation of the algorithm, the value $\delta > 0$ chosen equal to the grid step $h > 0$.

Table 2 shows the comparative results of the calculation by the considered maximum principle method (MPM) for the first four iterations of improving the functional with an index $s \geq 0$, starting from the above starting control, with the known [20] calculation data by the global method (GlM) and the gradient method (GrM). For an adequate comparison of the methods, the ε-neighborhood of zero was determined by the value $\varepsilon = 0.001$, specified in [20].

Table 2. The comparative results of the calculation by the considered maximum principle method for the first four iterations.

Number s	$\Phi(u^s)$ GlM	$\Phi(u^s)$ GrM	$\Phi(v^s)$ MPM
0	0.7681	0.7681	0.7680
1	0.1401	0.6911	0.5714
2	0.0040	0.6107	0.3612
3	0.0021	0.5421	0.1904
4	0.0015	0.4913	0.1380

In [20], the final calculated value of the functional $\Phi^* \approx 0.000952$, obtained by the global method at the ninth iteration of control improvement, is indicated. In this case, a singular section of the final control, determined according to the rules [20], is the interval $[0.0667, t_1]$.

Figure 3 shows the final computational control $v3(t), t \in T$, obtained by the MPM method, with the achieved value of the functional $\Phi^* \approx 0.000989$ and the number of control improvement iterations equal to 18. A singular section of the final control with the discretization grid accuracy is $[0.0693, 1.4717]$.

Figure 4 shows the final computational control $v4(t), t \in T$, obtained by the MPM method from the initial approximation $v^0 = u1$, which was obtained by the PPM1 method in example 1. In this case, the value of the functional is $\Phi^* \approx 0.000907$ with the number of control improvement iterations equal to 7. A singular section of the final control is approximately equal to $[0.0698, 1.4609]$.

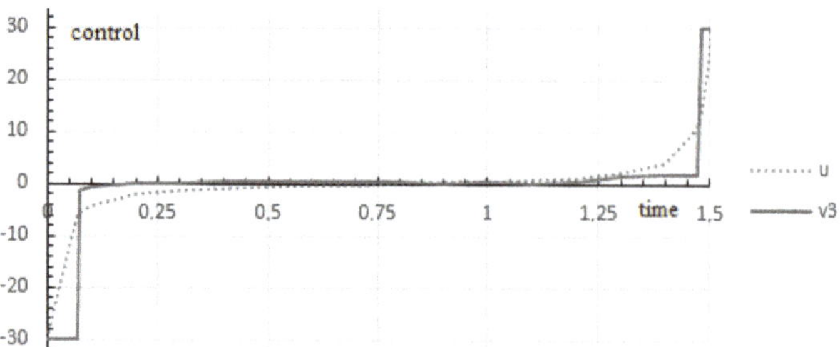

Figure 3. u—starting control, $v3$—computational control obtained by the MPM method.

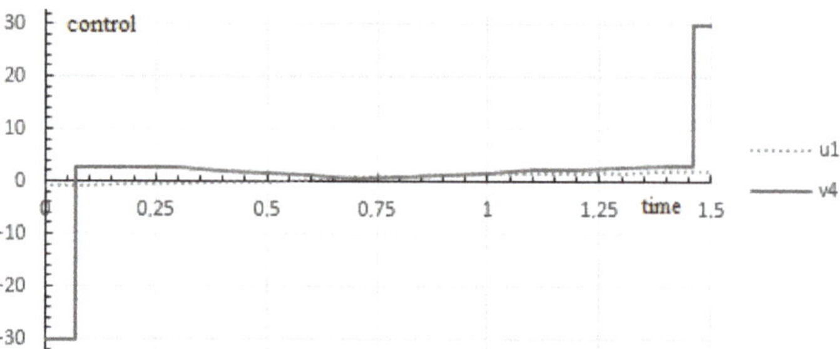

Figure 4. $u1$—starting control; $v4$—computational control obtained by the MPM method.

Figure 5 shows the final computational control $v5(t)$, $t \in T$, obtained by the MPM method from the initial approximation $v^0 = u2$, which was obtained by the PPM2 method in example 1. In this case, the value of the functional is $\Phi^* \approx 0.000620$, with the number of control improvement iterations equal to 3. A singular section of the final control is approximately equal to $[0.0751, 1.4512]$.

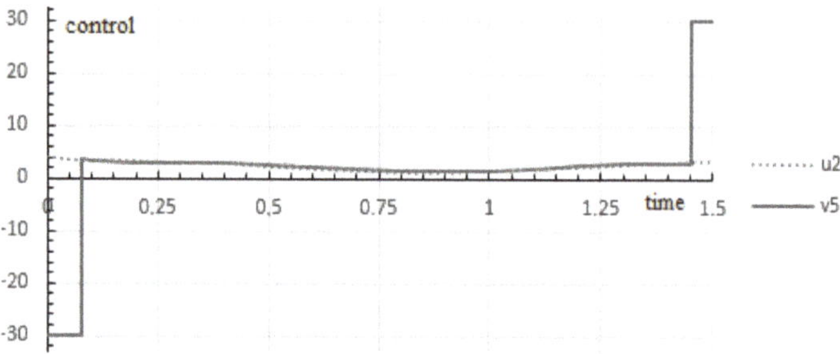

Figure 5. $u2$—starting control; $v5$—computational control obtained by the MPM method.

The calculations performed within the framework of the model problem show a high quantitative and qualitative efficiency of the implicit projection method of the maximum principle (20), which makes it possible to accurately calculate complex singular sections of extreme controls, which are typical in optimal control problems for quantum systems of the class under consideration. The main feature of this method, which is important for increasing efficiency, is the solution at each iteration of the Cauchy problems with a special uniquely defined and continuous right-hand side, in contrast to the global Krotov method.

6. Conclusions

In the considered class of optimal control problems for quantum systems, new operator forms of the maximum principle are proposed in the form of fixed-point problems in the control space, which make it possible to effectively apply and modify the well-known apparatus of the theory and methods of fixed points for constructing iterative algorithms to find extremal controls.

The developed iterative operator methods for searching for extremal controls are characterized by the following properties:

1. computational stability, in contrast to standard methods for solving the boundary value problem of the maximum principle;
2. nonlocality of successive control approximations;
3. the absence of a laborious procedure of needle or convex variation of the control in a small neighborhood of the considered approximation, which is typical for gradient methods;
4. the numerical solution of the Cauchy problems with a continuous and uniquely defined right-hand side at each iteration of the constructed projection methods, in contrast to the well-known global Krotov method.

The indicated properties of the proposed methods for searching for extremal controls are important factors for increasing the efficiency of the numerical solution of optimal control problems for quantum systems of the class under consideration.

In quantum systems with multidimensional control, the structures of the proposed operator methods of the maximum principle and the well-known global Krotov method remain the same, but the advantage of the indicated properties of the proposed projection methods of the maximum principle increases significantly in comparison with the global method.

Author Contributions: Investigation, A.B. and I.K.; Methodology, A.B. and I.K. All authors have read and agreed to the published version of the manuscript.

Funding: This work was supported by the Russian Foundation for Basic Research, project 18-41-030005, and Buryat State University, the project of the 2021 year.

Institutional Review Board Statement: Not applicable.

Informed Consent Statement: Not applicable.

Data Availability Statement: Not applicable.

Conflicts of Interest: The authors declare no conflict of interest.

References

1. Butkovskiy, A.; Samoilenko, Y. *Control of Quantum-Mechanical Processes and Systems*; Mathematics and its Applications; Kluwer Academic Publishers: Dordrecht, The Netherlands, 1990; pp. 1–232.
2. D'Alessandro, D. *Introduction to Quantum Control and Dynamics*; Chapman and Hall/CRC Appl. Math. Nonlinear Sci. Ser.; Chapman and Hall/CRC: Boca Raton, FL, USA, 2008; pp. 1–343.
3. Pechen, A. Some mathematical problems of control of quantum systems. *J. Math. Sci.* **2019**, *241*, 185–190. [CrossRef]
4. Stefanatos, D.; Paspalakis, E. A shortcut tour of quantum control methods for modern quantum technologies. *EPL (Europhys. Lett.)* **2020**, *132*, 60001. [CrossRef]
5. Mulero-Martínez, J.; Molina-Vilaplana, J. Quantum Pontryagin principle under continuous measurements. *J. Math. Phys.* **2020**, *61*, 102203. [CrossRef]
6. Boscain, U.; Sigalotti, M.; Sugny, D. Introduction to the Pontryagin maximum principle for quantum optimal control. *PRX Quantum* **2021**, *2*, 030203. [CrossRef]
7. Krotov, V. Control of quantum systems and some ideas of optimal control theory. *Autom. Remote Control* **2009**, *70*, 357–365. [CrossRef]
8. Krotov, V.; Bulatov, A.; Baturina, O. Optimization of linear systems with controlled coefficients. *Autom. Remote Control* **2011**, *72*, 1199–1212. [CrossRef]
9. Gurman, V. Turnpike solutions in optimal control problems for quantum-mechanical systems. *Autom. Remote Control* **2011**, *72*, 1248–1257. [CrossRef]
10. Pontryagin, L.; Boltyanskii, V.; Gamkrelidze, R.; Mishchenko, E. *The Mathematical Theory of Optimal Processes*; Interscience Publishers: New York, NY, USA; London, UK, 1962; pp. 1–360.
11. Vasiliev, O. *Optimization Methods*; World Federation Publishers Company INC.: Atlanta, GA, USA, 1996; pp. 1–276.
12. Krotov, V. *Global Methods in Optimal Control*; Marcel Dekker: New York, NY, USA, 1996; pp. 1–384.

13. Buldaev, A.; Khishektueva, I.-K. The fixed point method for the problems of nonlinear systems optimization on the managing functions and parameters. *Bull. Irkutsk State Univ. Ser. Math.* **2017**, *19*, 89–104. [CrossRef]
14. Buldaev, A.; Kazmin, I. On one method of optimization of quantum systems based on the search for fixed points. In *OPTIMA 2021: Advances in Optimization and Applications*; CCIS; Olenev, N., Evtushenko, Y., Jacimovic, M., Khachay, M., Malkova, V., Eds.; Springer: Cham, Switzerland, 2021; Volume 1514, pp. 67–81.
15. Buldaev, A. Operator forms of the maximum principle and iterative algorithms in optimal control problems. In *OPTIMA 2020: Advances in Optimization and Applications*; CCIS; Olenev, N., Evtushenko, Y., Khachay, M., Malkova, V., Eds.; Springer: Cham, Switzerland, 2020; Volume 1340, pp. 101–112.
16. Kirk, W.; Sims, B. (Eds.) *Handbook of Metric Fixed Point Theory*; Kluwer Academic: Dordrecht, The Netherlands, 2001; pp. 1–704.
17. Buldaev, A. Operator equations and maximum principle algorithms in optimal control problems. *Bull. Buryat State Univ. Math. Inform.* **2020**, *1*, 35–53. [CrossRef]
18. Krylov, I.; Chernous'ko, F. On a method of successive approximations for the solution of problems of optimal control. *USSR Comput. Math. Math. Phys.* **1963**, *2*, 1371–1382. [CrossRef]
19. Bartenev, O. *Fortran for Professionals. IMSL Mathematical Library. Part 2*; Dialog-MIFI: Moscow, Russia, 2001; pp. 1–320.
20. Baturina, O.; Morzhin, O. Optimal control of the spin system based on the global improvement method. *Autom. Remote Control* **2011**, *72*, 1213–1220. [CrossRef]

 *mathematics*

 MDPI

Article

Uniform Persistence and Global Attractivity in a Delayed Virus Dynamic Model with Apoptosis and Both Virus-to-Cell and Cell-to-Cell Infections

Meng Li, Ke Guo and Wanbiao Ma *

School of Mathematics and Physics, University of Science and Technology Beijing, Beijing 100083, China; limeng_4444x@163.com (M.L.); ke__guo@163.com (K.G.)
* Correspondence: wanbiao_ma@ustb.edu.cn

Abstract: In this paper, we study the global dynamics of a delayed virus dynamics model with apoptosis and both virus-to-cell and cell-to-cell infections. When the basic reproduction number $R_0 > 1$, we obtain the uniform persistence of the model, and give some explicit expressions of the ultimate upper and lower bounds of any positive solution of the model. In addition, by constructing the appropriate Lyapunov functionals, we obtain some sufficient conditions for the global attractivity of the disease-free equilibrium and the chronic infection equilibrium of the model. Our results extend existing related works.

Keywords: virus dynamic model; delay; uniform persistence; global attractivity; Lyapunov functional

MSC: 34K25; 92B05

Citation: Li, M.; Guo, K.; Ma, W. Uniform Persistence and Global Attractivity in a Delayed Virus Dynamic Model with Apoptosis and Both Virus-to-Cell and Cell-to-Cell Infections. *Mathematics* **2022**, *10*, 975. https://doi.org/10.3390/math10060975

Academic Editor: Quanxin Zhu

Received: 7 February 2022
Accepted: 14 March 2022
Published: 18 March 2022

Publisher's Note: MDPI stays neutral with regard to jurisdictional claims in published maps and institutional affiliations.

1. Introduction

It is well known that human health and safety have been seriously threatened by known or emerging new viral infections, such as human immunodeficiency virus (HIV), severe acute respiratory syndrome coronavirus 2 (SARS-CoV-2), etc. The mechanisms of transmission of viral infections have become increasingly complex, due to the mutation and evolution of viruses caused by changes in the physical environment, the use of drugs, and other factors. Since the 1980s, differential equations have been widely used in the study of important issues, such as the transmission mechanisms and control strategies of virus infection, and have gradually developed and established the important interdisciplinary research branch of virus dynamics [1–5]. In particular, in [1,2], the authors proposed the following classical viral infection dynamics model, describing the interactions among uninfected cells, infected cells and free viruses:

$$\begin{cases} \dot{x}(t) = s - dx(t) - \beta x(t)v(t), \\ \dot{y}(t) = \beta x(t)v(t) - py(t), \\ \dot{v}(t) = ky(t) - uv(t), \end{cases} \tag{1}$$

where $x(t)$, $y(t)$ and $v(t)$ denote the concentrations of uninfected cells, infected cells and free viruses at time t, respectively. The constant $s > 0$ is the rate at which new uninfected cells are generated. The constant $d > 0$ is the death rate of uninfected cells. The constant $\beta \geq 0$ is the characterizing infection of the cells. The constant $p > 0$ is the death rate of infected cells. The infected cells produce virus particles at the constant rate $k \geq 0$, and the constant $u > 0$ is the rate at which the virus is cleared. The term $\beta x(t)v(t)$ denotes the rate at which uninfected cells become infected cells through their contact with free viruses.

Based on model (1), many scholars have extended the linear growth rate $s - dx(t)$ of uninfected cells to classical logistic growth or more general nonlinear functions, and extended the bilinear functional response function $\beta x(t)v(t)$ to the following classical forms,

$\beta x(t)v(t)/(1+v(t))$, $\beta x(t)v(t)/(1+ax(t)+bv(t))$, $\beta x(t)v(t)/(1+av(t))$, and more general nonlinear functions (see, for example, [6–17] and the references therein). Additionally, important factors, such as delay and immune response, were considered, and a series of important results on global stability and existence of Hopf bifurcations were obtained [16,18–23].

In addition, recent studies have also shown that a large number of viral particles can also be transferred from infected cells to uninfected cells through the formation of virally induced structures termed virological synapses [24,25]. The direct fusion between infected cells and uninfected cells can also lead to cell infection, which is also known as cell-to-cell infection [25–27]. Based on this important fact, many scholars have proposed several important classes of viral dynamics models, considering the more general nonlinear growth rate of uninfected cells (which can include linear and logistic growth), while introducing important factors such as virus-to-cell infection, cell-to-cell infection, and immune response and delay, and have thoroughly investigated the local and global dynamics of equilibria and the existence of Hopf bifurcations. For details, see, for example, [18–21,25–31] and the references cited therein. In particular, based on the studies in [7,25,26], the authors [28] proposed and studied the following virus infection dynamic model:

$$
\begin{cases}
\dot{x}(t) = rx(t)\left(1 - \dfrac{x(t)+\alpha y(t)}{K}\right) - \beta_1 x(t)y(t) - \beta_2 x(t)v(t), \\
\dot{y}(t) = \beta_1 x(t)y(t) + \beta_2 x(t)v(t) - py(t), \\
\dot{v}(t) = ky(t) - uv(t).
\end{cases}
\tag{2}
$$

In model (2), the constant $K > 0$ denotes the effective carrying capacity of the environment of uninfected cells and infected cells. The term $rx(1 - (x+\alpha y)/K)$ indicates that the growth of uninfected cells conforms to the logistic growth function and takes into account the effect of infected cells on the maximum carrying capacity of the environment, the constant $r > 0$ is the growth rate, and $\alpha \geq 0$ is a constant; the constant $\beta_1 \geq 0$ is the cell-to-cell infection rate, and the constant $\beta_2 \geq 0$ is the virus-to-cell infection rate. All other parameters in model (2) have the same biological meaning as that in model (1). In [28], for model (2), the authors obtained the local stability of the equilibria, uniform persistence and the existence of Hopf bifurcations caused by the cell-to-cell infection rate β_1 or the virus-to-cell infection rate β_2.

HIV gene expression products can produce toxicity, which directly or indirectly induces apoptosis in uninfected cells [32]. Studies have shown that viral proteins interact with uninfected cells and induce an apoptotic signal, which induces the death of uninfected cells [33]. In [34], the authors considered the following virus infection dynamic model with delay:

$$
\begin{cases}
\dot{x}(t) = s - dx(t) - cx(t)y(t) - \beta x(t)v(t), \\
\dot{y}(t) = \delta x(t-\tau)v(t-\tau) - py(t), \\
\dot{v}(t) = ky(t) - uv(t),
\end{cases}
\tag{3}
$$

where the constant $\delta = \beta e^{-m_0\tau} > 0$ denotes the surviving rate of infected cells before it becomes productively infected, $m_0 \geq 0$ is a constant, and $\tau \geq 0$ is a delay. The constant $c \geq 0$ is the rate of apoptosis at which infected cells induce uninfected cells [32,33]. All other parameters in model (3) have the same biological meaning as that in model (1). Based on models (2) and (3), in [22,23,35], the authors further considered virus dynamics models with the logistic growth of uninfected cells, nonlinear infection rate, cell-to-cell infection, virus-to-cell infection and delay, and investigated the permanence, the global stability of the disease-free equilibrium, the local stability of the chronic infection equilibrium and the existence of Hopf bifurcations.

In this paper, based on [18–23,25,28,30,34,35], etc., we continue to consider the following delayed virus dynamic model with apoptosis and both virus-to-cell and cell-to-cell infections:

$$\begin{cases} \dot{x}(t) =s - dx(t) + rx(t)\left(1 - \dfrac{x(t) + \alpha y(t)}{K}\right) - cx(t)y(t) - \beta_1 x(t)y(t) - \beta_2 x(t)v(t), \\ \dot{y}(t) =\beta_1 e^{-m_1\tau_1}x(t-\tau_1)y(t-\tau_1) + \beta_2 e^{-m_2\tau_2}x(t-\tau_2)v(t-\tau_2) - py(t), \\ \dot{v}(t) =ky(t) - uv(t). \end{cases} \tag{4}$$

In model (4), the delay $\tau_1 \geq 0$ represents the time between infected cells spreading viruses into uninfected cells and the production of new free viruses; the delay $\tau_2 \geq 0$ represents the time between viral entry into an uninfected cell and the production of new free viruses. $m_1 \geq 0$ and $m_2 \geq 0$ are constants, and $\delta_1:=\beta_1 e^{-m_1\tau_1}$ and $\delta_2:=\beta_2 e^{-m_2\tau_2}$ denote the survival rates of uninfected cells during successful infection with infected cells and free viruses, respectively. All other parameters have the same biological meaning as that in models (1)–(3).

Let $\tau = \max\{\tau_1, \tau_2\}$. The initial condition of model (4) is given as follows, $x(\theta) = \phi_1(\theta)$, $y(\theta) = \phi_2(\theta)$, $v(\theta) = \phi_3(\theta)$ ($\theta \in [-\tau, 0]$), where $\phi = (\phi_1, \phi_2, \phi_3) \in C^+ := \{\phi \in C \mid \phi_i \geq 0, i = 1, 2, 3\}$, $C = C([-\tau, 0], \mathbb{R}^3)$ is the Banach space of continuous functions from $[-\tau, 0]$ to $\mathbb{R}^3$ equipped with the supremum norm.

By using the standard theory of functional differential equations (see [36–39]), it is easy to show that the solution $(x(t), y(t), v(t))$ of model (4) with the above initial condition is existent, unique, non-negative on $[0, +\infty)$, and satisfies

$$\limsup_{t\to\infty} x(t) \leq x_0, \quad \limsup_{t\to\infty} y(t) \leq M_2, \quad \limsup_{t\to\infty} v(t) \leq M_3, \tag{5}$$

where

$$x_0 = \frac{K}{2r}\left(r - d + \sqrt{(d-r)^2 + \frac{4rs}{K}}\right), \quad M_2 = \frac{2n_0(e^{-m_1\tau_1} + e^{-m_2\tau_2})}{\bar{p}}, \quad M_3 = \frac{kM_2}{u},$$

$$n_0 = \max_{x\in[0,x_0]}\left(s - dx + rx\left(1 - \frac{x}{K}\right)\right), \quad \bar{p} = \min\left\{p, \frac{n_0}{x_0}\right\}.$$

Obviously, model (4) always has a disease-free equilibrium (boundary equilibrium) $E_0 = (x_0, 0, 0)$. We can easily derive the expression of the basic reproduction number of model (4) as

$$R_0 = \frac{x_0(u\beta_1 e^{-m_1\tau_1} + k\beta_2 e^{-m_2\tau_2})}{pu} = \frac{x_0(u\delta_1 + k\delta_2)}{pu}$$

by the method of the next generation matrix [40,41]. The basic reproduction number R_0 is positively correlated with respect to the cell-to-cell infection rate β_1 and the virus-to-cell infection rate β_2. Hence, when only one route of infection is considered, the evolution of the disease infection may be underestimated.

The function is defined as

$$\Gamma(z) = s - dz + rz\left(1 - \frac{z}{K}\right) = \frac{r}{K}(z + x_1)(x_0 - z) \ (z \geq 0), \quad x_1 = -\frac{K}{2r}\left(r - d - \sqrt{(d-r)^2 + \frac{4rs}{K}}\right) > 0.$$

Note that if $R_0 = \frac{x_0}{x^*} > 1$, then model (4) has a unique chronic infection equilibrium (positive equilibrium) $E^* = (x^*, y^*, v^*)$, where

$$x^* = \frac{pu}{u\delta_1 + k\delta_2}, \quad y^* = \frac{\Gamma(x^*)}{x^*\zeta} = \frac{r(x^* + x_1)(x_0 - x^*)}{Kx^*\zeta} > 0, \quad v^* = \frac{ky^*}{u}, \quad \zeta = \frac{r\alpha}{K} + c + \beta_1 + \frac{k\beta_2}{u}. \tag{6}$$

It is noted that the apoptosis rate c has effects in reducing the loads of both infected cells and free viruses. In addition, it is easy to show that the set $G := \{\phi = (\phi_1, \phi_2, \phi_3) \in C^+ \mid 0 \le \phi_1 \le x_0\}$ is attractive and positively invariant with respect to model (4).

For the global asymptotic stability (global attractivity) of the disease-free equilibrium E_0 of model (4), using the method similar to that in [18,23,34,35], the following conclusion can be obtained (the proof is omitted): if $R_0 < 1$ ($R_0 = 1$), then the disease-free equilibrium E_0 is globally asymptotically stable (globally attractive) in G.

As far as we know, the global attractivity of the chronic infection equilibrium E^* of model (4) is still a difficult mathematical question and worthy of further study. This paper has the following two main purposes. First of all, we study the uniform persistence of model (4) in Section 2, and give explicit expressions of the ultimate upper and lower bounds of any positive solution of model (4). Second, by constructing some appropriate Lyapunov functionals and combining inequality analysis, some sufficient conditions for the global attractivity of the chronic infection equilibrium E^* of model (4) are given in Section 3. The brief summary of the conclusions of this paper is given in Section 4.

2. Uniform Persistence

In this section, we assume that $R_0 > 1$. It is not difficult to find that the function

$$f_1(z) := \frac{K}{2r}\left(z + \sqrt{z^2 + \frac{4rs}{K}}\right)$$

is strictly monotonically increasing with respect to z on $\mathbb{R}$. According to the first equation of model (4), x^* can be rewritten as $x^* = f_1(l_0)$, where $l_0 = r - d - \left(\frac{r\alpha}{K} + c + \beta_1\right)y^* - \beta_2 v^*$. For convenience, let us define the following parameters:

$$v_1 = f_1(l_1), \quad x_1^* = f_1(l_2), \quad \widehat{x_1^*} = \frac{K}{2r}\left(l_2 - \sqrt{l_2^2 + \frac{4rs}{K}}\right),$$

$$l_1 = r - d - \left(\frac{r\alpha}{K} + c + \beta_1\right)M_2 - \beta_2 M_3, \quad l_2 = r - d - \frac{1}{2}\left(\frac{r\alpha}{K} + c + \beta_1\right)y^* - \beta_2 v^*.$$

Note that $x_0 = f_1(r - d)$ and $r - d > l_2 > l_0 > l_1$, we can obtain $x_0 > x_1^* > x^* > v_1 > 0$. For the uniform persistence of model (4), we have the following main results.

Theorem 1. *If $R_0 > 1$, then model (4) is uniformly persistent in $X^+ := \{\phi = (\phi_1, \phi_2, \phi_3) \in C^+ \mid \phi_2(0) > 0, \phi_3(0) > 0\}$, and the solution $(x(t), y(t), v(t))$ of model (4) with any $\phi \in X^+$ satisfies*

$$\liminf_{t \to \infty} x(t) \ge v_1, \quad \liminf_{t \to \infty} y(t) \ge \frac{y^*}{2}e^{-p\omega} \equiv v_2, \quad \liminf_{t \to \infty} v(t) \ge \frac{ky^*}{2u}e^{-p\omega} = \frac{k}{u}v_2 \equiv v_3, \tag{7}$$

where $\omega = T_0 + T_1 + T_2 + \tau_1 + \tau_2$,

$$T_0 = -\frac{1}{u}\ln\left(\frac{y^*}{2M_2}\right) > 0, \quad T_1 = \frac{-K}{r(x_1^* - \widehat{x_1^*})}\ln\left[\left(\frac{x_1^* - x^0}{x^0 - \widehat{x_1^*}}\right)\left(\frac{\gamma v_1 - \widehat{x_1^*}}{x_1^* - \gamma v_1}\right)\right] > 0,$$

$$T_2 = \frac{q}{u(1-q)} > 0, \quad \gamma \in (0,1), \quad x^0 \in (x^*, x_1^*), \quad q = \frac{x^*}{x^0} < 1.$$

Proof. Let $(x(t), y(t), v(t))$ be any solution of model (4) with any $\phi \in X^+$. By (5), for any $\varepsilon > 0$, there exists a sufficiently large $\widehat{t} > \tau$ such that, for $t > \widehat{t}$, $y(t) \le M_2 + \varepsilon$ and $v(t) \le M_3 + \varepsilon$. From the first equation of model (4), we have, for $t > \widehat{t}$,

$$\dot{x}(t) \geq s - dx(t) + rx(t)\left(1 - \frac{x(t) + \alpha(M_2 + \varepsilon)}{K}\right) - (c + \beta_1)(M_2 + \varepsilon)x(t) - \beta_2(M_3 + \varepsilon)x(t)$$

$$= -\frac{r}{K}(x(t) - v_1(\varepsilon))(x(t) - \widehat{v_1}(\varepsilon)),$$

where

$$v_1(\varepsilon) = \frac{K}{2r}\left(l_1(\varepsilon) + \sqrt{l_1^2(\varepsilon) + \frac{4rs}{K}}\right) > 0, \quad \widehat{v_1}(\varepsilon) = \frac{K}{2r}\left(l_1(\varepsilon) - \sqrt{l_1^2(\varepsilon) + \frac{4rs}{K}}\right) < 0,$$

$$l_1(\varepsilon) = r - d - \left(\frac{r\alpha}{K} + c + \beta_1\right)(M_2 + \varepsilon) - \beta_2(M_3 + \varepsilon).$$

Using the arbitrariness of ε, we have $\liminf_{t\to\infty} x(t) \geq v_1(0) = v_1$.

Next, using a method similar to that in [34,35], let us prove that $\liminf_{t\to\infty} y(t) \geq v_2$. Note that $\gamma \in (0,1)$, then there exists a sufficiently large $T > 0$ such that for $t \geq T$,

$$x(t) > \gamma v_1, \quad v(t) \leq \frac{kM_2}{u} + \frac{ky^*}{2u}.$$

Let us first claim that, for any $t_0 \geq T$, when $t \geq t_0$, the inequality $y(t) \leq \frac{y^*}{2}$ cannot always hold.

If this claim is not true, then there exists a $t_0 \geq T$ such that $y(t) \leq \frac{y^*}{2}$ for all $t \geq t_0$. Then, from the third equation of model (4), we have for $t \geq t_0$, $\dot{v}(t) \leq \frac{1}{2}ky^* - uv(t)$, which implies that, for $t \geq t_0$,

$$v(t) \leq \frac{ky^*}{2u} + \left(v(t_0) - \frac{ky^*}{2u}\right)e^{-u(t-t_0)} \leq \frac{ky^*}{2u} + \frac{kM_2}{u}e^{-u(t-t_0)}.$$

Hence, we have, for $t \geq t_0 + T_0$, $v(t) \leq \frac{k}{u}y^* = v^*$. Then, from the first equation of model (4), we have, for $t \geq t_0 + T_0$,

$$\dot{x}(t) \geq s - dx(t) + rx(t)\left(1 - \frac{x(t)}{K}\right) - \frac{r\alpha y^*}{2K}x(t) - \frac{cy^*}{2}x(t) - \frac{\beta_1 y^*}{2}x(t) - \beta_2 v^* x(t)$$

$$= -\frac{r}{K}(x(t) - x_1^*)(x(t) - \widehat{x_1^*}),$$

which implies that, for $t \geq t_0 + T_0$,

$$x(t) \geq \frac{x_1^* - \widehat{x_1^*}\left(\frac{x(t_0+T_0)-x_1^*}{x(t_0+T_0)-\widehat{x_1^*}}\right)e^{-\frac{r}{K}(x_1^*-\widehat{x_1^*})(t-t_0-T_0)}}{1 - \left(\frac{x(t_0+T_0)-x_1^*}{x(t_0+T_0)-\widehat{x_1^*}}\right)e^{-\frac{r}{K}(x_1^*-\widehat{x_1^*})(t-t_0-T_0)}} > \frac{x_1^* + \widehat{x_1^*}\left(\frac{x_1^*-\gamma v_1}{\gamma v_1-\widehat{x_1^*}}\right)e^{-\frac{r}{K}(x_1^*-\widehat{x_1^*})(t-t_0-T_0)}}{1 + \left(\frac{x_1^*-\gamma v_1}{\gamma v_1-\widehat{x_1^*}}\right)e^{-\frac{r}{K}(x_1^*-\widehat{x_1^*})(t-t_0-T_0)}}.$$

Hence, we have, for $t \geq t_0 + T_0 + T_1$,

$$x(t) > \frac{x_1^* + \widehat{x_1^*}\left(\frac{x_1^*-\gamma v_1}{\gamma v_1-\widehat{x_1^*}}\right)e^{-\frac{r}{K}(x_1^*-\widehat{x_1^*})T_1}}{1 + \left(\frac{x_1^*-\gamma v_1}{\gamma v_1-\widehat{x_1^*}}\right)e^{-\frac{r}{K}(x_1^*-\widehat{x_1^*})T_1}} = x^0 > x^*. \tag{8}$$

Define $m = \min\{\bar{y}, \frac{u\bar{v}}{k}\} > 0$, where

$$\bar{y} = \min_{\theta\in[-(\tau_1+\tau_2),0]} y(T_* + \theta) > 0, \quad \bar{v} = \min_{\theta\in[-(\tau_1+\tau_2),0]} v(T_* + \theta) > 0, \quad T_* = t_0 + T_0 + T_1 + \tau_1 + \tau_2.$$

Next, we show that $y(t) \geq m$ for $t \geq t_0 + T_0 + T_1$. In fact, otherwise, there exists a $\hat{T}_1 \geq 0$ such that $y(t) \geq m$ for $t_0 + T_0 + T_1 \leq t \leq T_* + \hat{T}_1$, $y(T_* + \hat{T}_1) = m$ and $\dot{y}(T_* + \hat{T}_1) \leq 0$. Then, from the third equation of model (4), we have, for $t_0 + T_0 + T_1 \leq t \leq T_* + \hat{T}_1$, $\dot{v}(t) \geq km - uv(t)$, which implies that, for $t_0 + T_0 + T_1 \leq t \leq T_* + \hat{T}_1$,

$$
\begin{aligned}
v(t) &\geq \left(v(t_0 + T_0 + T_1) - \frac{km}{u} \right) e^{-u(t - t_0 - T_0 - T_1)} + \frac{km}{u} \\
&\geq (v(t_0 + T_0 + T_1) - \bar{v}) e^{-u(t - t_0 - T_0 - T_1)} + \frac{km}{u} \\
&\geq \frac{km}{u}.
\end{aligned}
\tag{9}
$$

Therefore, from (8) and (9), we have, for $t = T_* + \hat{T}_1$,

$$
\begin{aligned}
\dot{y}(t) &= \delta_1 x(t - \tau_1) y(t - \tau_1) + \delta_2 x(t - \tau_2) v(t - \tau_2) - pm \\
&\geq \delta_1 x^0 m + \delta_2 x^0 \frac{km}{u} - pm \\
&= pm \left(\frac{x^0}{x^*} - 1 \right) > 0.
\end{aligned}
$$

This is a contradiction to $\dot{y}(T_* + \hat{T}_1) \leq 0$. This shows that for $t \geq t_0 + T_0 + T_1$, $y(t) \geq m$.

Using the derivation completely similar to (9), we have, for $t \geq t_0 + T_0 + T_1$, $v(t) \geq \frac{km}{u}$. Consider the following auxiliary function:

$$
V(t) = y(t) + \frac{\delta_2}{u} x^* v(t) + \delta_1 \int_{t-\tau_1}^{t} x(\theta) y(\theta) d\theta + \delta_2 \int_{t-\tau_2}^{t} x(\theta) v(\theta) d\theta.
$$

Then, we have, for $t \geq t_0 + T_0 + T_1$,

$$
\dot{V}(t) = \delta_1 (x(t) - x^*) y(t) + \delta_2 (x(t) - x^*) v(t) \geq m \left(\delta_1 + \frac{k}{u} \delta_2 \right) (x^0 - x^*) > 0,
$$

which leads to, for $t \geq t_0 + T_0 + T_1$,

$$
V(t) \geq V(t_0 + T_0 + T_1) + m \left(\delta_1 + \frac{k}{u} \delta_2 \right) (x^0 - x^*)(t - t_0 - T_0 - T_1),
$$

which implies that $V(t) \to +\infty (t \to +\infty)$. This is a contradiction with the boundedness of $V(t)$. Therefore, the claim is proved.

Below, there are two remaining cases that need to be discussed.

(i) $y(t) \geq \frac{y^*}{2}$ for sufficiently large t. (ii) $y(t)$ oscillates about $\frac{y^*}{2}$ for sufficiently large t. For the case (i), it clearly has $\liminf_{t \to +\infty} y(t) \geq v_2$.

For the case (ii), let t_1 and t_2 be sufficiently large such that $y(t_1) = y(t_2) = \frac{y^*}{2}$, $y(t) < \frac{y^*}{2}$ $(t_1 < t < t_2)$.

If $t_2 - t_1 \leq \omega$, from the second equation of model (4), we have, for $t_1 \leq t \leq t_2$, $\dot{y}(t) \geq -py(t)$, which implies that for $t_1 \leq t \leq t_2$,

$$
y(t) \geq y(t_1) e^{-p(t - t_1)} \geq \frac{y^*}{2} e^{-p\omega} = v_2.
$$

If $t_2 - t_1 > \omega$, it is easily obtained that, for $t_1 \leq t \leq t_1 + \omega$, $y(t) \geq v_2$. Further, we prove that for $t_1 + \omega \leq t \leq t_2$, $y(t) \geq v_2$.

If not, there exists a $\hat{T}_2 \geq 0$ such that for $t_1 \leq t \leq t_1 + \omega + \hat{T}_2(< t_2)$, $y(t) \geq v_2$, $y(t_1 + \omega + \hat{T}_2) = v_2$ and $\dot{y}(t_1 + \omega + \hat{T}_2) \leq 0$. Using the derivation method similar to (8), treating t_1 as t_0, we have, for $t_1 + T_0 + T_1 \leq t \leq t_1 + \omega + \hat{T}_2$, $x(t) > x^0 > x^*$. Then, let us prove that there exists $\bar{t} \in [t_1, t_1 + T_0 + T_1 + T_2]$ such that $v(\bar{t}) \geq \frac{qkv_2}{u}$. If not, then we have, for $t_1 \leq t \leq t_1 + T_0 + T_1 + T_2$, $v(t) < \frac{qkv_2}{u}$. From the third equation of model (4), we have, for $t_1 \leq t \leq t_1 + T_0 + T_1 + T_2$,

$$\dot{v}(t) \geq kv_2 - uv(t) \geq kv_2(1 - q),$$

which implies that, for $t = t_1 + T_0 + T_1 + T_2$,

$$v(t) \geq v(t_1) + kv_2(1 - q)(t - t_1) > v(t_1) + kv_2(1 - q)T_2 > \frac{qkv_2}{u},$$

which is a contradiction. Hence, we conclude that there exists $\bar{t} \in [t_1, t_1 + T_0 + T_1 + T_2]$ such that $v(\bar{t}) \geq \frac{qkv_2}{u}$.

Note that, for $\bar{t} \leq t \leq t_1 + \omega + \hat{T}_2$, $\dot{v}(t) \geq qkv_2 - uv(t)$, which implies that, for $\bar{t} \leq t \leq t_1 + \omega + \hat{T}_2$,

$$v(t) \geq \left(v(\bar{t}) - \frac{qkv_2}{u} \right) e^{-u(t - t_1)} + \frac{qkv_2}{u} \geq \frac{qkv_2}{u}.$$

Hence, we have from the second equation of model (4) that for $t = t_1 + \omega + \hat{T}_2$,

$$\begin{aligned}
\dot{y}(t) &= \delta_1 x(t - \tau_1)y(t - \tau_1) + \delta_2 x(t - \tau_2)v(t - \tau_2) - pv_2 \\
&\geq pv_2 \left(\frac{\delta_1}{p} x(t - \tau_1) + \frac{qk\delta_2}{up} x(t - \tau_2) - 1 \right) \\
&> pv_2 \left(\frac{qx^0}{x^*} - 1 \right) = 0.
\end{aligned} \tag{10}$$

This is a contradiction to $\dot{y}(t_1 + \omega + \hat{T}_2) \leq 0$. Based on the above analysis, we have $y(t) \geq v_2$ for $t \in [t_1, t_2]$. Since this kind of interval $[t_1, t_2]$ is chosen in an arbitrary way, we conclude that $y(t) \geq v_2$ for any sufficiently large t. Thus, $\liminf_{t \to \infty} y(t) \geq v_2$.

Finally, according to the third equation of model (4), we have $\liminf_{t \to \infty} v(t) \geq \frac{kv_2}{u} = v_3$. $\square$

3. Global Attractivity of the Chronic Infection Equilibrium

Now, we continue to discuss global attractivity of the chronic infection equilibrium E^*. The following lemma is used.

Lemma 1. *(Barbalat's lemma [42]) Let $q(t)$ be a real valued differentiable function defined on some half line $[\vartheta, +\infty)$, $\vartheta \in (-\infty, +\infty)$. If $\lim_{t \to +\infty} q(t) = \vartheta_1$ ($|\vartheta_1| < +\infty$) and $\dot{q}(t)$ is uniformly continuous for $t > \vartheta$, then $\lim_{t \to +\infty} \dot{q}(t) = 0$.*

Due to the technical requirements of the proof, we assume that $m_1\tau_1 = m_2\tau_2$.
For any sufficient small $0 < \varepsilon < v_1$, we define

$$\nu_1(\varepsilon) = \nu_1 - \varepsilon, \quad x_0(\varepsilon) = x_0 + \varepsilon, \quad M_2(\varepsilon) = M_2 + \varepsilon, \quad M_3(\varepsilon) = M_3 + \varepsilon, \quad \Lambda_1 = \frac{r\alpha}{K} + c + \beta_1,$$

$$Y_1(\varepsilon) = d - r + \frac{r}{K}(x_0(\varepsilon) + x^*) + \Lambda_1 y^* + \beta_2 v^*, \quad Y_2(\varepsilon) = \beta_1 x_0(\varepsilon) + \beta_1 y^* + \beta_2 x_0(\varepsilon) + \beta_2 v^* + p e^{m_1 \tau_1},$$

$$\Psi_1(\varepsilon) = \frac{1}{2}\beta_1[x_0(\varepsilon)e^{-m_1\tau_1}Y_2(\varepsilon) + M_2(\varepsilon)(Y_1(\varepsilon) + \Lambda_1 x_0(\varepsilon) + \beta_2 x_0(\varepsilon))],$$

$$\Psi_2(\varepsilon) = \frac{1}{2}\beta_2[M_3(\varepsilon)(Y_1(\varepsilon) + \Lambda_1 x_0(\varepsilon) + \beta_2 x_0(\varepsilon)) + (k+u)x_0(\varepsilon)],$$

$$\Psi_3(\varepsilon) = \frac{1}{2}\beta_1^2 x_0(\varepsilon)y^*(1 + e^{-m_1\tau_1}), \quad \Psi_4(\varepsilon) = \frac{1}{2}\beta_1\beta_2 x_0(\varepsilon)v^*(1 + e^{-m_1\tau_1}),$$

$$\Psi_5(\varepsilon) = \frac{1}{2}Y_1(\varepsilon)\beta_1 M_2(\varepsilon)(1 + e^{m_1\tau_1}), \quad \Psi_6(\varepsilon) = \frac{1}{2}\beta_1^2 x_0^2(\varepsilon)(1 + e^{-m_1\tau_1}),$$

$$\Psi_7(\varepsilon) = \frac{1}{2}p\beta_1 x_0(\varepsilon)(1 + e^{m_1\tau_1}) + \frac{1}{2}\Lambda_1 x_0(\varepsilon)\beta_1 M_2(\varepsilon)(1 + e^{m_1\tau_1}), \quad \Psi_8(\varepsilon) = \frac{1}{2}\beta_1\beta_2 x_0^2(\varepsilon)(1 + e^{-m_1\tau_1}),$$

$$\Psi_9(\varepsilon) = \frac{1}{2}\beta_2 x_0(\varepsilon)\beta_1 M_2(\varepsilon)(1 + e^{m_1\tau_1}), \quad \Psi_{10}(\varepsilon) = \frac{1}{2}Y_1(\varepsilon)\beta_2 M_3(\varepsilon)(1 + e^{m_1\tau_1}),$$

$$\Psi_{11}(\varepsilon) = \frac{1}{2}k\beta_2 x_0(\varepsilon)(1 + e^{m_1\tau_1}) + \frac{1}{2}\Lambda_1 x_0(\varepsilon)\beta_2 M_3(\varepsilon)(1 + e^{m_1\tau_1}),$$

$$\Psi_{12}(\varepsilon) = \frac{1}{2}u\beta_2 x_0(\varepsilon)(1 + e^{m_1\tau_1}) + \frac{1}{2}\beta_2 x_0(\varepsilon)\beta_2 M_3(\varepsilon)(1 + e^{m_1\tau_1}).$$

Let us define the real symmetric matrices as follows,

$$J(\varepsilon) = \begin{pmatrix} A_{11}(\varepsilon) & -A_{12}(\varepsilon) & 0 \\ -A_{12}(\varepsilon) & A_{22}(\varepsilon) & -A_{23}(\varepsilon) \\ 0 & -A_{23}(\varepsilon) & A_{33}(\varepsilon) \end{pmatrix},$$

where

$$A_{11}(\varepsilon) = \frac{1}{x_0(\varepsilon)}\left[d - r + \frac{rx^*}{K} + \left(\frac{r\alpha}{K} + c\right)y^*\right] + \frac{r}{K} + \theta_1\left\{d - r + \frac{r}{K}(\nu_1(\varepsilon) + x^*) + \left(c + \frac{r\alpha}{K}\right)y^*\right.$$

$$\left. - [(\Psi_1(\varepsilon) + \Psi_3(\varepsilon) + \Psi_4(\varepsilon) + \Psi_5(\varepsilon))\tau_1 + (\Psi_2(\varepsilon) + \Psi_{10}(\varepsilon))\tau_2]\right\},$$

$$A_{12}(\varepsilon) = \frac{1}{2}\left(c + \frac{r\alpha}{K}\right) + \frac{1}{2}\theta_1\left\{e^{m_1\tau_1}\left[d - r + \frac{r}{K}(x_0(\varepsilon) + x^*) + \left(c + \frac{r\alpha}{K}\right)y^* + p\right] + \left(c + \frac{r\alpha}{K}\right)x_0(\varepsilon)\right\},$$

$$A_{22}(\varepsilon) = \theta_1\left\{e^{m_1\tau_1}\left[e^{m_1\tau_1}p + \left(c + \frac{r\alpha}{K}\right)\nu_1(\varepsilon)\right]\right.$$

$$\left. - [(\Psi_1(\varepsilon)e^{m_1\tau_1} + \Psi_6(\varepsilon) + \Psi_7(\varepsilon))\tau_1 + (\Psi_2(\varepsilon)e^{m_1\tau_1} + \Psi_{11}(\varepsilon))\tau_2]\right\},$$

$$A_{23}(\varepsilon) = \frac{1}{2}\theta_2 k, \quad A_{33}(\varepsilon) = \theta_2 u - \theta_1[(\Psi_8(\varepsilon) + \Psi_9(\varepsilon))\tau_1 + \Psi_{12}(\varepsilon)\tau_2],$$

where θ_1 and θ_2 are arbitrary positive constants.

Theorem 2. *If $R_0 > 1$, $d - r + \frac{rx^*}{K} \geq 0$, $m_1\tau_1 = m_2\tau_2$ and matrix $J(0)$ is positive definite, then the chronic infection equilibrium E^* is globally attractive in X^+.*

Proof. Let $(x(t), y(t), v(t)) \in X^+ (t \geq 0)$ be any solution of model (4). If $J(0)$ is positive definite, then $J(\varepsilon)$ is also positive definite for any sufficiently small $0 < \varepsilon < \nu_1$. For the above ε, there exists a sufficiently large $T(\varepsilon) > \tau_1 + \tau_2$ such that, for $t > T(\varepsilon)$,

$$0 < \nu_1(\varepsilon) < x(t) < x_0(\varepsilon), \quad y(t) < M_2(\varepsilon), \quad v(t) < M_3(\varepsilon).$$

Let $g(z) = z - 1 - \ln z$ $(z > 0)$. Clearly, $g(z) \geq 0$ $(z > 0)$, and $g(z) = 0$ if and only if $z = 1$. Define

$$
\begin{aligned}
U_1 =\; & x^* g\left(\frac{x(t)}{x^*}\right) + e^{m_1 \tau_1} y^* g\left(\frac{y(t)}{y^*}\right) + \frac{\beta_2 x^* v^*}{k y^*} v^* g\left(\frac{v(t)}{v^*}\right) \\
& + \beta_1 x^* y^* \int_{t-\tau_1}^{t} g\left(\frac{x(s)y(s)}{x^* y^*}\right) ds + \beta_2 x^* v^* \int_{t-\tau_2}^{t} g\left(\frac{x(s)v(s)}{x^* v^*}\right) ds.
\end{aligned}
$$

Hence, U_1 is positive definite with respect to the chronic infection equilibrium $E^* = (x^*, y^*, v^*)$. Similar to the calculation in [18,19], for $t \geq 0$, the derivative along the solution of model (4) satisfies

$$
\begin{aligned}
\frac{dU_1}{dt} =\; & -\frac{1}{x(t)}\left(d - r + \frac{rx^*}{K}\right)(x(t) - x^*)^2 - \frac{r}{K}(x(t) - x^*)^2 \\
& + \left(\frac{r\alpha}{K} + c\right)\left(1 - \frac{x^*}{x(t)}\right)(x^* y^* - x(t)y(t)) \\
& + \beta_1 x^* y^* \left\{ -g\left(\frac{x(t-\tau_1)y(t-\tau_1)}{x^* y(t)}\right) - g\left(\frac{x^*}{x(t)}\right)\right\} \\
& + \beta_2 x^* v^* \left\{ -g\left(\frac{x(t-\tau_2)v(t-\tau_2)y^*}{x^* v^* y(t)}\right) - g\left(\frac{x^*}{x(t)}\right) - g\left(\frac{y(t)v^*}{y^* v(t)}\right)\right\} \\
=\; & -\frac{1}{x(t)}\left[d - r + \frac{rx^*}{K} + \left(\frac{r\alpha}{K} + c\right)y^*\right](x(t) - x^*)^2 - \frac{r}{K}(x(t) - x^*)^2 \\
& - \left(\frac{r\alpha}{K} + c\right)(x(t) - x^*)(y(t) - y^*) \\
& + \beta_1 x^* y^* \left\{ -g\left(\frac{x(t-\tau_1)y(t-\tau_1)}{x^* y(t)}\right) - g\left(\frac{x^*}{x(t)}\right)\right\} \\
& + \beta_2 x^* v^* \left\{ -g\left(\frac{x(t-\tau_2)v(t-\tau_2)y^*}{x^* v^* y(t)}\right) - g\left(\frac{x^*}{x(t)}\right) - g\left(\frac{y(t)v^*}{y^* v(t)}\right)\right\}.
\end{aligned} \tag{11}
$$

It is worth mentioning that if $\alpha = c = 0$ and $d - r + \frac{rx^*}{K} \geq 0$, then $\frac{dU_1}{dt} \leq 0$, which leads to E^* is stable (see [37,38]). Then, it follows from [19] that E^* is globally attractive. Thus, E^* is globally asymptotically stable.

If α and c are not 0 at the same time, inspired by [43,44], we define

$$
U_2 = \frac{1}{2}[(x(t) - x^*) + e^{m_1 \tau_1}(y(t) - y^*)]^2.
$$

From model (4), we have, for $t > T(\varepsilon)$,

$$
\begin{aligned}
\dot{x}(t) + e^{m_1 \tau_1}\dot{y}(t) =\; & s - dx(t) + rx(t)\left(1 - \frac{x(t)}{K}\right) - \left(c + \frac{r\alpha}{K}\right)x(t)y(t) - e^{m_1 \tau_1}py(t) \\
& + \beta_1 x(t-\tau_1)y(t-\tau_1) - \beta_1 x(t)y(t) + \beta_2 x(t-\tau_2)v(t-\tau_2) - \beta_2 x(t)v(t) \\
=\; & -\left[d - r + \frac{r}{K}(x(t) + x^*) + \left(c + \frac{r\alpha}{K}\right)y^*\right](x(t) - x^*) \\
& - \left[e^{m_1 \tau_1}p + \left(c + \frac{r\alpha}{K}\right)x(t)\right](y(t) - y^*) \\
& + \beta_1 x(t-\tau_1)(y(t-\tau_1) - y(t)) + \beta_1 y(t)(x(t-\tau_1) - x(t)) \\
& + \beta_2 x(t-\tau_2)(v(t-\tau_2) - v(t)) + \beta_2 v(t)(x(t-\tau_2) - x(t)).
\end{aligned}
$$

Further, we have, for $t > T(\varepsilon)$,

$$
\begin{aligned}
\frac{dU_2}{dt} =& [(x(t) - x^*) + e^{m_1\tau_1}(y(t) - y^*)](\dot{x}(t) + e^{m_1\tau_1}\dot{y}(t)) \\
=& - \left[d - r + \frac{r}{K}(x(t) + x^*) + \left(c + \frac{r\alpha}{K}\right)y^*\right](x(t) - x^*)^2 \\
& - e^{m_1\tau_1}\left[e^{m_1\tau_1}p + \left(c + \frac{r\alpha}{K}\right)x(t)\right](y(t) - y^*)^2 \\
& - \left\{e^{m_1\tau_1}\left[d - r + \frac{r}{K}(x(t) + x^*) + \left(c + \frac{r\alpha}{K}\right)y^* + p\right] + \left(c + \frac{r\alpha}{K}\right)x(t)\right\} \\
& \times (x(t) - x^*)(y(t) - y^*) + \Gamma_1(t) + \Gamma_2(t) + \Gamma_3(t) + \Gamma_4(t),
\end{aligned}
\tag{12}
$$

where

$$
\Gamma_1(t) = - \beta_1 x(t - \tau_1)[(x(t) - x^*) + e^{m_1\tau_1}(y(t) - y^*)] \int_{t-\tau_1}^{t} \dot{y}(s)ds,
$$

$$
\Gamma_2(t) = - \beta_1 y(t)[(x(t) - x^*) + e^{m_1\tau_1}(y(t) - y^*)] \int_{t-\tau_1}^{t} \dot{x}(s)ds,
$$

$$
\Gamma_3(t) = - \beta_2 x(t - \tau_2)[(x(t) - x^*) + e^{m_1\tau_1}(y(t) - y^*)] \int_{t-\tau_2}^{t} \dot{v}(s)ds,
$$

$$
\Gamma_4(t) = - \beta_2 v(t)[(x(t) - x^*) + e^{m_1\tau_1}(y(t) - y^*)] \int_{t-\tau_2}^{t} \dot{x}(s)ds.
$$

Since E^* is a positive equilibrium of model (4), $\dot{x}(t)$, $\dot{y}(t)$ and $\dot{v}(t)$ can be rewritten as

$$
\begin{aligned}
\dot{x}(t) =& - \left[d - r + \frac{r}{K}(x(t) + x^*)\right](x(t) - x^*) + \Lambda_1 x^* y^* - \Lambda_1 x(t)y(t) + \beta_2 x^* v^* - \beta_2 x(t)v(t) \\
=& - \left[d - r + \frac{r}{K}(x(t) + x^*)\right](x(t) - x^*) + \Lambda_1(x^* - x(t))y^* + \Lambda_1 x(t)(y^* - y(t)) \\
& + \beta_2(x^* - x(t))v^* + \beta_2 x(t)(v^* - v(t)) \\
=& - \widetilde{Y_1}(t)(x(t) - x^*) + \Lambda_1 x(t)(y^* - y(t)) + \beta_2 x(t)(v^* - v(t)), \\
& \text{where } \widetilde{Y_1}(t) = d - r + \tfrac{r}{K}(x(t) + x^*) + \Lambda_1 y^* + \beta_2 v^*,
\end{aligned}
\tag{13}
$$

$$
\begin{aligned}
\dot{y}(t) =& \beta_1 e^{-m_1\tau_1}x(t - \tau_1)(y(t - \tau_1) - y^*) + \beta_1 e^{-m_1\tau_1}y^*(x(t - \tau_1) - x^*) \\
& + \beta_2 e^{-m_2\tau_2}x(t - \tau_2)(v(t - \tau_2) - v^*) + \beta_2 e^{-m_2\tau_2}v^*(x(t - \tau_2) - x^*) + p(y^* - y(t)),
\end{aligned}
\tag{14}
$$

$$
\dot{v}(t) = k(y(t) - y^*) + u(v^* - v(t)).
\tag{15}
$$

From (14), we have, for $t > T(\varepsilon) + 2(\tau_1 + \tau_2)$,

$$
\begin{aligned}
|\Gamma_1(t)| =&\, \beta_1 x(t-\tau_1)|[(x(t)-x^*)+e^{m_1\tau_1}(y(t)-y^*)]| \\
&\times \left| \int_{t-\tau_1}^{t} \{\beta_1 e^{-m_1\tau_1} x(s-\tau_1)(y(s-\tau_1)-y^*)+\beta_1 e^{-m_1\tau_1} y^*(x(s-\tau_1)-x^*)\right. \\
&\quad \left. +\beta_2 e^{-m_2\tau_2} x(s-\tau_2)(v(s-\tau_2)-v^*)+\beta_2 e^{-m_2\tau_2} v^*(x(s-\tau_2)-x^*)+p(y^*-y(s))\}ds \right| \\
\leq&\, \beta_1 x_0(\varepsilon)\int_{t-\tau_1}^{t} \left\{ \frac{\beta_1 e^{-m_1\tau_1} x_0(\varepsilon)}{2}[(x(t)-x^*)^2+(y(s-\tau_1)-y^*)^2] \right. \\
&+\frac{\beta_1 x_0(\varepsilon)}{2}[(y(t)-y^*)^2+(y(s-\tau_1)-y^*)^2] \\
&+\frac{\beta_1 e^{-m_1\tau_1} y^*}{2}[(x(t)-x^*)^2+(x(s-\tau_1)-x^*)^2]+\frac{\beta_1 y^*}{2}[(y(t)-y^*)^2+(x(s-\tau_1)-x^*)^2] \\
&+\frac{\beta_2 e^{-m_2\tau_2} x_0(\varepsilon)}{2}[(x(t)-x^*)^2+(v(s-\tau_2)-v^*)^2]+\frac{\beta_2 x_0(\varepsilon)}{2}[(y(t)-y^*)^2+(v(s-\tau_2)-v^*)^2] \\
&+\frac{\beta_2 e^{-m_2\tau_2} v^*}{2}[(x(t)-x^*)^2+(x(s-\tau_2)-x^*)^2]+\frac{\beta_2 v^*}{2}[(y(t)-y^*)^2+(x(s-\tau_2)-x^*)^2] \\
&\left. +\frac{p}{2}[(x(t)-x^*)^2+(y(s)-y^*)^2]+\frac{pe^{m_1\tau_1}}{2}[(y(t)-y^*)^2+(y(s)-y^*)^2] \right\}ds \\
=&\, \frac{1}{2}\beta_1 x_0(\varepsilon)e^{-m_1\tau_1}Y_2(\varepsilon)(x(t)-x^*)^2\tau_1+\frac{1}{2}\beta_1 x_0(\varepsilon)Y_2(\varepsilon)(y(t)-y^*)^2\tau_1 \\
&+\frac{1}{2}\beta_1^2 x_0^2(\varepsilon)(1+e^{-m_1\tau_1})\int_{t-\tau_1}^{t}(y(s-\tau_1)-y^*)^2ds \\
&+\frac{1}{2}\beta_1^2 x_0(\varepsilon)y^*(1+e^{-m_1\tau_1})\int_{t-\tau_1}^{t}(x(s-\tau_1)-x^*)^2ds \\
&+\frac{1}{2}\beta_1\beta_2 x_0^2(\varepsilon)(1+e^{-m_1\tau_1})\int_{t-\tau_1}^{t}(v(s-\tau_2)-v^*)^2ds \\
&+\frac{1}{2}\beta_1\beta_2 x_0(\varepsilon)v^*(1+e^{-m_1\tau_1})\int_{t-\tau_1}^{t}(x(s-\tau_2)-x^*)^2ds \\
&+\frac{1}{2}p\beta_1 x_0(\varepsilon)(1+e^{m_1\tau_1})\int_{t-\tau_1}^{t}(y(s)-y^*)^2ds.
\end{aligned}
$$

Similarly, from (13) and (15), we have, for $t>T(\varepsilon)+2(\tau_1+\tau_2)$,

$$
\begin{aligned}
|\Gamma_2(t)| \leq&\, \beta_1 M_2(\varepsilon)|[(x(t)-x^*)+e^{m_1\tau_1}(y(t)-y^*)]| \\
&\times \int_{t-\tau_1}^{t}\left\{ Y_1(\varepsilon)|(x(s)-x^*)|+\Lambda_1 x_0(\varepsilon)|(y(s)-y^*)|+\beta_2 x_0(\varepsilon)|(v(s)-v^*)| \right\}ds \\
\leq&\, \frac{1}{2}\beta_1 M_2(\varepsilon)(Y_1(\varepsilon)+\Lambda_1 x_0(\varepsilon)+\beta_2 x_0(\varepsilon))(x(t)-x^*)^2\tau_1 \\
&+\frac{1}{2}\beta_1 M_2(\varepsilon)e^{m_1\tau_1}(Y_1(\varepsilon)+\Lambda_1 x_0(\varepsilon)+\beta_2 x_0(\varepsilon))(y(t)-y^*)^2\tau_1 \\
&+\frac{1}{2}Y_1(\varepsilon)\beta_1 M_2(\varepsilon)(1+e^{m_1\tau_1})\int_{t-\tau_1}^{t}(x(s)-x^*)^2ds \\
&+\frac{1}{2}\Lambda_1 x_0(\varepsilon)\beta_1 M_2(\varepsilon)(1+e^{m_1\tau_1})\int_{t-\tau_1}^{t}(y(s)-y^*)^2ds \\
&+\frac{1}{2}\beta_2 x_0(\varepsilon)\beta_1 M_2(\varepsilon)(1+e^{m_1\tau_1})\int_{t-\tau_1}^{t}(v(s)-v^*)^2ds,
\end{aligned}
$$

$$
\begin{aligned}
|\Gamma_3(t)| \leq&\, \beta_2 x_0(\varepsilon)[|(x(t)-x^*)|+e^{m_1\tau_1}|(y(t)-y^*)|]\int_{t-\tau_2}^{t}[k|(y(s)-y^*)|+u|(v^*-v(s))|]ds, \\
\leq&\, \frac{1}{2}\beta_2 x_0(\varepsilon)(k+u)(x(t)-x^*)^2\tau_2+\frac{1}{2}\beta_2 x_0(\varepsilon)(k+u)e^{m_1\tau_1}(y(t)-y^*)^2\tau_2 \\
&+\frac{1}{2}k\beta_2 x_0(\varepsilon)(1+e^{m_1\tau_1})\int_{t-\tau_2}^{t}(y(s)-y^*)^2ds+\frac{1}{2}u\beta_2 x_0(\varepsilon)(1+e^{m_1\tau_1})\int_{t-\tau_2}^{t}(v(s)-v^*)^2ds,
\end{aligned}
$$

$$|\Gamma_4(t)| \leq \beta_2 M_3(\varepsilon)[|(x(t) - x^*)| + e^{m_1\tau_1}|(y(t) - y^*)|]$$

$$\times \int_{t-\tau_2}^{t} \left\{ Y_1(\varepsilon)|(x(s) - x^*)| + \Lambda_1 x_0(\varepsilon)|(y(s) - y^*)| + \beta_2 x_0(\varepsilon)|(v(s) - v^*)| \right\} ds$$

$$\leq \frac{1}{2}\beta_2 M_3(\varepsilon)(Y_1(\varepsilon) + \Lambda_1 x_0(\varepsilon) + \beta_2 x_0(\varepsilon))(x(t) - x^*)^2 \tau_2$$

$$+ \frac{1}{2}\beta_2 M_3(\varepsilon)e^{m_1\tau_1}(Y_1(\varepsilon) + \Lambda_1 x_0(\varepsilon) + \beta_2 x_0(\varepsilon))(y(t) - y^*)^2 \tau_2$$

$$+ \frac{1}{2}Y_1(\varepsilon)\beta_2 M_3(\varepsilon)(1 + e^{m_1\tau_1}) \int_{t-\tau_2}^{t} (x(s) - x^*)^2 ds$$

$$+ \frac{1}{2}\Lambda_1 x_0(\varepsilon)\beta_2 M_3(\varepsilon)(1 + e^{m_1\tau_1}) \int_{t-\tau_2}^{t} (y(s) - y^*)^2 ds$$

$$+ \frac{1}{2}\beta_2 x_0(\varepsilon)\beta_2 M_3(\varepsilon)(1 + e^{m_1\tau_1}) \int_{t-\tau_2}^{t} (v(s) - v^*)^2 ds.$$

Hence, we have, for $t > T(\varepsilon) + 2(\tau_1 + \tau_2)$,

$$\Gamma(t) := |\Gamma_1(t)| + |\Gamma_2(t)| + |\Gamma_3(t)| + |\Gamma_4(t)|$$

$$\leq \frac{1}{2}\beta_1[x_0(\varepsilon)e^{-m_1\tau_1}Y_2(\varepsilon) + M_2(\varepsilon)(Y_1(\varepsilon) + \Lambda_1 x_0(\varepsilon) + \beta_2 x_0(\varepsilon))]\tau_1(x(t) - x^*)^2$$

$$+ \frac{1}{2}\beta_1[x_0(\varepsilon)Y_2(\varepsilon) + e^{m_1\tau_1}M_2(\varepsilon)(Y_1(\varepsilon) + \Lambda_1 x_0(\varepsilon) + \beta_2 x_0(\varepsilon))]\tau_1(y(t) - y^*)^2$$

$$+ \frac{1}{2}\beta_2[M_3(\varepsilon)(Y_1(\varepsilon) + \Lambda_1 x_0(\varepsilon) + \beta_2 x_0(\varepsilon)) + x_0(\varepsilon)(k + u)]\tau_2(x(t) - x^*)^2$$

$$+ \frac{1}{2}\beta_2 e^{m_1\tau_1}[M_3(\varepsilon)(Y_1(\varepsilon) + \Lambda_1 x_0(\varepsilon) + \beta_2 x_0(\varepsilon)) + x_0(\varepsilon)(k + u)]\tau_2(y(t) - y^*)^2$$

$$+ \frac{1}{2}\beta_1^2 x_0(\varepsilon)y^*(1 + e^{-m_1\tau_1}) \int_{t-\tau_1}^{t} (x(s - \tau_1) - x^*)^2 ds$$

$$+ \frac{1}{2}\beta_1\beta_2 x_0(\varepsilon)v^*(1 + e^{-m_1\tau_1}) \int_{t-\tau_1}^{t} (x(s - \tau_2) - x^*)^2 ds$$

$$+ \frac{1}{2}Y_1(\varepsilon)\beta_1 M_2(\varepsilon)(1 + e^{m_1\tau_1}) \int_{t-\tau_1}^{t} (x(s) - x^*)^2 ds$$

$$+ \frac{1}{2}\beta_1^2 x_0^2(\varepsilon)(1 + e^{-m_1\tau_1}) \int_{t-\tau_1}^{t} (y(s - \tau_1) - y^*)^2 ds$$

$$+ \left[\frac{1}{2}p\beta_1 x_0(\varepsilon)(1 + e^{m_1\tau_1}) + \frac{1}{2}\Lambda_1 x_0(\varepsilon)\beta_1 M_2(\varepsilon)(1 + e^{m_1\tau_1})\right] \int_{t-\tau_1}^{t} (y(s) - y^*)^2 ds$$

$$+ \frac{1}{2}\beta_1\beta_2 x_0^2(\varepsilon)(1 + e^{-m_1\tau_1}) \int_{t-\tau_1}^{t} (v(s - \tau_2) - v^*)^2 ds$$

$$+ \frac{1}{2}\beta_2 x_0(\varepsilon)\beta_1 M_2(\varepsilon)(1 + e^{m_1\tau_1}) \int_{t-\tau_1}^{t} (v(s) - v^*)^2 ds$$

$$+ \frac{1}{2}Y_1(\varepsilon)\beta_2 M_3(\varepsilon)(1 + e^{m_1\tau_1}) \int_{t-\tau_2}^{t} (x(s) - x^*)^2 ds$$

$$+ \left[\frac{1}{2}k\beta_2 x_0(\varepsilon)(1 + e^{m_1\tau_1}) + \frac{1}{2}\Lambda_1 x_0(\varepsilon)\beta_2 M_3(\varepsilon)(1 + e^{m_1\tau_1})\right] \int_{t-\tau_2}^{t} (y(s) - y^*)^2 ds$$

$$+ \left[\frac{1}{2}u\beta_2 x_0(\varepsilon)(1 + e^{m_1\tau_1}) + \frac{1}{2}\beta_2 x_0(\varepsilon)\beta_2 M_3(\varepsilon)(1 + e^{m_1\tau_1})\right] \int_{t-\tau_2}^{t} (v(s) - v^*)^2 ds$$

$$= \Psi_1(\varepsilon)\tau_1(x(t) - x^*)^2 + \Psi_1(\varepsilon)e^{m_1\tau_1}\tau_1(y(t) - y^*)^2 + \Psi_2(\varepsilon)\tau_2(x(t) - x^*)^2 + \Psi_2(\varepsilon)e^{m_1\tau_1}\tau_2(y(t) - y^*)^2$$

$$+ \Psi_3(\varepsilon) \int_{t-\tau_1}^{t} (x(s - \tau_1) - x^*)^2 ds + \Psi_4(\varepsilon) \int_{t-\tau_1}^{t} (x(s - \tau_2) - x^*)^2 ds + \Psi_5(\varepsilon) \int_{t-\tau_1}^{t} (x(s) - x^*)^2 ds$$

$$+ \Psi_6(\varepsilon) \int_{t-\tau_1}^{t} (y(s - \tau_1) - y^*)^2 ds + \Psi_7(\varepsilon) \int_{t-\tau_1}^{t} (y(s) - y^*)^2 ds$$

$$+ \Psi_8(\varepsilon) \int_{t-\tau_1}^{t} (v(s - \tau_2) - v^*)^2 ds + \Psi_9(\varepsilon) \int_{t-\tau_1}^{t} (v(s) - v^*)^2 ds$$

$$+ \Psi_{10}(\varepsilon) \int_{t-\tau_2}^{t} (x(s) - x^*)^2 ds + \Psi_{11}(\varepsilon) \int_{t-\tau_2}^{t} (y(s) - y^*)^2 ds + \Psi_{12}(\varepsilon) \int_{t-\tau_2}^{t} (v(s) - v^*)^2 ds.$$

For $t > T(\varepsilon) + 2(\tau_1 + \tau_2)$, we define

$$
\begin{aligned}
U_3 =& \Psi_3(\varepsilon)\left[\int_{t-\tau_1}^{t}\int_{\theta}^{t}(x(s-\tau_1)-x^*)^2 ds d\theta + \tau_1\int_{t-\tau_1}^{t}(x(s)-x^*)^2 ds\right]\\
&+ \Psi_4(\varepsilon)\left[\int_{t-\tau_1}^{t}\int_{\theta}^{t}(x(s-\tau_2)-x^*)^2 ds d\theta + \tau_1\int_{t-\tau_2}^{t}(x(s)-x^*)^2 ds\right] + \Psi_5(\varepsilon)\int_{t-\tau_1}^{t}\int_{\theta}^{t}(x(s)-x^*)^2 ds d\theta\\
&+ \Psi_6(\varepsilon)\left[\int_{t-\tau_1}^{t}\int_{\theta}^{t}(y(s-\tau_1)-y^*)^2 ds d\theta + \tau_1\int_{t-\tau_1}^{t}(y(s)-y^*)^2 ds\right] + \Psi_7(\varepsilon)\int_{t-\tau_1}^{t}\int_{\theta}^{t}(y(s)-y^*)^2 ds d\theta\\
&+ \Psi_8(\varepsilon)\left[\int_{t-\tau_1}^{t}\int_{\theta}^{t}(v(s-\tau_2)-v^*)^2 ds d\theta + \tau_1\int_{t-\tau_2}^{t}(v(s)-v^*)^2 ds\right] + \Psi_9(\varepsilon)\int_{t-\tau_1}^{t}\int_{\theta}^{t}(v(s)-v^*)^2 ds d\theta\\
&+ \Psi_{10}(\varepsilon)\int_{t-\tau_2}^{t}\int_{\theta}^{t}(x(s)-x^*)^2 ds d\theta + \Psi_{11}(\varepsilon)\int_{t-\tau_2}^{t}\int_{\theta}^{t}(y(s)-y^*)^2 ds d\theta\\
&+ \Psi_{12}(\varepsilon)\int_{t-\tau_2}^{t}\int_{\theta}^{t}(v(s)-v^*)^2 ds d\theta.
\end{aligned}
$$

Computing the derivative of U_3, we have, for $t > T(\varepsilon) + 2(\tau_1 + \tau_2)$,

$$
\begin{aligned}
\frac{dU_3}{dt} =& \Psi_3(\varepsilon)\left[-\int_{t-\tau_1}^{t}(x(s-\tau_1)-x^*)^2 ds + \tau_1(x(t)-x^*)^2\right]\\
&+ \Psi_4(\varepsilon)\left[-\int_{t-\tau_1}^{t}(x(s-\tau_2)-x^*)^2 ds + \tau_1(x(t)-x^*)^2\right] + \Psi_5(\varepsilon)\left[-\int_{t-\tau_1}^{t}(x(s)-x^*)^2 ds + \tau_1(x(t)-x^*)^2\right]\\
&+ \Psi_6(\varepsilon)\left[-\int_{t-\tau_1}^{t}(y(s-\tau_1)-y^*)^2 ds + \tau_1(y(t)-y^*)^2\right] + \Psi_7(\varepsilon)\left[-\int_{t-\tau_1}^{t}(y(s)-y^*)^2 ds + \tau_1(y(t)-y^*)^2\right]\\
&+ \Psi_8(\varepsilon)\left[-\int_{t-\tau_1}^{t}(v(s-\tau_2)-v^*)^2 ds + \tau_1(v(t)-v^*)^2\right] + \Psi_9(\varepsilon)\left[-\int_{t-\tau_1}^{t}(v(s)-v^*)^2 ds + \tau_1(v(t)-v^*)^2\right]\\
&+ \Psi_{10}(\varepsilon)\left[-\int_{t-\tau_2}^{t}(x(s)-x^*)^2 ds + \tau_2(x(t)-x^*)^2\right] + \Psi_{11}(\varepsilon)\left[-\int_{t-\tau_2}^{t}(y(s)-y^*)^2 ds + \tau_2(y(t)-y^*)^2\right]\\
&+ \Psi_{12}(\varepsilon)\left[-\int_{t-\tau_2}^{t}(v(s)-v^*)^2 ds + \tau_2(v(t)-v^*)^2\right].
\end{aligned}
$$

Hence, we have, for $t > T(\varepsilon) + 2(\tau_1 + \tau_2)$,

$$
\begin{aligned}
\frac{dU_3}{dt} + \Gamma(t) \leq &[(\Psi_1(\varepsilon) + \Psi_3(\varepsilon) + \Psi_4(\varepsilon) + \Psi_5(\varepsilon))\tau_1 + (\Psi_2(\varepsilon) + \Psi_{10}(\varepsilon))\tau_2](x(t)-x^*)^2\\
&+ [(\Psi_1(\varepsilon)e^{m_1\tau_1} + \Psi_6(\varepsilon) + \Psi_7(\varepsilon))\tau_1 + (\Psi_2(\varepsilon)e^{m_1\tau_1} + \Psi_{11}(\varepsilon))\tau_2](y(t)-y^*)^2\\
&+ [(\Psi_8(\varepsilon) + \Psi_9(\varepsilon))\tau_1 + \Psi_{12}(\varepsilon)\tau_2](v(t)-v^*)^2.
\end{aligned}
$$

Further, we have, for $t > T(\varepsilon) + 2(\tau_1 + \tau_2)$,

$$
\begin{aligned}
\frac{dU_2}{dt} + \frac{dU_3}{dt} \leq & -\left\{d - r + \frac{r}{K}(v_1(\varepsilon) + x^*) + \left(c + \frac{r\alpha}{K}\right)y^*\right.\\
&\left. - [(\Psi_1(\varepsilon) + \Psi_3(\varepsilon) + \Psi_4(\varepsilon) + \Psi_5(\varepsilon))\tau_1 + (\Psi_2(\varepsilon) + \Psi_{10}(\varepsilon))\tau_2]\right\}(x(t)-x^*)^2\\
&- \left\{e^{m_1\tau_1}\left[e^{m_1\tau_1}p + \left(c + \frac{r\alpha}{K}\right)v_1(\varepsilon)\right]\right.\\
&\left. - [(\Psi_1(\varepsilon)e^{m_1\tau_1} + \Psi_6(\varepsilon) + \Psi_7(\varepsilon))\tau_1 + (\Psi_2(\varepsilon)e^{m_1\tau_1} + \Psi_{11}(\varepsilon))\tau_2]\right\}(y(t)-y^*)^2\\
&+ \left\{e^{m_1\tau_1}\left[d - r + \frac{r}{K}(x_0(\varepsilon) + x^*) + \left(c + \frac{r\alpha}{K}\right)y^* + p\right] + \left(c + \frac{r\alpha}{K}\right)x_0(\varepsilon)\right\}|(x(t)-x^*)(y(t)-y^*)|\\
&+ [(\Psi_8(\varepsilon) + \Psi_9(\varepsilon))\tau_1 + \Psi_{12}(\varepsilon)\tau_2](v(t)-v^*)^2.
\end{aligned}
\tag{16}
$$

Define

$$U_4 = \frac{1}{2}(v(t) - v^*)^2,$$

then we have, for $t > T(\varepsilon) + 2(\tau_1 + \tau_2)$,

$$\frac{dU_4}{dt} = (v(t) - v^*)[k(y(t) - y^*) + u(v^* - v(t))] = k(y(t) - y^*)(v(t) - v^*) - u(v(t) - v^*)^2. \tag{17}$$

Finally, we define

$$U = U_1 + \theta_1(U_2 + U_3) + \theta_2 U_4.$$

From (11), (16) and (17), we have, for $t > T(\varepsilon) + 2(\tau_1 + \tau_2)$,

$$\begin{aligned}
\frac{dU}{dt} =& \frac{dU_1}{dt} + \theta_1\left(\frac{dU_2}{dt} + \frac{dU_3}{dt}\right) + \theta_2\frac{dU_4}{dt} \\
\leq& -A_{11}(\varepsilon)(x(t) - x^*)^2 - A_{22}(\varepsilon)(y(t) - y^*)^2 - A_{33}(\varepsilon)(v(t) - v^*)^2 \\
& + 2A_{12}(\varepsilon)|x(t) - x^*||y(t) - y^*| + 2A_{23}(\varepsilon)|y(t) - y^*||v(t) - v^*| \\
=& -(|x(t) - x^*|, |y(t) - y^*|, |v(t) - v^*|)J(\varepsilon)(|x(t) - x^*|, |y(t) - y^*|, |v(t) - v^*|)^T.
\end{aligned} \tag{18}$$

Since $J(\varepsilon)$ is positive definite, then using the classic Barbalat's lemma [42], we have

$$\lim_{t\to+\infty}|x(t) - x^*| = \lim_{t\to+\infty}|y(t) - y^*| = \lim_{t\to+\infty}|v(t) - v^*| = 0.$$

Thus, the chronic infection equilibrium E^* is globally attractive. $\square$

4. Conclusions

In this paper, we mainly study the uniform persistence and global attractivity of chronic infection equilibrium E^* of model (4). For the uniform persistence of model (4), Theorem 1 and (5) give explicit expressions of the ultimate upper and lower bounds of any positive solution of model (4). In fact, the global attractivity of the chronic infection equilibrium E^* of model (4) is still a question worthy of further discussion. Using standard analytical methods, it is not difficult to find that when delay τ_1 or τ_2 changes, Hopf bifurcations can appear near the chronic infection equilibrium E^*. Under the condition $m_1\tau_1 = m_2\tau_2$, Theorem 2 gives a class of sufficient conditions to determine the global attractivity of the chronic infection equilibrium E^*. Of course, it is also easy to see that the verification of these sufficient conditions are complex and highly conservative, relying strongly on the construction of the Lyapunov functional $U = U_1 + \theta_1(U_2 + U_3) + \theta_2 U_4$.

To illustrate the feasibility of the application of Theorem 2, we specifically choose the following parameter values: $s = 1$, $d = 1$, $r = 0.01$, $K = 1000$, $c = 0.01$, $p = 2$, $k = 1$, $u = 1$, $\alpha = 2$, $\beta_1 = 1$, $\beta_2 = 1.06$, $\tau_1 = 0.015$, $\tau_2 = 0.01$, $m_1 = 0.02$, $m_2 = 0.03$. Then, we have $R_0 \approx 1.040081 > 1$, $m_1\tau_1 = m_2\tau_2$, and model (4) has a unique chronic infection equilibrium $E^* \approx (0.971165, 0.0191695, 0.0191695)$. We further choose $\theta_1 = 1$ and $\theta_2 = 0.5$, then we can obtain $A_{11}(0) \approx 1.660181 > 0$, $A_{12}(0) \approx 1.505625$, $A_{22}(0) \approx 1.637131 > 0$, $A_{23}(0) = 0.25$, and $A_{33}(0) \approx 0.3623425 > 0$. It is easy to verify that $J(0)$ is positive definite and $d - r + \frac{rx^*}{K} \approx 0.99001 > 0$. Therefore, the conditions of Theorem 2 are satisfied, and chronic infection equilibrium E^* is globally attractive.

Author Contributions: Writing—original draft preparation, M.L., K.G. and W.M.; methodology, M.L., K.G. and W.M.; writing—review and editing, K.G., W.M. All authors have read and agreed to the published version of the manuscript.

Funding: This paper is supported by Beijing Natural Science Foundation (No. 1202019) and National Natural Science Foundation of China (No. 11971055).

Institutional Review Board Statement: Not applicable.

Informed Consent Statement: Not applicable.

Data Availability Statement: Not applicable.

Acknowledgments: We are grateful to the editor and anonymous reviewers for their valuable comments.

Conflicts of Interest: The authors declare no conflict of interest.

References

1. Nowak, M.A.; May, R.M. *Virus Dynamics: Mathematics Principles of Immunology and Virology*; Oxford University Press: London, UK, 2000.
2. Nowak, M.A.; Bangham, C.R.M. Population dynamics of immune responses to persistent viruses. *Science* **1996**, *272*, 74–79. [CrossRef] [PubMed]
3. Herz, A.V.; Bonhoeffer, S.; Anderson, R.M.; May, R.M.; Nowak, M.A. Viral dynamics in vivo: Limitations on estimates of intracellular delay and virus decay. *Proc. Natl. Acad. Sci. USA* **1996**, *93*, 7247–7251. [CrossRef] [PubMed]
4. Perelson, A.S.; Neumann, A.U.; Markowitz, M.; Leonard, J.M.; Ho, D.D. HIV-1 dynamics in vivo: virion clearance rate, infected cell life-span, and viral generation time. *Science* **1996**, *271*, 1582–1586. [CrossRef] [PubMed]
5. Mittler, J.E.; Sulzer, B.; Neumann, A.U.; Perelson, A.S. Influence of delayed viral production on viral dynamics in HIV-1 infected patients. *Math. Biosci.* **1998**, *152*, 143–163. [CrossRef]
6. Perelson, A.S.; Kirschner, D.E.; De Boer, R. Dynamics of HIV infection of CD4+ T cells. *Math. Biosci.* **1993**, *114*, 81–125. [CrossRef]
7. De Boer, R.J.; Perelson, A.S. Target cell limited and immune control models of HIV infection: A comparison. *J. Theor. Biol.* **1998**, *190*, 201–214. [CrossRef] [PubMed]
8. Perelson, A.S.; Nelson, P.W. Mathematical analysis of HIV-1 dynamics in vivo. *SIAM Rev.* **1999**, *41*, 3–44. [CrossRef]
9. Wang, L.; Li, M.Y. Mathematical analysis of the global dynamics of a model for HIV infection of CD4+ T cells. *Math. Biosci.* **2006**, *200*, 44–57. [CrossRef]
10. Li, D.; Ma, W. Asymptotic properties of a HIV-1 infection model with time delay. *J. Math. Anal. Appl.* **2007**, *335*, 683–691. [CrossRef]
11. Song, X.; Neumann, A.U. Global stability and periodic solution of the viral dynamics. *J. Math. Anal. Appl.* **2007**, *329*, 281–297. [CrossRef]
12. Huang, G.; Takeuchi, Y.; Ma, W. Lyapunov functionals for delay differential equations model of viral infections. *SIAM J. Appl. Math.* **2010**, *70*, 2693–2708. [CrossRef]
13. Huang, G.; Ma, W.; Takeuchi, Y. Global analysis for delay virus dynamics model with Beddington-DeAngelis functional response. *Appl. Math. Lett.* **2011**, *24*, 1199–1203. [CrossRef]
14. Xu, R. Global stability of an HIV-1 infection model with saturation infection and intracellular delay. *J. Math. Anal. Appl.* **2011**, *375*, 75–81. [CrossRef]
15. Li, J.; Wang, K.; Yang, Y. Dynamical behaviors of an HBV infection model with logistic hepatocyte growth. *Math. Comput. Model.* **2011**, *54*, 704–711. [CrossRef]
16. Pawelek, K.A.; Liu, S.; Pahlevani, F.; Rong, L. A model of HIV-1 infection with two time delays: Mathematical analysis and comparison with patient. data. *Math. Biosci.* **2012**, *235*, 98–109. [CrossRef] [PubMed]
17. Zhou, X.; Zhang, L.; Zheng, T.; Li, H.; Teng, Z. Global stability for a delayed HIV reactivation model with latent infection and Beddington-DeAngelis incidence. *Appl. Math. Lett.* **2021**, *117*, 107047. [CrossRef]
18. Yang, Y.; Zou, L.; Ruan, S. Global dynamics of a delayed within-host viral infection model with both virus-to-cell and cell-to-cell transmissions. *Math. Biosci.* **2015**, *270*, 183–191. [CrossRef]
19. Chen, S.S.; Cheng, C.Y.; Takeuchi, Y. Stability analysis in delayed within-host viral dynamics with both viral and cellular infections. *J. Math. Anal. Appl.* **2016**, *442*, 642–672. [CrossRef]
20. Alshorman, A.; Wang, X.; Meyer, M.J.; Rong, L. Analysis of HIV models with two time delays. *J. Biol. Dyn.* **2017**, *11*, 40–64. [CrossRef]
21. She, Z.; Jiang, X. Threshold dynamics of a general delayed within-host viral infection model with humoral immunity and two modes of virus transmission. *Discrete Cont. Dyn. Syst. B* **2021**, *26*, 3835–3861. [CrossRef]
22. Ji, Y.; Ma, W.; Song, K. Modeling inhibitory effect on the growth of uninfected T cells caused by infected T cells: Stability and Hopf bifurcation. *Comput. Math. Method Med.* **2018**, *2018*, 3176893. [CrossRef] [PubMed]
23. Zhang, T.; Wang, J.; Song, Y.; Jiang, Z. Dynamical analysis of a delayed HIV virus dynamic model with cell-to-cell transmission and apoptosis of bystander cells. *Complexity* **2020**, *2020*, 2313102. [CrossRef]
24. Huebner, W.; Mcnemey, G.P.; Chen, P.; Dale, B.M.; Gordon, R.E.; Chuang, F.Y.S.; Li, X.; Asmuth, D.M.; Huser, T.; Chen, B.K. Quantitative 3D video microscopy of HIV transfer across T cell virological synapses. *Science* **2009**, *323*, 1743–1747. [CrossRef] [PubMed]
25. Lai, X.; Zou, X. Modeling HIV-1 virus dynamics with both virus-to-cell infection and cell-to-cell transmission. *SIAM J. Appl. Math.* **2014**, *74*, 898–917. [CrossRef]
26. Culshaw, R.V.; Ruan, S.; Webb, G. A mathematical model of cell-to-cell spread of HIV-1 that includes a time delay. *J. Math. Biol.* **2003**, *46*, 425–444. [CrossRef] [PubMed]
27. Sattentau, Q. Avoiding the void: Cell-to-cell spread of human viruses. *Nat. Rev. Microbiol.* **2008**, *6*, 815–826. [CrossRef]
28. Lai, X.; Zou, X. Modeling cell-to-cell spread of HIV-1 with logistic target cell growth. *J. Math. Anal. Appl.* **2015**, *426*, 563–584. [CrossRef]

29. Li, F.; Ma, W. Dynamics analysis of an HTLV-1 infection model with mitotic division of actively infected cells and delayed CTL immune response. *Math. Meth. Appl. Sci.* **2018**, *41*, 3000–3017. [CrossRef]
30. Shu, H.; Chen, Y.; Wang, L. Impacts of the cell-free and cell-to-cell infection modes on viral dynamics. *J. Dyn. Differ. Equ.* **2018**, *30*, 1817–1836. [CrossRef]
31. Pan, X.; Chen, Y.; Shu, H. Rich dynamics in a delayed HTLV-1 infection model: stability switch, multiple stable cycles, and torus. *J. Math. Anal. Appl.* **2019**, *479*, 2214–2235. [CrossRef]
32. Finkel, T.H.; Tudor-Williams, G.; Banda, N.K.; Cotton, M.F.; Curiel, T.; Monks, C.; Baba, T.W.; Ruprecht, R.M.; Kupfer, A. Apoptosis occurs predominantly in bystander cells and not in productively infected cells of HIV-and SIV-infected lymph nodes. *Nat. Med.* **1995**, *1*, 129–134. [CrossRef]
33. Selliah, N.; Finkel, T.H. Biochemical mechanisms of HIV induced T cell apoptosis. *Cell Death Differ.* **2001**, *8*, 127–136. [CrossRef] [PubMed]
34. Cheng, W.; Ma, W.; Guo, S. A class of virus dynamic model with inhibitory effect on the growth of uninfected T cells caused by infected T cells and its stability analysis. *Commun. Pure Appl. Anal* **2016**, *15*, 795–806.
35. Guo, S.; Ma, W. Global behavior of delay differential equations model of HIV infection with apoptosis. *Discrete Cont. Dyn. Syst. B* **2016**, *21*, 103–119. [CrossRef]
36. Li, S.; Wen, L. *Functional Differential Equations*; Hunan Science and Technology Press: Changsha, China, 1987.
37. Hale, J.K.; Verduyn Lunel, S.M. *Introduction to Functional Differential Equations*; Springer: New York, NY, USA, 1993.
38. Kuang, Y. *Delay Differential Equations with Applications in Population Dynamics*; Academic Press: Boston, MA, USA, 1993.
39. Zheng, Z. *Theorey of Functional Differential Equations*; Anhui Education Press: Hefei, China, 1994.
40. Diekmann, O.; Heesterbeek, J.A.; Metz, J.A. On the definition and the computation of the basic reproduction ratio R_0 in models for infectious diseases in heterogeneous populations. *J. Math. Biol.* **1990**, *28*, 365–382. [CrossRef] [PubMed]
41. van den Dreessche, P.; Watmough, J. Reproduction numbers and sub-threshold endemic equilibria for compartmental models of disease transmission. *Math. Biosci.* **2002**, *180*, 29–48. [CrossRef]
42. Barbălat, I. Systèmes d'équations différentielles d'oscillations non lineairés. *Rev. Roumaine Math. Pures Appl.* **1959**, *4*, 267–270.
43. Guo, K.; Ma, W.; Qiang, R. Global dynamics analysis of a time-delayed dynamic model of Kawasaki disease pathogenesis. *Discrete Cont. Dyn. Syst. B* **2022**, *27*, 2367–2400. [CrossRef]
44. Guo, K.; Ma, W. Global dynamics of an SI epidemic model with nonlinear incidence rate, feedback controls and time delays. *Math. Biosci. Eng.* **2021**, *18*, 643–672. [CrossRef]

 *mathematics*

Article

Adaptive Evolutionary Computation for Nonlinear Hammerstein Control Autoregressive Systems with Key Term Separation Principle

Faisal Altaf [1], Ching-Lung Chang [2], Naveed Ishtiaq Chaudhary [3,*], Muhammad Asif Zahoor Raja [3], Khalid Mehmood Cheema [4], Chi-Min Shu [5] and Ahmad H. Milyani [6]

[1] Graduate School of Engineering Science and Technology, National Yunlin University of Science and Technology, 123 University Road, Section 3, Douliou, Yunlin 64002, Taiwan; d10910221@yuntech.edu.tw
[2] Department of Computer Science and Information Engineering, National Yunlin University of Science and Technology, 123 University Road, Section 3, Douliou, Yunlin 64002, Taiwan; chang@yuntech.edu.tw
[3] Future Technology Research Center, National Yunlin University of Science and Technology, 123 University Road, Section 3, Douliou, Yunlin 64002, Taiwan; rajamaz@yuntech.edu.tw
[4] Department of Electrical Engineering, Khwaja Fareed University of Engineering & Information Technology, Rahim Yar Khan 64200, Pakistan; km.cheema@kfueit.edu.pk
[5] Department of Safety, Health, and Environmental Engineering, National Yunlin University of Science and Technology, 123 University Road, Section 3, Douliou, Yunlin 64002, Taiwan; shucm@yuntech.edu.tw
[6] Department of Electrical and Computer Engineering, King Abdulaziz University, Jeddah 21589, Saudi Arabia; ahmilyani@kau.edu.sa
*   Correspondence: chaudni@yuntech.edu.tw

Citation: Altaf, F.; Chang, C.-L.; Chaudhary, N.I.; Raja, M.A.Z.; Cheema, K.M.; Shu, C.-M.; Milyani, A.H. Adaptive Evolutionary Computation for Nonlinear Hammerstein Control Autoregressive Systems with Key Term Separation Principle. *Mathematics* 2022, 10, 1001. https://doi.org/10.3390/math10061001

Academic Editor: Simeon Reich

Received: 17 February 2022
Accepted: 18 March 2022
Published: 21 March 2022

Publisher's Note: MDPI stays neutral with regard to jurisdictional claims in published maps and institutional affiliations.

Abstract: The knacks of evolutionary and swarm computing paradigms have been exploited to solve complex engineering and applied science problems, including parameter estimation for nonlinear systems. The population-based computational heuristics applied for parameter identification of nonlinear systems estimate the redundant parameters due to an overparameterization problem. The aim of this study was to exploit the key term separation (KTS) principle-based identification model with adaptive evolutionary computing to overcome the overparameterization issue. The parameter estimation of Hammerstein control autoregressive (HC-AR) systems was conducted through integration of the KTS idea with the global optimization efficacy of genetic algorithms (GAs). The proposed approach effectively estimated the actual parameters of the HC-AR system for noiseless as well as noisy scenarios. The simulation results verified the accuracy, convergence, and robustness of the proposed scheme. While consistent accuracy and reliability of the designed approach was validated through statistical assessments on multiple independent trials.

Keywords: Hammerstein nonlinear systems; parameter estimation; bioinspired computing; genetic algorithms

MSC: 93C10; 93B30

1. Introduction

Parameter estimation is an essential and fundamental step for solving various engineering and applied science problems [1–3]. Parameter estimation and control of nonlinear systems is a challenging task and has been explored in various studies [4–7]. Nonlinear systems/processes can be modeled through block structure representation, i.e., Hammerstein, Wiener, and Hammerstein–Wiener models [8–10]. The Hammerstein model representation given in Figure 1 consists of two blocks where the first block normally represents the static nonlinearity, while the second block is a linear dynamical subsystem [11]. The Hammerstein structure has been used to model different nonlinear processes. For instance, joint stiffness dynamics [12], heating process [13], cascade water tanks [14], geochemical problems [15], pneumatic muscle actuator [16], financial analysis [17], electric load forecasting [18], and muscle dynamics [19,20].

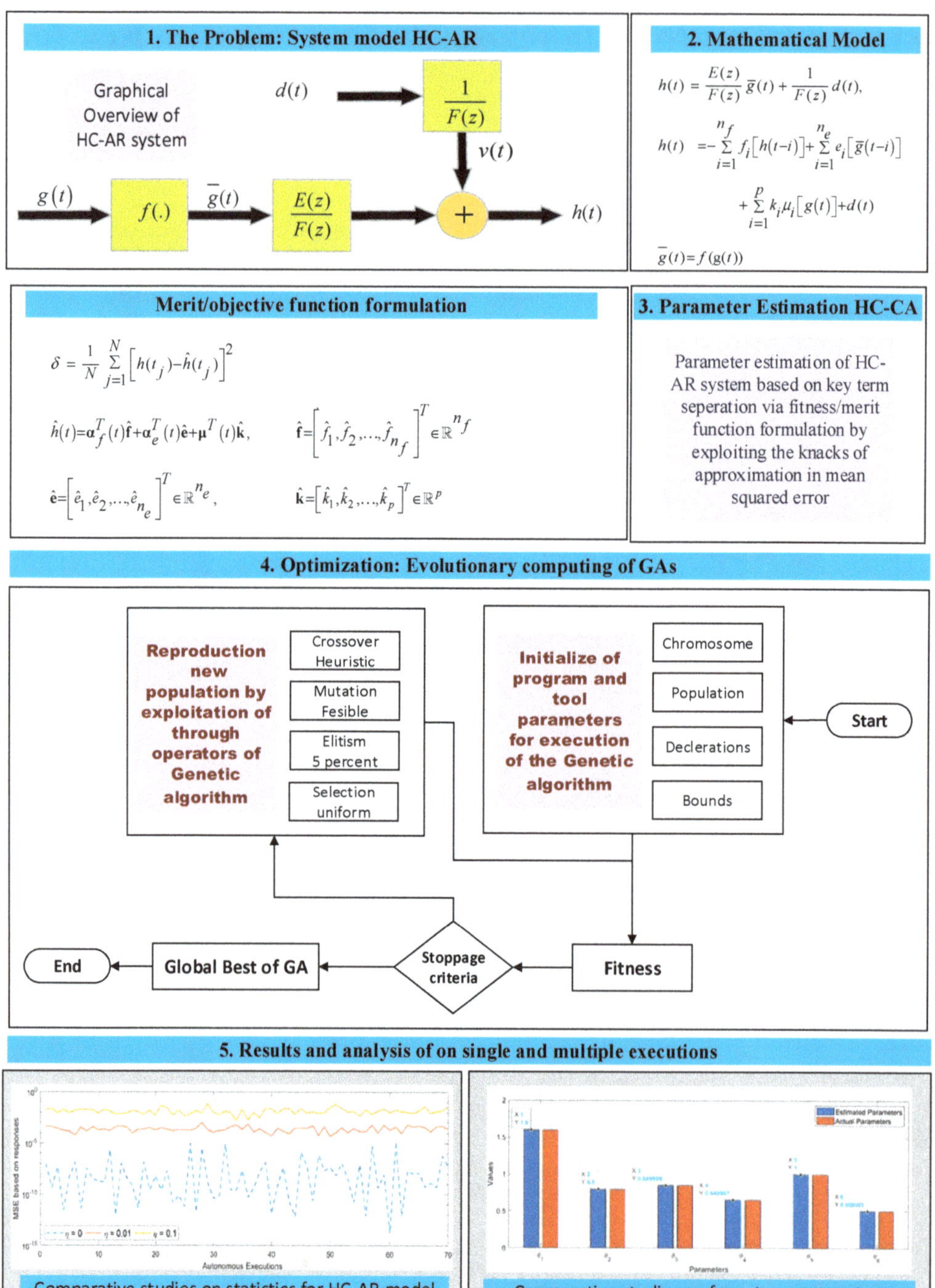

Figure 1. Workflow of methodology for HC-AR systems with evolutionary heuristics of GAs.

The research community proposed various algorithms/methods for parameter estimation for the Hammerstein model owing to its significance in modeling different nonlinear systems: for example, gradient/least squares iterative methods [21–25], fractional gradient based adaptive strategies [26–29], Newton iterative scheme [30], Kalman filtering [31], reframed model [32], filtering technique [33], separable block approach [34], Levenberg–Marquardt optimization [35], orthogonal matching pursuit technique [36], and the maximum likelihood scheme [37]. The biological/nature-inspired computations through evolutionary/swarm optimization were also explored for Hammerstein system identification. For instance, Mehmood et al. exploited the strength of genetic algorithms (GA), differential evolution, pattern search, simulated annealing, and backtracking search optimization heuristics for Hammerstein structure identification [38–40]. Tariq et al. exploited the maximum likelihood-based adaptive DE for nonlinear system identification [41]. Raja et al. presented a detailed study of applying GAs to the Hammerstein control autoregressive (HC-AR) structure [42]. In [42], the identification of the HC-AR system through GAs was done through an overparameterization approach by making the system linear in parameters which causes the estimation of redundant parameters rather than identifying only the actual parameters of the HC-AR system.

In order to avoid the redundant parameters involved in the overparameterization identification approach used in genetic algorithms, we integrated the key term separation (KTS) principle with the evolutionary computing paradigm of a GA that allowed us to estimate only the actual parameters of the HC-AR system. The KTS principle identifies and separates the key term in the HC-AR identification model [43] and then exploits the global search competency of GAs to estimate only the actual parameters of the system. The performance of the proposed KTS-based scheme was assessed in terms of accuracy, convergence, robustness, consistency, and reliability for varying parameters of the proposed scheme. The main contributions of the proposed study are as follows:

- A global search identification scheme through the integration of key term separation, KTS principle identification model with the evolutionary computing algorithm of GA is presented for parameter estimation of Hammerstein nonlinear systems.
- The proposed scheme avoids identifying redundant parameters and effectively estimates only the actual parameters of Hammerstein control autoregressive (HC-AR) systems through minimizing the mean square error-based criterion function.
- The accuracy, robustness, and convergence of the proposed approach is established through optimal values of estimation-error-based evaluation metrics.
- The stability and reliability of the designed approach is ascertained through statistical inferences obtained after executing multiple independent trials of the scheme.

The remaining article is organized as follows: Section 2 provides the proposed key term separation-based identification model for HC-AR systems. Section 3 presents the evolutionary computing approach of GAs for the KTS-based identification model of HC-AR systems. Section 4 gives the results of numerical experimentation with elaborative discussion. Section 5 concludes the findings of the study and lists future research directions.

2. Key Term Separation Identification Model

The block diagram of the HC-AR system is given in Figure 1 while mathematically represented as [43,44]

$$h(t) = \frac{E(z)}{F(z)}\overline{g}(t) + \frac{1}{F(z)}d(t) \tag{1}$$

where $h(t)$, $g(t)$, and $d(t)$ represent input, output, and disturbance signal, respectively, while $\overline{g}(t)$ is a nonlinear function of known basis and written as

$$\overline{g}(t) = k_1\mu_1[g(t)] + k_2\mu_2[g(t)] + \ldots + k_p\mu_p[g(t)] \tag{2}$$

$E(z)$ and $F(z)$ are defined as

$$E(z) = e_0 + e_1 z^{-1} + e_2 z^{-2} +, \ldots, + e_{n_e} z^{-n_e}, \tag{3}$$

$$F(z) = 1 + f_1 z^{-1} + f_2 z^{-2} +, \ldots, + f_{n_f} z^{-n_f} \tag{4}$$

Rearrange Equation (1) as

$$h(t) = (1 - F(z))h(t) + E(z)\overline{g}(t) + d(t) \tag{5}$$

while using Equations (2)–(4) in Equation (5) and assuming $e_0 = 1$. Apply the key term separation (KTS) principle by considering $\overline{g}(t)$ as a key term

$$
\begin{aligned}
h(t) \quad &= -\sum_{i=1}^{n_f} f_i [h(t-i)] + \sum_{i=0}^{n_e} e_i [\overline{g}(t-i)] + d(t) \\
&= -\sum_{i=1}^{n_f} f_i [h(t-i)] + e_0 [\overline{g}(t)] + \sum_{i=1}^{n_e} e_i [\overline{g}(t-i)] + d(t) \\
&= -\sum_{i=1}^{n_f} f_i [h(t-i)] + \sum_{i=1}^{n_e} e_i [\overline{g}(t-i)] + \sum_{i=1}^{p} k_i \mu_i [g(t)] + d(t)
\end{aligned}
\tag{6}
$$

Write Equation (6) in terms of information and parameter vectors as

$$h(t) = \boldsymbol{\alpha}_f^T(t)\mathbf{f} + \boldsymbol{\alpha}_e^T(t)\mathbf{e} + \boldsymbol{\mu}^T(t)\mathbf{k} + d(t) \tag{7}$$

where the information vectors are defined as

$$\boldsymbol{\alpha}_f(t) = \left[-h(t-1), -h(t-2), \ldots, -h\left(t-n_f\right)\right]^T \in \mathbb{R}^{n_f}, \tag{8}$$

$$\boldsymbol{\alpha}_e(t) = [\overline{g}(t-1), \overline{g}(t-2), \ldots, \overline{g}(t-n_e)]^T \in \mathbb{R}^{n_e}, \tag{9}$$

$$\boldsymbol{\mu}(t) = [\mu_1[g(t)], \mu_2[g(t)], \ldots, \mu_p[g(t)]]^T \in \mathbb{R}^{p}, \tag{10}$$

and the corresponding parameter vectors are

$$\mathbf{f} = \left[f_1, f_2, \ldots, f_{n_f}\right]^T \in \mathbb{R}^{n_f}, \tag{11}$$

$$\mathbf{e} = [e_1, e_2, \ldots, e_{n_e}]^T \in \mathbb{R}^{n_e}, \tag{12}$$

$$\mathbf{k} = [k_1, k_2, \ldots, k_p]^T \in \mathbb{R}^{p}. \tag{13}$$

Equations (7)–(13) represent the KTS identification model for HC-AR systems that avoids the estimation of redundant parameters due to the overparameterization approach.

3. Proposed Methodology for KTS System Model

The proposed methodology for parameter estimation of the KTS-based identification model of HC-AR systems was developed in two phases. First, the objective/fitness function was formulated for the KTS model of the HC-AR system presented in Section 2. Second, the HC-AR system was identified through estimating the actual parameters of the HC-AR system using optimization knacks of the evolutionary computing paradigm of a GA. The overall flow diagram of the proposed study in terms of fundamental compartments is provided in Figure 1.

3.1. Fitness Function Formulation

The iterative and recursive identification approaches for parameter estimation of nonlinear systems develop the identification model by expressing the system output as a product of information and parameter vectors [23]. However, the population-based stochastic computing techniques have no such requirement. The fitness function for a GA based on an evolutionary computing paradigm is formulated by exploiting the strength of approximation theory in mean square error sense as

$$\delta = \frac{1}{N}\sum_{j=1}^{N}\left[h(t_j) - \hat{h}(t_j)\right]^2, \tag{14}$$

where N represents the number of samples involved in the parameter identification of HC-AR systems. The desired response h is calculated using Equation (7) while the estimated response is given by the following:

$$\hat{h}(t) = \boldsymbol{\alpha}_f^T(t)\hat{\mathbf{f}} + \boldsymbol{\alpha}_e^T(t)\hat{\mathbf{e}} + \boldsymbol{\mu}^T(t)\hat{\mathbf{k}}. \tag{15}$$

The estimated parameter is written as

$$\hat{\boldsymbol{\theta}} = [\hat{\mathbf{f}}, \hat{\mathbf{e}}, \hat{\mathbf{k}}], \tag{16}$$

where

$$\hat{\mathbf{f}} = \left[\hat{f}_1, \hat{f}_2, \ldots, \hat{f}_{n_f}\right]^T \in \mathbb{R}^{n_f}, \tag{17}$$

$$\hat{\mathbf{e}} = \left[\hat{e}_1, \hat{e}_2, \ldots, \hat{e}_{n_e}\right]^T \in \mathbb{R}^{n_e}, \tag{18}$$

$$\hat{\mathbf{k}} = \left[\hat{k}_1, \hat{k}_2, \ldots, \hat{k}_p\right]^T \in \mathbb{R}^{p}. \tag{19}$$

Now the objective was to estimate the parameters of the HC-AR system through minimizing the fitness of Equation (14) using a GA-based evolutionary computing approach such that the desired response given by Equation (7) approached the estimate calculated from Equation (15).

3.2. Optimization Procedure: Evolutionary Computing Paradigm

The legacy of global optimization knacks of genetic algorithms (GAs) belongs to a class of evolutionary computational paradigm that is narrated here which is used for learning the parameters of the HC-AR system as portrayed in the fitness function in Equation (14).

The GAs were introduced in a pioneer work conducted by Holland to mimic an optimization task [45]. Normally, the adaptive performance of GAs to find the appropriate candidate solution in a large search dimension is controlled by a reproduction mechanism consisting of the feasible selection of individuals in the nest population, viable crossover operation for the offspring generation, and the diversity maintenance procedure of mutation. GAs were implemented since their introduction in a variety of research domains such as the viable optimization of closed-loop supply chain design [46], optimization of the weights of neural networks representing the nonlinear singular prediction differential system [47], optimization of electroless NiB coating model [48], optimization of the solar selective absorber design [49], and the crack sensitivity control system for nickel-based laser coating [50]. We were motivated/inspired from these significant applications of GA-based evolutionary computing and used GAs for parameter identification of the HC-AR system.

The process flow structure, in terms of the fundamental steps the Gas used for the optimization of the HC-AR system is shown in Figure 2, i.e., representation of the population, fitness-based ranking, selection of the matting pair, crossover procedure, and mutation. A generic process workflow in the form of a block structure is portrayed in Figure 2 for the GAs that were used for the optimization mechanism of the HC-AR system. The simulation and experimentation of GAs was conducted with the help of the invoking routines/program/tools of optimization available in the MATLAB toolbox for optimization while Windows 10 was used as an operating system. The necessary details of GAs with their implementation procedure is given in pseudocode as provided in Algorithm 1.

Algorithm 1: Pseudocode of evolutionary computing with GAs for HC-AR system identification.

Start: Evolutionary computing of genetic algorithms (GAs)
Inputs: Chromosomes or individual representation as follows:

$$\theta = [\theta_f, \theta_e, \theta_k] = [(f_1, f_2, \ldots, f_{n_f})\ (e_1, e_2, \ldots, e_{n_e})\ (k_1, k_2, \ldots, k_{n_k})]$$

Population for an ensemble of chromosomes or individuals is given as

$$P = \begin{bmatrix} \theta_1 \\ \theta_2 \\ \vdots \\ \theta_l \end{bmatrix} = \begin{bmatrix} (f_{1,1}, f_{2,1}, \ldots, f_{n_f,1}) & (e_{1,1}, e_{2,1}, \ldots, e_{n_e,1}) & (k_{1,1}, k_{2,1}, \ldots, k_{n_k,1}) \\ (f_{1,2}, f_{2,2}, \ldots, f_{n_f,2}) & (e_{1,2}, e_{2,2}, \ldots, e_{n_e,2}) & (k_{1,2}, k_{2,2}, \ldots, k_{n_k,2}) \\ \vdots & \vdots & \vdots \\ (f_{1,l}, f_{2,l}, \ldots, f_{n_f,l}) & (e_{1,l}, e_{2,l}, \ldots, e_{n_e,l}) & (k_{1,l}, k_{2,l}, \ldots, k_{n_k,l}) \end{bmatrix},$$

for l members in θ in P
Output: Global Best θ in P
Begin GAs
    //Initialize
    Arbitrarily formulate θ with bounded pseudo real numbers.
    A group of l number of θ represents initial P.
    //Termination/Stoppage Criteria
    Set stoppage of execution of GAs for the following conditions:
        Desire fitness attained i.e., $\delta \rightarrow 10^{-16}$,
        Fitness function-Tolerance attained i.e., TolFun $\rightarrow 10^{-20}$,
        Constrained-Tolerance attained, i.e., TolCon $\rightarrow 10^{-20}$,
        Set total number of generations = 600,
        Other default of GA routine in optimization toolbox
    //Main loop of GA
    While {until termination conditions attained} do %
        //Fitness calculation step
        Evaluate δ using Expression (14) and repeat the procedure for each θ in P
        //Check for termination requirements
        If any of termination level attained then go out of the while loop
        else continues
        //Ranking of individual step
        Rank each θ on the basis of quality of fitness θ achieved.
        //Reproduction step through GA operators
        Appropriate/suitable invoking for
        selection (Stochastic uniform via routine '@selectionstochunif'),
        crossover (heuristics via routine '@crossoverheuristic'),
        mutations (adaptive feasible via routine '@mutationadaptfeasible')
        Elitism operations up to 5%, i.e., elitism count set as 26 best ranking
        individuals in the population P
        Modify/update P and go to fitness calculation step
    End
    //Storage step of GAs outcomes
    Store the global best θ with credentials of fitness attained, time spent,
    generations executed and fitness function counts of the algorithm.
End GAs
Statistical Analysis:
Dataset generation for the statistical observation by repetition of GAs for a sufficiently large number of multiple execution to identify the parameters of the HC-AR and analysis of these datasets was performed for exhaustive assessments.

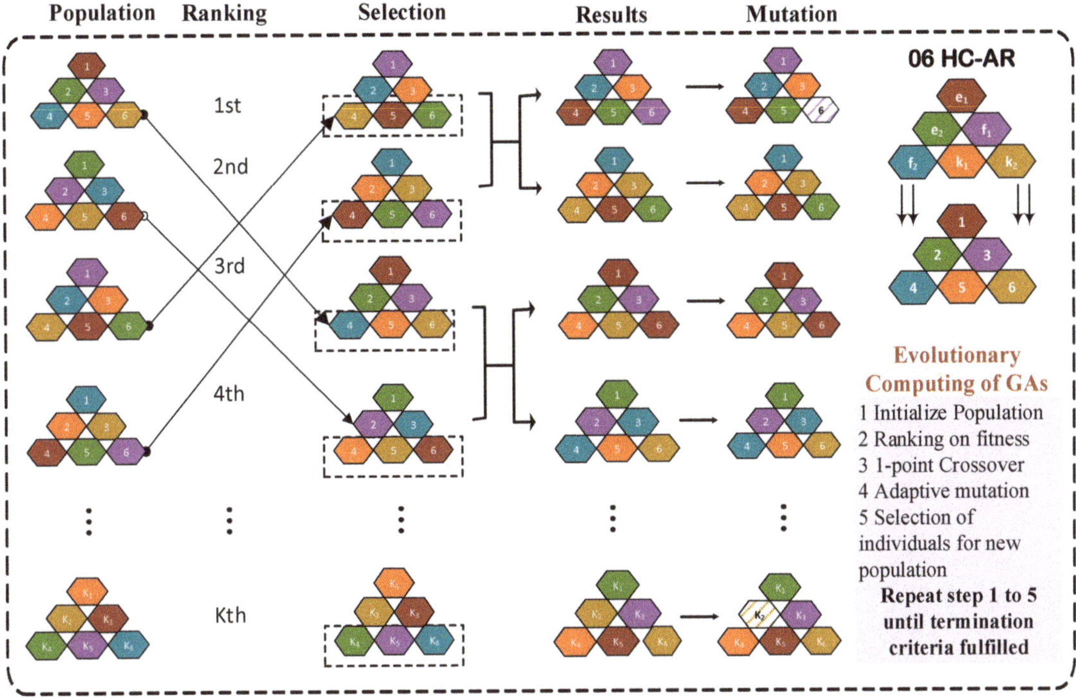

Figure 2. Overview of reproduction operators of GAs representing HC-AR systems.

3.3. Evaluation Metrics

In order to assess the performance of the evolutionary computing paradigm for parameter estimation of nonlinear systems through the KTS-based identification model of HC-AR systems, we defined three evaluation metrics. The formulated assessment criterions are mean square error based on the difference between the responses, i.e., $(\text{MSE})_h$; as given in Equation (14), mean square error based on the difference between the desired and the estimated parameters, i.e., $(\text{MSE})_\theta$; and the normalized parameter deviation, i.e., NPD.

$$(\text{MSE})_\theta = \text{mean}\left(\theta - \hat{\theta}\right)^2, \tag{20}$$

$$\text{NPD} = \frac{\|\theta - \hat{\theta}\|}{\|\theta\|} \tag{21}$$

where $\|\cdot\|$ denote the 2-norm of a vector.

4. Results of Numerical Experimentation with Discussion

The results of the numerical experimentation for parameter estimation for two HC-AR systems are presented in this section. In problem 1, a standard HC-AR system was considered, while in problem 2, a practical application of an HC-AR system representing the dynamics of stimulated muscle model was considered.

4.1. Problem 1

In Problem 1, the HC-AR system was considered with the following parameters, as taken from recent relevant studies to demonstrate the effectiveness of the proposed schemes:

$$h(t) = \frac{E(z)}{F(z)}\overline{g}(t) + \frac{1}{F(z)}d(t),$$

$$F(z) = 1 + 1.6_1 z^{-1} + 0.8z^{-2},$$

$$E(z) = 0.85z^{-1} + 0.65z^{-2},$$

$$\overline{g}(t) = k_1\mu_1[g(t)] + k_2\mu_2[g(t)] = 1.0g(t) + 0.5g^2(t)$$

The actual parameters of the HC-AR system were

$$\begin{aligned}
\theta = [\mathbf{f}, \mathbf{e}, \mathbf{k}]^T &= [f_1,\ f_2,\ e_1,\ e_2\ k_1,\ k_2]^T \\
&= [\theta_1,\ \theta_2,\ \theta_3,\ \theta_4,\ \theta_5,\ \theta_6]^T \\
&= [1.6,\ 0.8,\ 0.85,\ 0.65,\ 1,\ 0.5]^T
\end{aligned} \tag{22}$$

Simulations were performed in MATLAB 2020b running on an Asuspro Laptop core i7 with 16GB RAM. The input g was randomly generated with characteristics of zero-mean and unit variance. The disturbance signal was generated with characteristics of Gaussian distribution having zero-mean and constant variance. The robustness of the proposed scheme was assessed for three disturbance levels, i.e., 0, 0.01, and 0.1. The parameter settings of the GA used in the simulations are given in Algorithm 1. The performance of the proposed scheme was deeply investigated through the results of executing a single random run, the statistics through multiple autonomous trials, and evaluating the results for the three different evaluation metrics described in Section 3.3.

The results of the proposed scheme generated for a single random run based on evaluation criteria from Equation (14) in terms of learning curve, best individual scores (best, worst, and mean), and average distance between individuals are provided in Figures 3–5 for 0, 0.01, and 0.1 noise levels, respectively. The results indicated that the proposed identification scheme accurately estimated the parameters of the HC-AR system by optimizing the cost function through minimizing the error between the desired and the estimated responses.

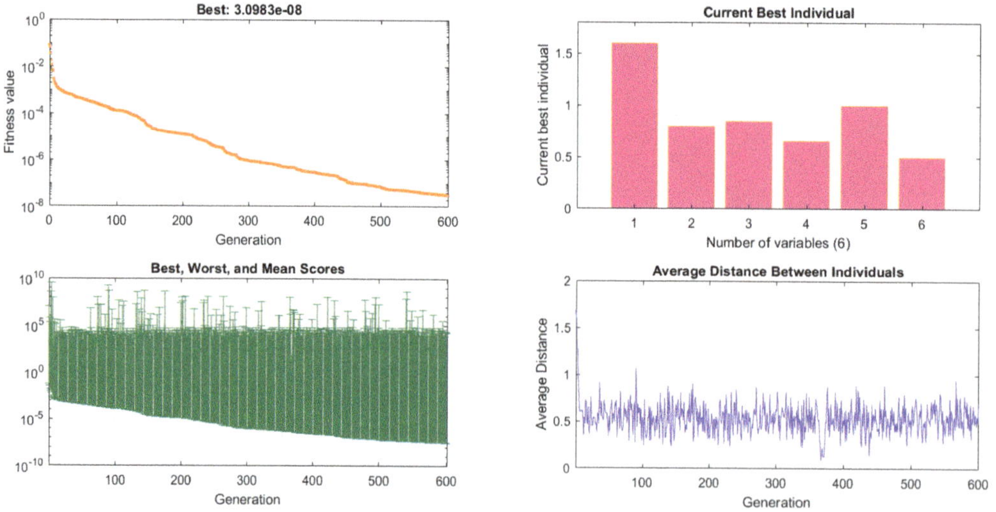

Figure 3. Results of Problem 1 in terms of learning curve, best individual scores, and average distance for no noise scenario.

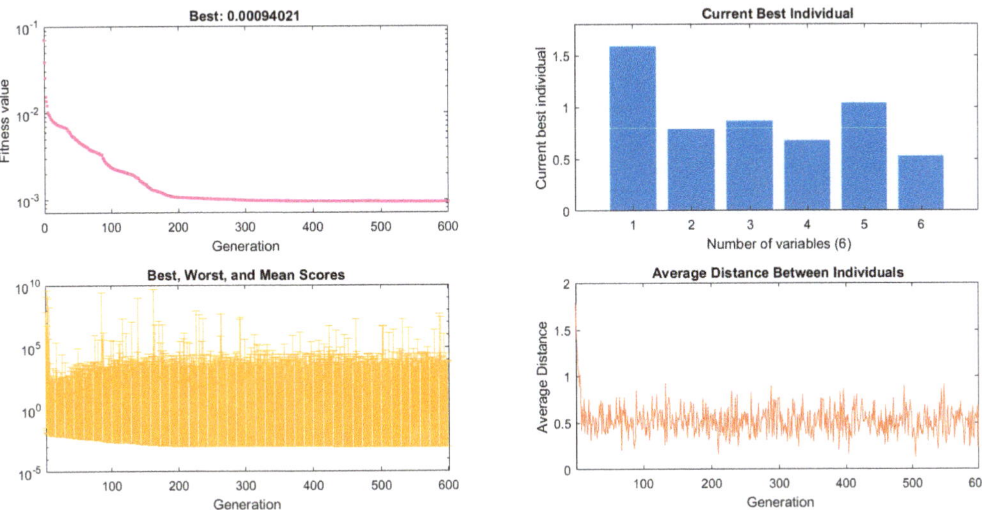

Figure 4. Results of Problem 1 in terms of learning curve, best individual scores, and average distance for 0.01 noise level.

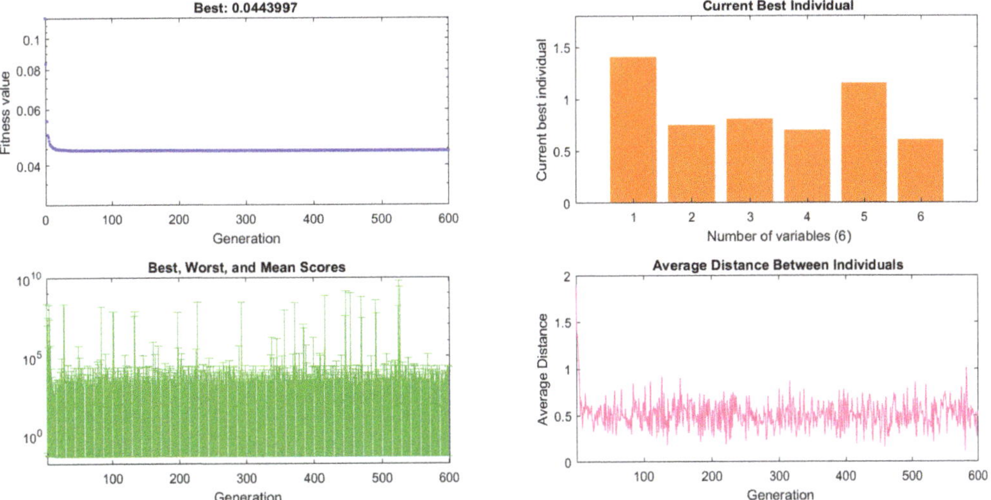

Figure 5. Results of Problem 1 in terms of learning curve, best individual scores, and average distance for 0.1 noise level.

The one good run of the evolutionary approach does not guarantee consistently accurate performance. The identification of the HC-AR system through the proposed scheme was also investigated for multiple autonomous executions, and the results are given in Figures 6 and 7 for standard and ascending order, respectively, in the case of all three evaluation metrics. The results verified the consistently accurate performance of the proposed methodology for 70 autonomous trials in the case of all three evaluation metrics given in Equations (14), (20) and (21).

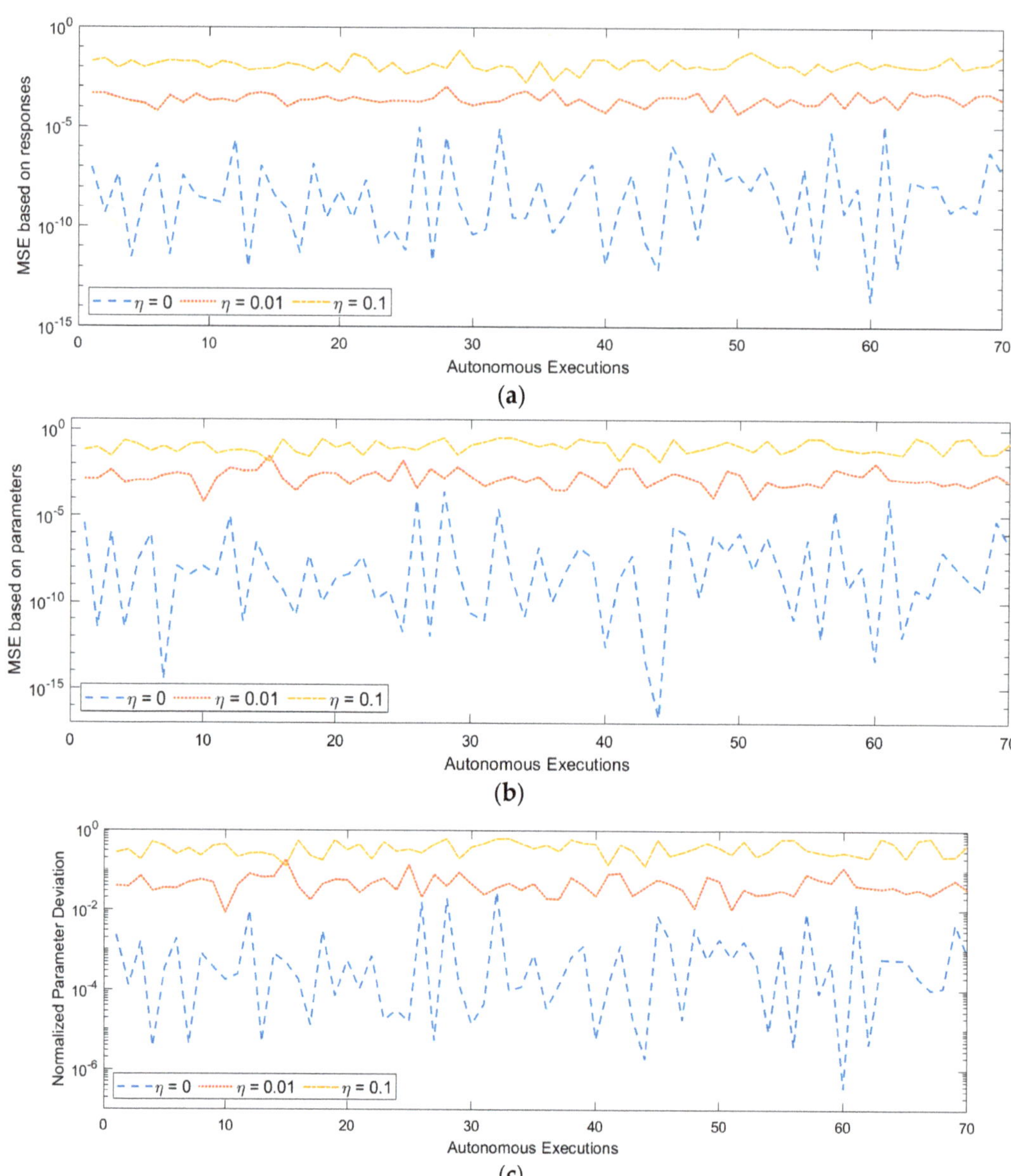

Figure 6. Results of autonomous executions through different evaluation metrics for Problem 1. (**a**) MSE through estimated response (**b**) MSE through estimated parameters (**c**) Normalized parameter deviation.

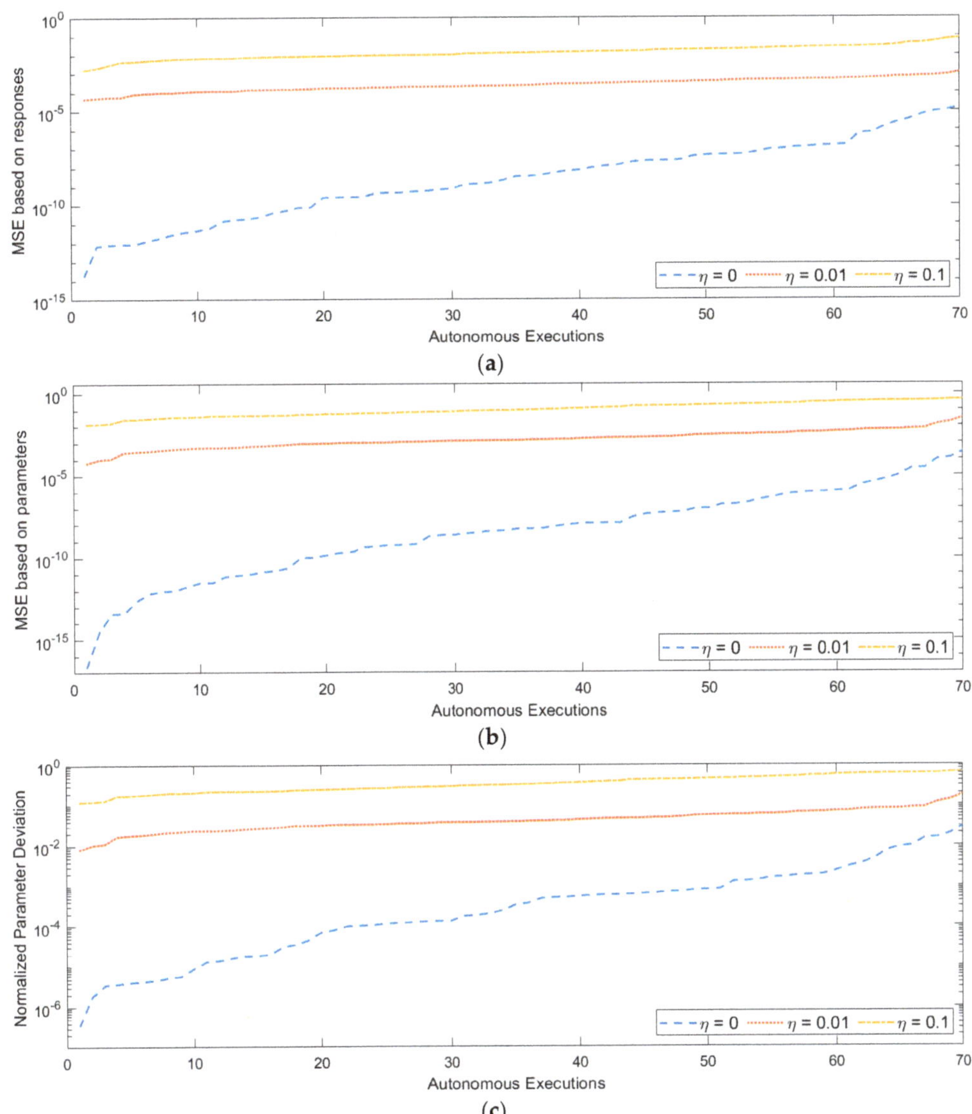

Figure 7. Result of autonomous executions in ascending order through different evaluation metrics for Problem 1. (**a**) MSE (ascending order) through estimated response (**b**) MSE (ascending order) through estimated parameters (**c**) Normalized (ascending order) parameter deviation.

The stability of the design approach was assessed through statistical measurements of the best, mean, and standard deviation. The results of the statistical indices are presented in Table 1 for all considered disturbances and evaluation metrics. The mean values for evaluation criteria (14) were 7.4405×10^{-7}, 3.0590×10^{-4}, and 1.7138×10^{-2} for disturbance level 0, 0.001, and 0.1, respectively, while the respective mean values in the case of evaluation measures (20) and (21) were 8.0612×10^{-6}, 2.8981×10^{-3}, 1.4764×10^{-1} and 2.0806×10^{-3}, 4.7970×10^{-2}, 3.6962×10^{-1}, respectively. For a better interpretation, the statistical results are also given in Figure 13. It was witnessed that the proposed scheme consistently provided the accurate results for all considered disturbance levels in the HC-AR system (22). However, the precision level decreased with an increase in disturbance level. The statistical results endorsed the stability, consistently accurate performance, robustness, and reliability of the proposed scheme.

Table 1. Results of statistical indices for different evaluation metrics in Problem 1.

Noise	Statistical Indices	MSE Responses	MSE Parameters	NPD
0	Minimum	1.8583×10^{-14}	2.1427×10^{-17}	3.5541×10^{-7}
	Mean	7.4405×10^{-17}	8.0612×10^{-6}	2.0806×10^{-3}
	Standard Deviation	2.6148×10^{-6}	3.6142×10^{-5}	5.0126×10^{-3}
0.01	Minimum	4.9615×10^{-5}	6.2525×10^{-5}	8.1884×10^{-3}
	Mean	3.0590×10^{-4}	2.8981×10^{-3}	4.7970×10^{-2}
	Standard Deviation	2.0307×10^{-4}	4.3865×10^{-3}	2.8608×10^{-2}
0.1	Minimum	1.7040×10^{-3}	1.4920×10^{-2}	1.2649×10^{-1}
	Mean	1.7138×10^{-2}	1.4764×10^{-1}	3.6962×10^{-1}
	Standard Deviation	1.3315×10^{-2}	1.1119×10^{-1}	1.4840×10^{-1}

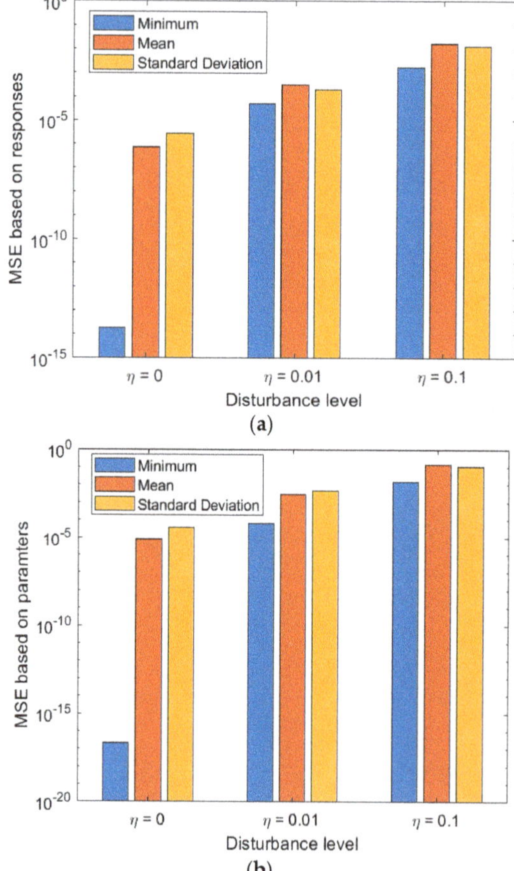

Figure 8. *Cont.*

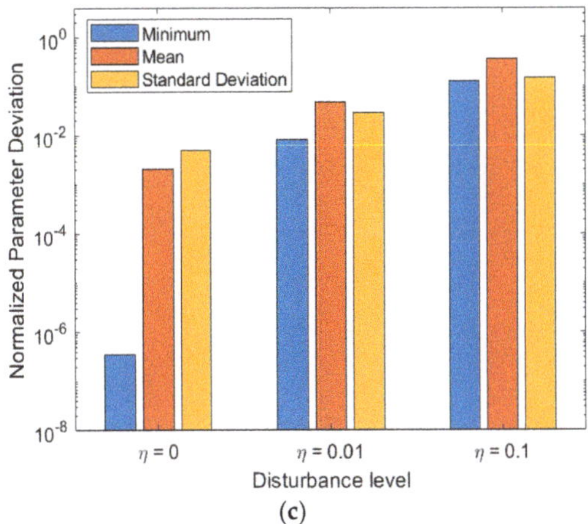

Figure 8. Graphic interpretation of statistics for different evaluation metrics in the case of Problem 1. (**a**) MSE through estimated responses (**b**) MSE through estimated parameters (**c**) Normalized parameter deviation.

The comparison of the actual parameters of the HC-AR system (22) with the estimated parameters through the proposed scheme was conducted, and the results are presented in Figure 9 and Table 2 along with the actual system parameters. The results validated the accurate and convergent performance of the proposed scheme in estimating the parameters of the HC-AR system (22) for different evaluation measurements based on mean square error of the responses (14), mean square error of the parameters (20), and normalized parameter deviation (21).

Table 2. Comparison of the estimated parameter values with the actual parameters of Problem 1.

Metric	Noise	θ_1	θ_2	θ_3	θ_4	θ_5	θ_6	Metric Value
MSE	0	1.6000	0.8000	0.8500	0.6500	1.0000	0.5000	2.14×10^{-17}
	0.01	1.5934	0.7972	0.8534	0.6614	1.0101	0.5089	6.25×10^{-5}
	0.1	1.7468	0.9828	0.9678	0.7525	1.0791	0.5626	1.49×10^{-2}
NWD	0	1.6000	0.8000	0.8500	0.6500	1.0000	0.5000	3.55×10^{-7}
	0.01	1.5934	0.7972	0.8534	0.6614	1.0101	0.5089	8.19×10^{-3}
	0.1	1.7468	0.9828	0.9678	0.7525	1.0791	0.5626	1.26×10^{-1}
DW		1.6000	0.8000	0.8500	0.6500	1.0000	0.5000	0

While comparing the proposed scheme with the conventional evolutionary approaches [42], the KTS-based GA was more efficient than the conventional GA presented in [42] for the HC-AR identification in the sense that it avoided the estimation of redundant parameters and estimated only the actual parameters of the HC-AR system, thus, making it computationally more efficient than the conventional GA.

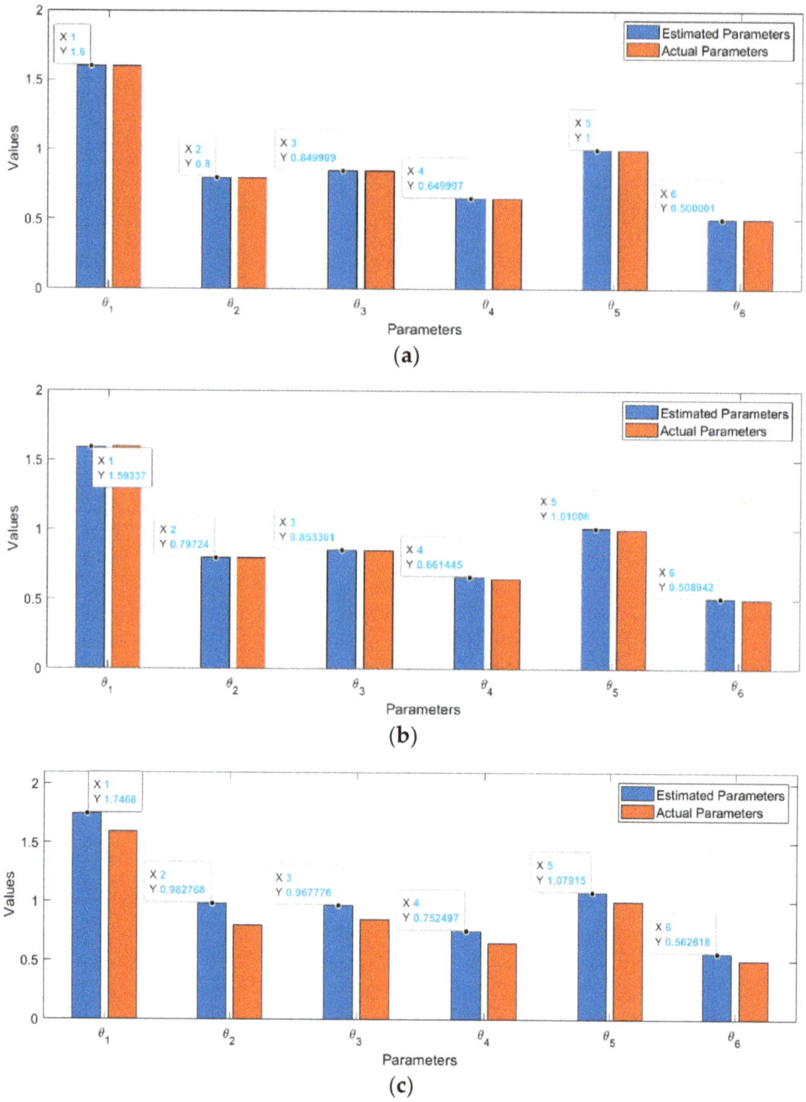

Figure 9. Results of estimated parameters in comparison with actual HC-AR parameters considered in Problem 1. (**a**) MSE through estimated responses (**b**) MSE through estimated parameters (**c**) Normalized parameter deviation.

4.2. Problem 2

In Problem 2, a practical application of an HC-AR system representing the muscle dynamics required to restore the functional use of paralyzed muscles through automatically controlled stimulations was considered by taking the actual parameters from the real time experimentations performed in the rehabilitation center of the Southampton University [51].

$$h(t) = \frac{E(z)}{F(z)}\overline{g}(t) + \frac{1}{F(z)}d(t),$$

$$F(z) = 1 - z^{-1} + 0.8z^{-2},$$

$$E(z) = 2.8z^{-1} - 4.8z^{-2},$$

$$\overline{g}(t) = k_1\mu_1[g(t)] + k_2\mu_2[g(t)] = 1.68g(t) - 2.88g^2(t) + 3.42g^3(t)$$

The actual parameters of the HC-AR system representing the dynamics of the stimulated muscle model are

$$\begin{aligned} \theta = [\mathbf{f}, \mathbf{e}, \mathbf{k}]^T &= [f_1,\ f_2,\ e_1,\ e_2\ k_1,\ k_2,\ k_3]^T \\ &= [\theta_1,\ \theta_2,\ \theta_3,\ \theta_4,\ \theta_5,\ \theta_6,\ \theta_7]^T \\ &= [-1.0,\ 0.8,\ 2.8,\ -4.8,\ 1.68,\ -2.88,\ 3.42]^T \end{aligned} \tag{23}$$

In this problem, the same input and disturbance signal were considered as taken from Problem 1. The robustness of the proposed scheme in Problem 2 was assessed for three disturbance levels, i.e., 0, 0.001, and 0.01.

The results of the proposed scheme for Problem 2 of the HC-AR system (23) generated from a single random run based on the evaluation criteria off Equation (14) in terms of learning curve, best individual scores (best, worst, and mean), and average distance between individuals are provided in Figures 10–12 for 0, 0.001, and 0.01 noise levels, respectively. The results indicated that the proposed identification scheme accurately estimated the parameters of the HC-AR system (23) by optimizing the cost function through minimizing the error between the desired and the estimated responses.

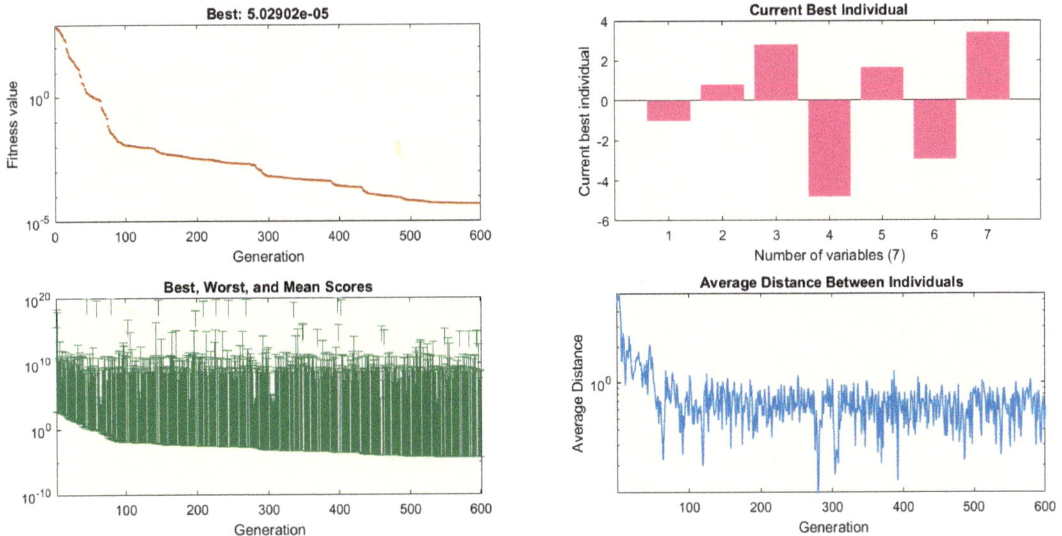

Figure 10. Results of Problem 2 in terms of learning curve, best individual scores, and average distance for 0 noise level.

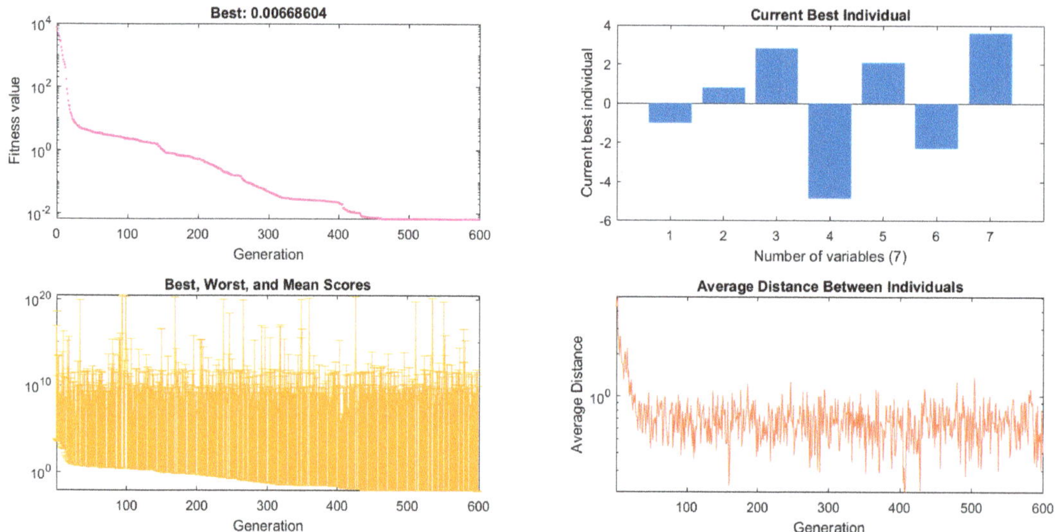

Figure 11. Results of Problem 2 in terms of learning curve, best individual scores, and average distance for 0.001 noise level.

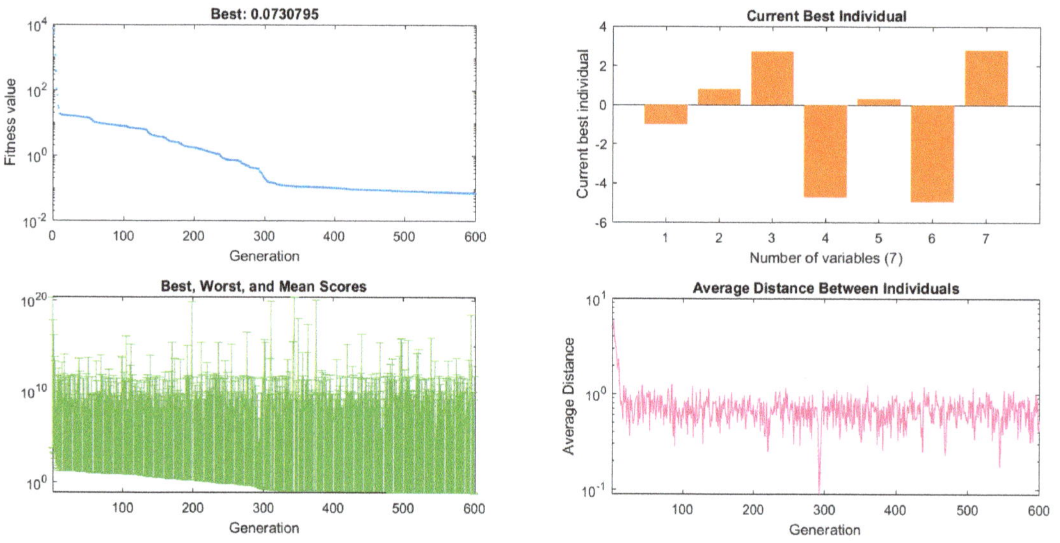

Figure 12. Results of Problem 2 in terms of learning curve, best individual scores, and average distance for 0.01 noise level.

The comparison of the actual parameters of the HC-AR system (23) with the estimated parameters through the proposed scheme was conducted, and the results based on the best run are presented in Figure 13 and Table 3 along with the actual system parameters. The results validated the accurate and convergent performance of the proposed scheme in estimating the parameters of the muscle model represented through the HC-AR system (23) for different evaluation measures based on the mean square error of the responses (14), the mean square error of the parameters (20), and the normalized parameter deviation (21). This case study presented a KTS-based GA approach for parameter estimation of an HC-AR system representing the parameters of muscle dynamics, while the details for the real rehabilitation procedure can be seen in [51].

Table 3. Comparison of the estimated parameter values with the actual parameters of Problem 2.

Metric	Noise	θ_1	θ_2	θ_3	θ_4	θ_5	θ_6	θ_7	Value
MSE	0	-1.0001	0.8000	2.7942	-4.7928	1.6636	-2.9094	3.4155	4.88×10^{-5}
	0.001	-0.9999	0.8001	2.7836	-4.7786	1.7292	-2.8884	3.4251	4.64×10^{-4}
	0.01	-1.0001	0.8000	2.7912	-4.7903	1.6224	-2.9880	3.3795	2.40×10^{-3}
NWD	0	-1.0001	0.8000	2.7942	-4.7928	1.6636	-2.9094	3.4155	4.73×10^{-3}
	0.001	-0.9999	0.8001	2.7836	-4.7786	1.7292	-2.8884	3.4251	7.66×10^{-3}
	0.01	-1.0001	0.8000	2.7912	-4.7903	1.6224	-2.9880	3.3795	1.74×10^{-2}
DW		-1.0000	0.8000	2.8000	-4.8000	1.6800	-2.8800	3.4200	0

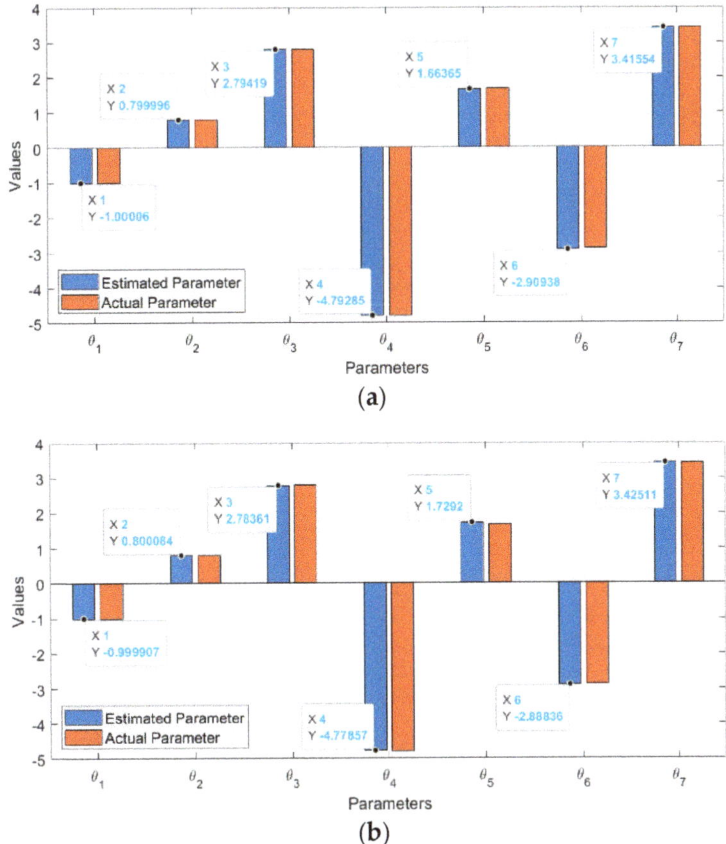

Figure 13. *Cont.*

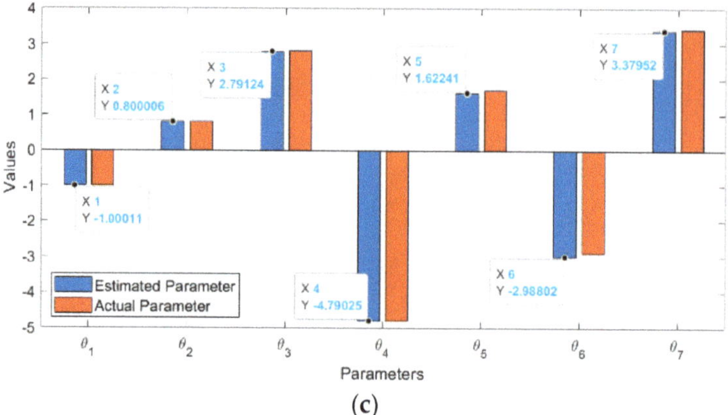

Figure 13. Results of estimated parameters in comparison with actual HC-AR parameters considered in Problem 2. (**a**) MSE through estimated responses (**b**) MSE through estimated parameters (**c**) Normalized parameter deviation.

5. Conclusions

The conclusions drawn from the study are

- The integration of an evolutionary computing paradigm of genetic algorithms, GA, with a key term separation-based identification model was presented for parameter estimation of Hammerstein control autoregressive (HC-AR) systems.
- The proposed identification scheme effectively estimated only the actual parameters of the HC-AR system without estimating the redundant parameters due to an overparameterization approach.
- The accurate and convergent behavior of the proposed strategy was ascertained through achieving an optimal value of different evaluation metrics based on response error and parameter estimation error.
- The results of the Monte Carlo simulations and statistical indices established the consistent accuracy of the proposed scheme.
- The accurate estimation of HC-AR parameters representing the dynamics of a muscle model for the rehabilitation of paralyzed muscles further endorsed the efficacy of the design approach.

The proposed KTS-based evolutionary optimization scheme seems to be an attractive alternative to be exploited for solving complex nonlinear problems [52–56].

Author Contributions: Conceptualization, N.I.C. and M.A.Z.R.; methodology, N.I.C. and M.A.Z.R.; software, F.A.; validation, M.A.Z.R. and N.I.C.; resources, C.-L.C. and C.-M.S.; writing—original draft preparation, F.A.; writing—review and editing, N.I.C. and M.A.Z.R.; project administration, C.-L.C., K.M.C., C.-M.S. and A.H.M.; funding acquisition, K.M.C. and A.H.M. All authors have read and agreed to the published version of the manuscript.

Funding: This research received no external funding.

Institutional Review Board Statement: Not applicable.

Informed Consent Statement: Not applicable.

Data Availability Statement: Not applicable.

Conflicts of Interest: The authors declare no conflict of interest.

References

1. Bock, H.G.; Carraro, T.; Jäger, W.; Körkel, S.; Rannacher, R.; Schlöder, J.P. (Eds.) *Model Based Parameter Estimation: Theory and Applications*; Springer Science & Business Media: Berlin/Heidelberg, Germany, 2013; Volume 4.
2. Chaudhary, N.I.; Latif, R.; Raja, M.A.Z.; Machado, J.T. An innovative fractional order LMS algorithm for power signal parameter estimation. *Appl. Math. Model.* **2020**, *83*, 703–718. [CrossRef]
3. Ćalasan, M.; Micev, M.; Ali, Z.M.; Zobaa, A.F.; Abdel Aleem, S.H. Parameter Estimation of Induction Machine Single-Cage and Double-Cage Models Using a Hybrid Simulated Annealing—Evaporation Rate Water Cycle Algorithm. *Mathematics* **2020**, *8*, 1024. [CrossRef]
4. Billings, S.A.; Leontaritis, I.J. Parameter estimation techniques for nonlinear systems. *IFAC Proc. Vol.* **1982**, *15*, 505–510. [CrossRef]
5. Shao, K.; Zheng, J.; Wang, H.; Wang, X.; Liang, B. Leakage-type adaptive state and disturbance observers for uncertain nonlinear systems. *Nonlinear Dyn.* **2021**, *105*, 2299–2311. [CrossRef]
6. Fei, J.; Feng, Z. Fractional-order finite-time super-twisting sliding mode control of micro gyroscope based on double-loop fuzzy neural network. *IEEE Trans. Syst. Man Cybern. Syst.* **2020**, *51*, 7692–7706. [CrossRef]
7. Fei, J.; Wang, Z.; Liang, X.; Feng, Z.; Xue, Y. Fractional sliding mode control for micro gyroscope based on multilayer recurrent fuzzy neural network. *IEEE Trans. Fuzzy Syst.* **2021**. [CrossRef]
8. Morales, J.Y.R.; Mendoza, J.A.B.; Torres, G.O.; Vázquez, F.D.J.S.; Rojas, A.C.; Vidal, A.F.P. Fault-Tolerant Control implemented to Hammerstein–Wiener model: Application to Bio-ethanol dehydration. *Fuel* **2022**, *308*, 121836. [CrossRef]
9. Janczak, A. *Identification of Nonlinear Systems Using Neural Networks and Polynomial Models: A Block-Oriented Approach*; Springer Science & Business Media: Berlin/Heidelberg, Germany, 2004; Volume 310.
10. Billings, S.A. *Nonlinear System Identification: NARMAX Methods in the Time, Frequency, and Spatio-Temporal Domains*; John Wiley & Sons: Hoboken, NJ, USA, 2013.
11. Dong, S.; Liu, T.; Wang, Q.G. Identification of Hammerstein systems with time delay under load disturbance. *IET Control Theory Appl.* **2018**, *12*, 942–952. [CrossRef]
12. Tehrani, E.S.; Golkar, M.A.; Guarin, D.L.; Jalaleddini, K.; Kearney, R.E. Methods for the identification of time-varying hammerstein systems with applications to the study of dynamic joint stiffness. *IFAC-PapersOnLine* **2015**, *48*, 1023–1028. [CrossRef]
13. Hammar, K.; Djamah, T.; Bettayeb, M. Identification of fractional Hammerstein system with application to a heating process. *Nonlinear Dyn.* **2019**, *96*, 2613–2626. [CrossRef]
14. Janjanam, L.; Saha, S.K.; Kar, R.; Mandal, D. Improving the Modelling Efficiency of Hammerstein System using Kalman Filter and its Parameters Optimised using Social Mimic Algorithm: Application to Heating and Cascade Water Tanks. *J. Frankl. Inst.* **2022**, *359*, 1239–1273. [CrossRef]
15. Piao, H.; Cheng, D.; Chen, C.; Wang, Y.; Wang, P.; Pan, X. A High Accuracy CO_2 Carbon Isotope Sensing System Using Subspace Identification of Hammerstein Model for Geochemical Application. *IEEE Trans. Instrum. Meas.* **2021**, *71*, 1–9. [CrossRef]
16. Ai, Q.; Peng, Y.; Zuo, J.; Meng, W.; Liu, Q. Hammerstein model for hysteresis characteristics of pneumatic muscle actuators. *Int. J. Intell. Robot. Appl.* **2019**, *3*, 33–44. [CrossRef]
17. Tissaoui, K. Forecasting implied volatility risk indexes: International evidence using Hammerstein-ARX approach. *Int. Rev. Financ. Anal.* **2019**, *64*, 232–249. [CrossRef]
18. Zhang, Z.; Hong, W.C. Electric load forecasting by complete ensemble empirical mode decomposition adaptive noise and support vector regression with quantum-based dragonfly algorithm. *Nonlinear Dyn.* **2019**, *98*, 1107–1136. [CrossRef]
19. Mehmood, A.; Zameer, A.; Chaudhary, N.I.; Raja, M.A.Z. Backtracking search heuristics for identification of electrical muscle stimulation models using Hammerstein structure. *Appl. Soft Comput.* **2019**, *84*, 105705. [CrossRef]
20. Mehmood, A.; Zameer, A.; Chaudhary, N.I.; Ling, S.H.; Raja, M.A.Z. Design of meta-heuristic computing paradigms for Hammerstein identification systems in electrically stimulated muscle models. *Neural Comput. Appl.* **2020**, *32*, 12469–12497. [CrossRef]
21. Mao, Y.; Ding, F.; Xu, L.; Hayat, T. Highly efficient parameter estimation algorithms for Hammerstein non-linear systems. *IET Control Theory Appl.* **2019**, *13*, 477–485. [CrossRef]
22. Ding, F.; Chen, H.; Xu, L.; Dai, J.; Li, Q.; Hayat, T. A hierarchical least squares identification algorithm for Hammerstein nonlinear systems using the key term separation. *J. Frankl. Inst.* **2018**, *355*, 3737–3752. [CrossRef]
23. Ding, F.; Ma, H.; Pan, J.; Yang, E. Hierarchical gradient-and least squares-based iterative algorithms for input nonlinear output-error systems using the key term separation. *J. Frankl. Inst.* **2021**, *358*, 5113–5135. [CrossRef]
24. Shen, B.; Ding, F.; Xu, L.; Hayat, T. Data filtering based multi-innovation gradient identification methods for feedback nonlinear systems. *Int. J. Control Autom. Syst.* **2018**, *16*, 2225–2234. [CrossRef]
25. Ding, J.; Cao, Z.; Chen, J.; Jiang, G. Weighted parameter estimation for Hammerstein nonlinear ARX systems. *Circuits Syst. Signal Process.* **2020**, *39*, 2178–2192. [CrossRef]
26. Chaudhary, N.I.; Raja, M.A.Z.; He, Y.; Khan, Z.A.; Machado, J.T. Design of multi innovation fractional LMS algorithm for parameter estimation of input nonlinear control autoregressive systems. *Appl. Math. Model.* **2021**, *93*, 412–425. [CrossRef]
27. Chaudhary, N.I.; Aslam, M.S.; Baleanu, D.; Raja, M.A.Z. Design of sign fractional optimization paradigms for parameter estimation of nonlinear Hammerstein systems. *Neural Comput. Appl.* **2020**, *32*, 8381–8399. [CrossRef]
28. Xu, C.; Mao, Y. Auxiliary Model-Based Multi-Innovation Fractional Stochastic Gradient Algorithm for Hammerstein Output-Error Systems. *Machines* **2021**, *9*, 247. [CrossRef]

29. Chaudhary, N.I.; Zubair, S.; Raja, M.A.Z.; Dedovic, N. Normalized fractional adaptive methods for nonlinear control autoregressive systems. *Appl. Math. Model.* **2019**, *66*, 457–471. [CrossRef]
30. Li, J. Parameter estimation for Hammerstein CARARMA systems based on the Newton iteration. *Appl. Math. Lett.* **2013**, *26*, 91–96. [CrossRef]
31. Ma, J.; Ding, F.; Xiong, W.; Yang, E. Combined state and parameter estimation for Hammerstein systems with time delay using the Kalman filtering. *Int. J. Adapt. Control Signal Process.* **2017**, *31*, 1139–1151. [CrossRef]
32. Wang, D. Hierarchical parameter estimation for a class of MIMO Hammerstein systems based on the reframed models. *Appl. Math. Lett.* **2016**, *57*, 13–19. [CrossRef]
33. Wang, D.; Zhang, Z.; Xue, B. Decoupled parameter estimation methods for Hammerstein systems by using filtering technique. *IEEE Access* **2018**, *6*, 66612–66620. [CrossRef]
34. Zhang, S.; Wang, D.; Liu, F. Separate block-based parameter estimation method for Hammerstein systems. *R. Soc. Open Sci.* **2018**, *5*, 172194. [CrossRef] [PubMed]
35. Li, J.; Zheng, W.X.; Gu, J.; Hua, L. Parameter estimation algorithms for Hammerstein output error systems using Levenberg–Marquardt optimization method with varying interval measurements. *J. Frankl. Inst.* **2017**, *354*, 316–331. [CrossRef]
36. Mao, Y.; Ding, F.; Liu, Y. Parameter estimation algorithms for Hammerstein time-delay systems based on the orthogonal matching pursuit scheme. *IET Signal Process.* **2017**, *11*, 265–274. [CrossRef]
37. Wang, D.Q.; Zhang, Z.; Yuan, J.Y. Maximum likelihood estimation method for dual-rate Hammerstein systems. *Int. J. Control Autom. Syst.* **2017**, *15*, 698–705. [CrossRef]
38. Mehmood, A.; Aslam, M.S.; Chaudhary, N.I.; Zameer, A.; Raja, M.A.Z. Parameter estimation for Hammerstein control autoregressive systems using differential evolution. *Signal Image Video Process.* **2018**, *12*, 1603–1610. [CrossRef]
39. Mehmood, A.; Chaudhary, N.I.; Zameer, A.; Raja, M.A.Z. Backtracking search optimization heuristics for nonlinear Hammerstein controlled auto regressive auto regressive systems. *ISA Trans.* **2019**, *91*, 99–113. [CrossRef] [PubMed]
40. Mehmood, A.; Chaudhary, N.I.; Zameer, A.; Raja, M.A.Z. Novel computing paradigms for parameter estimation in Hammerstein controlled auto regressive auto regressive moving average systems. *Appl. Soft Comput.* **2019**, *80*, 263–284. [CrossRef]
41. Tariq, H.B.; Chaudhary, N.I.; Khan, Z.A.; Raja MA, Z.; Cheema, K.M.; Milyani, A.H. Maximum-Likelihood-Based Adaptive and Intelligent Computing for Nonlinear System Identification. *Mathematics* **2021**, *9*, 3199. [CrossRef]
42. Raja, M.A.Z.; Shah, A.A.; Mehmood, A.; Chaudhary, N.I.; Aslam, M.S. Bio-inspired computational heuristics for parameter estimation of nonlinear Hammerstein controlled autoregressive system. *Neural Comput. Appl.* **2018**, *29*, 1455–1474. [CrossRef]
43. Chaudhary, N.I.; Raja, M.A.Z.; Khan, Z.A.; Cheema, K.M.; Milyani, A.H. Hierarchical Quasi-Fractional Gradient Descent Method for Parameter Estimation of Nonlinear ARX Systems Using Key Term Separation Principle. *Mathematics* **2021**, *9*, 3302. [CrossRef]
44. Giri, F.; Bai, E.W. *Block-Oriented Nonlinear System Identification*, 1st ed.; Springer: London, UK, 2010; Volume 1.
45. Holland, J.H. Genetic algorithms and the optimal allocation of trials. *SIAM J. Comput.* **1973**, *2*, 88–105. [CrossRef]
46. Yun, Y.; Chuluunsukh, A.; Gen, M. Sustainable closed-loop supply chain design problem: A hybrid genetic algorithm approach. *Mathematics* **2020**, *8*, 84. [CrossRef]
47. Sabir, Z.; Raja, M.A.Z.; Botmart, T.; Weera, W. A Neuro-Evolution Heuristic Using Active-Set Techniques to Solve a Novel Nonlinear Singular Prediction Differential Model. *Fractal Fract.* **2022**, *6*, 29. [CrossRef]
48. Vijayanand, M.; Varahamoorthi, R.; Kumaradhas, P.; Sivamani, S.; Kulkarni, M.V. Regression-BPNN modelling of surfactant concentration effects in electroless NiB coating and optimization using genetic algorithm. *Surf. Coat. Technol.* **2021**, *409*, 126878. [CrossRef]
49. Wang, Z.Y.; Hu, E.T.; Cai, Q.Y.; Wang, J.; Tu, H.T.; Yu, K.H.; Chen, L.Y.; Wei, W. Accurate design of solar selective absorber based on measured optical constants of nano-thin Cr Film. *Coatings* **2020**, *10*, 938. [CrossRef]
50. Yu, J.; Sun, W.; Huang, H.; Wang, W.; Wang, Y.; Hu, Y. Crack sensitivity control of nickel-based laser coating based on genetic algorithm and neural network. *Coatings* **2019**, *9*, 728. [CrossRef]
51. Le, F.; Markovsky, I.; Freeman, C.T.; Rogers, E. Recursive identification of Hammerstein systems with application to electrically stimulated muscle. *Control Eng. Pract.* **2012**, *20*, 386–396. [CrossRef]
52. Cao, W.; Zhu, Q. Stability of stochastic nonlinear delay systems with delayed impulses. *Appl. Math. Comput.* **2022**, *421*, 126950. [CrossRef]
53. Kong, F.; Zhu, Q.; Huang, T. New Fixed-Time Stability Criteria of Time-Varying Delayed Discontinuous Systems and Application to Discontinuous Neutral-Type Neural Networks. *IEEE Trans. Cybern.* **2021**, 1–10. [CrossRef]
54. Aadhithiyan, S.; Raja, R.; Zhu, Q.; Alzabut, J.; Niezabitowski, M.; Lim, C.P. A Robust Non-Fragile Control Lag Synchronization for Fractional Order Multi-Weighted Complex Dynamic Networks with Coupling Delays. *Neural Process. Lett.* **2022**, 1–22. [CrossRef]
55. Kong, F.; Zhu, Q.; Huang, T. Improved Fixed-Time Stability Lemma of Discontinuous System and Its Application. *IEEE Trans. Circuits Syst. I Regul. Pap.* **2021**, *69*, 835–842. [CrossRef]
56. Xiao, H.; Zhu, Q. Stability analysis of switched stochastic delay system with unstable subsystems. *Nonlinear Anal. Hybrid Syst.* **2021**, *42*, 101075. [CrossRef]

 *mathematics*

Article

Stability of Impulsive Stochastic Delay Systems with Markovian Switched Delay Effects

Wei Hu

School of Mathematics and Physics, Jiangsu University of Technology, Changzhou 213001, China; huw@jsut.edu.cn

Abstract: In this paper, we investigate the pth moment exponential stability of impulsive stochastic delay systems with Markovian switched delay effects. The model we consider here is rather different from the models in the existing literature. In particular, the delay is a Markov chain, which is quite different from the traditional deterministic delay. By using the Markov chain theory, stochastic analysis theory, Razumikhin technology and the Lyaponov method, we derive a criterion of pth moment exponential stability for the suggested system. Finally, an example is provided to illustrate the effectiveness of the obtained result.

Keywords: Markovian switched delay; impulsive stochastic delay system; moment exponential stability; Lyapunov approach; Razumikhin technique

MSC: 93D05; 93D23; 93E03; 93E15

Citation: Hu, W. Stability of Impulsive Stochastic Delay Systems with Markovian Switched Delay Effects. *Mathematics* **2022**, *10*, 1110. https://doi.org/10.3390/math10071110

Academic Editor: Dimplekumar N. Chalishajar

Received: 28 February 2022
Accepted: 23 March 2022
Published: 30 March 2022

Publisher's Note: MDPI stays neutral with regard to jurisdictional claims in published maps and institutional affiliations.

1. Introduction

Stochastic systems with Markovian switching is an important class of stochastic hybrid systems. In the real world, they are widely used in many fields such as automatic control, aircraft and air traffic control and electrics. So, the research on stochastic systems with Markovian switching has attracted an increasing amount of attention. Just as studying the stability for stochastic systems is important, the stability for stochastic systems with Markovian switching is also a significant issue. In the past several decades, a large number of works on this topic have been reported in the literature. For example, Mao in [1] considered the stability of stochastic differential equations with Markovian switching. Later, he generalized the above result to stochastic functional differential equations [2] and stochastic delay interval systems [3]. Recently, Ref. [4] considered the mean-square stability for a class of singular stochastic systems with Markovian switching. Ref. [5] derived the stability in distribution of stochastic differential delay equations with Markovian switching based on the pure probability method. By using the multiple Lyapunov functions method, Ref. [6] considered the asymptotic stability of stochastic differential equations with Markovian switching , which generalized the result in [7]. Ref. [8] applied a mode-dependent parameter approach to give a sufficient condition for the finite-time stability of Itô's stochastic systems with Markovian switching. Refs. [9–14] considered the stability and the related questions for the Markovian jump linear systems by using different approaches. In [15,16], the authors obtained some properties for stochastic systems with Markovian switching, and then they also applied these properties to study the networked control systems with delays. Ref. [17] established the exponential-m stability for stochastic switched systems. Refs. [18–21] considered the stability for stochastic systems with semi-Markovian switching. Ref. [22] obtained the pth moment exponential stability for stochastic pantograph systems with Markovian switching, which is an important class of stochastic hybrid systems with unbounded delays. In addition, the "dwell time" method has been used to investigate switching systems. In [23], Hespanha and Morse proved that a switched system is exponentially stable if all the subsystems are exponentially stable and its average dwell time is large enough. From then on, a large number of stability criteria have been reported by using "dwell time" method. For example, Ref. [24] studied the stochastic stability of continuous-time systems with random switching signals by using the Lyapunov approach, the LMI (i.e., linear matrix inequalities, see [25]) technique and the "dwell time" method. Ref. [26] used the "dwell time" and the non-convolution type multiple Lyapunov functionals to derive the almost sure exponential stability for switched delay systems with nonlinear stochastic perturbations. Ref. [27] applied the average dwell time approach to obtain the stability results for neutral stochastic switching delay systems. By using the average dwell time method and Lyapunov–Krasovskii functional theory, an H_∞ control problem for network-based stochastic systems with two additive delays was studied in [28,29]. The are also some other methods used in the stability analysis for switching systems. For example, Wu et al. in [30] applied Itô's formula and Dynkin's formula to investigate the stability on stochastic systems with state-dependent switching. Ref. [31] obtained some stability results for slowly switched systems by using the multiple discontinuous Lyapunov function approach. By using the comparison principle and the multiple Lyapunov functions method, Ref. [32] studied the stability of deterministic and stochastic switched systems. For other results with respect to the stability of stochastic differential equations with Markovian switching, please refer to Refs. [33–36] and references therein. For a survey of stability for stochastic hybrid systems, please refer to [37].

In most of the existing works, the switching signals are deterministic functions. However, in the real world, the delay switching is not usually determined. In other words, the delay switching may be random. Therefore, it is interesting and challenging to investigate the stability of randomly switched delay systems. To the best of our knowledge, there have been only a few results on this issue. For example, Ref. [38] studied the moment exponential stability of random delay systems with two-time-scale Markovian switching and the main tool based on the theory of two-time-scale Markov chains. However, the noise disturbance in [38] was ignored.

In addition, impulsive stochastic systems are also important systems in control engineering. There is also some important literature in the stability for impulsive stochastic systems. For example, Refs. [39,40] considered the exponential stability for impulsive stochastic delay differential systems. Ref. [41] established the exponential stability for neutral impulsive stochastic delay differential systems. Ref. [42] designed the impulsive controller for stochastic recurrent neural networks. Ref. [43] studied the stability for impulsive stochastic differential equations driven by G-Brownian motion. For the other stability analysis for impulsive stochastic systems, please refer to [44–46] and references therein.

Inspired by the above discussion, we can see that there is still no result for the pth moment exponential stability for impulsive stochastic functional differential equations with Markovian switched delay effects. Thus, in this paper, we will focus on this question. The systems combine the characteristics of the continuous-time systems and discrete-time systems, which leads the stability analysis for such systems being more complicated than in the case of the pure continuous-time systems or discrete-time systems. By applying Markov chain theory, stochastic analysis theory and the Razumikhin technique, we establish a criterion of pth moment exponential stability. Finally, an example is provided to verify the efficiency of the obtained result.

The rest of the paper is organized as follows. In Section 2, we introduce the model and some preliminaries. The main result and its proof will be presented in Section 3. An illustrative example is provided in Section 4. Conclusions are drawn in Section 5.

Notation 1. *Throughout this paper, we use the following notations. Let $(\Omega, \mathcal{F}, \{\mathcal{F}_t\}_{t \geq 0}, \mathbb{P})$ be a complete probability space with a natural filtration $\{\mathcal{F}_t\}_{t \geq 0}$ satisfying the usual condition (i.e., it is right continuous, and $\mathcal{F}_0$ contains all $\mathbb{P}$-null sets). For $r > 0$, the symbol $PC([-r, 0], \mathbb{R}^n)$ denotes the family of piecewise right continuous function φ from $[-r, 0]$ to $\mathbb{R}^n$ with the norm $\|\varphi\| = \sup_{-r \leq u \leq 0} |\varphi(u)|$. We use $PL^p_{\mathcal{F}_0}([-r, 0], \mathbb{R}^n)$ to denote the family of all $\mathcal{F}_0$-measurable, $PC([-r, 0], \mathbb{R}^n)$-valued random variables satisfying $\sup_{-r \leq u \leq 0} \mathbb{E}|\varphi(u)|^p < \infty$. We use $\mathbb{E}[\cdot]$ to denote the correspondent expectation* *operator with respect to the probability measure* $\mathbb{P}$. *Let* $B_t = B(t) = (B_1(t), B_2(t), \cdots, B_m(t))^T$ *be an m-dimensional Brownian motion defined on a complete probability space.*

2. Preliminaries

Let $\tau(t)$ be a right-continuous Markov chain on the probability space $(\Omega, \mathcal{F}, \mathbb{P})$ taking values in a finite state space $\mathcal{S} = \{r_1, r_2, \cdots, r_m\}$ with generator $Q = (q_{ij})_{m \times m}$ given by:

$$\mathbb{P}(\tau(t + \Delta t) = r_j | \tau(t) = r_i) = \begin{cases} q_{ij}\Delta t + o(\Delta t) & if \ i \neq j \\ 1 + q_{ii}\Delta t + o(\Delta t) & if \ i = j \end{cases}$$

where $\Delta t > 0$. Here, $q_{ij} \geq 0$ is the transition rate from r_i to r_j if $r_i \neq r_j$, while $q_{ii} = -\sum_{j \neq i} q_{ij} = -q_i$. Define $\zeta_0 = 0$, $\zeta_k = \inf\{t > \zeta_{k-1}; \tau(t) \neq \tau(\zeta_{k-1})\}$ and $r(n) = \tau(\zeta_n)$. From Markov chain theory, we know that $\{r(n), n \in Z\}$ is a Markov chain (called the embedded chain of $\tau(t)$). Its transition probability is $r_{ij}^{(1)} = (1 - \delta_{ij})\frac{q_{ij}}{q_i}$, where $\delta_{ij} = 1$ if $i = j$, and $\delta_{ij} = 0$ if $i \neq j$. We use $R = (R_{ij})$ to denote the transition probability matrix of embedded chain $\{r(n), n \in Z\}$. In this paper, we assume that the Markov chain $\{\tau(t), t \geq 0\}$ is independent of the Brownian motion $\{B(t), t \geq 0\}$. For the sake of simplicity, we denote $q_i \doteq q_{r_i}$. We denote $\lambda = \max\{q_i, i \in \mathcal{S}\}$, $v = \max_i\{1 - \frac{q_i}{\lambda}, i \in \mathcal{S}\}$ and $q = \max_{i,j} q_{ij}$.

Assumption 1. *In this paper, we always assume that* $v, q and \lambda$ *satisfy* $v \leq \frac{q}{\lambda}$.

We will consider the following impulsive stochastic delay differential equation with Markovian switched delay effects:

$$\begin{cases} dx(t) = f(t, x(t - \tau(t)), x(t))dt + g(t, x(t - \tau(t)), x(t))dB_t, t \neq t_k, \\ x(t_k) = I_k(t_k^- - \tau(t_k^-), x(t_k^- - \tau(t_k^-))), k = 1, 2, \cdots, t = t_k, \\ x_0 = \varphi, \end{cases} \tag{1}$$

where $\varphi = \{\varphi(u) : u \in [-r, 0]\} \in PL_{\mathcal{F}_0}^p([-r, 0], \mathbb{R}^n)$, $f : \mathbb{R}^+ \times \mathbb{R}^n \times \mathbb{R}^n \to \mathbb{R}^n$ and $g : \mathbb{R}^+ \times \mathbb{R}^n \times \mathbb{R}^n \to \mathbb{R}^{n \times m}$ are all Borel measurable. $r = \max\{r_1, r_2, \cdots, r_m\}$. $x(t_k^-)$ and $x(t_k^+)$ denote the left and right limits at t_k, respectively. Here, the impulse instants $\{t_k\}_{k=1}^{\infty}$ are all deterministic. $\Delta x(t_k) = x(t_k) - x(t_k^-)$, i.e., $x(t)$ is right-continuous at the impulse moment. We assume f, g and I_k satisfy the Lips conditions and the linear growth condition (see also [44]) in order to guarantee the existence and uniqueness of solutions $x(t)$ for system (1). We also assume that $f(t, 0, 0) = 0$, $g(t, 0, 0) = 0$ and $I_k(t, 0) = 0$ for all $k = 1, 2, \cdots$, which implies that the trivial solution of system (1) exists.

Now, we will provide a real example to show the usefulness of the model (1).

Example 1 (Electronic control systems). *On the road, if the vehicle flow reaches A, the red light is on. If the vehicle flow is below A, the green light is on. We define a two states of Markov chains* $\{\tau(t), t \geq 0\}$ *with generator:*

$$Q = \begin{bmatrix} q_{11} & q_{12} \\ q_{21} & q_{22} \end{bmatrix},$$

$\{the \ vehicle \ flow \ exceed \ A\} = \{the \ red \ light \ is \ on\} = \{\tau(t) = r_1\}$. $\{the \ vehicle \ flow \ notexceed \ A\} = \{the \ green \ light \ is \ on\} = \{\tau(t) = r_2\}$. The red light or green light sign will transfer to the electronic control system. For different sign lights, the sign transport delay is different. Moreover, the electronic control system is perturbed by noise, and the impulsive sign emerge in some determined instants. If the red light (or green light) is on, the electronic control system can be described by the following two systems, respectively:

$$\begin{cases} dx(t) = f(t, x(t), x(t - r_i))dt + g(t, x(t), x(t - r_i))dB_t, \\ x(t_k) = I_k(t_k^- - r_i, x(t_k^- - r_i)), \quad k = 1, 2, \cdots, \quad t = t_k, \end{cases}$$

where $i = 1,2$. However, the vehicle flow is random, so the electronic control system randomly switched during above two systems, and the switching law is governed by the Markov chain $\tau(t)$. In other words, we can describe the electronic control system by Equation (1). A natural question is how to consider its stability.

Definition 1. *The trivial solution of system (1) is said to be pth moment exponentially stable if there is a pair of positive constants C and β, such that:*

$$\mathbb{E}|x(t)|^p \leq C\mathbb{E}\|\varphi\|^p e^{-\beta t} \quad on \quad t \geq 0$$

for all $\varphi \in PL^p_{\mathcal{F}_0}([-r,0],\mathbb{R}^n)$.

Remark 1. *When $p = 2$, it is said to be mean square exponentially stable.*

In order to use Lyapunov's method, we need the following definition.

Definition 2. *The function $V = V(t,x) : \mathbb{R}^+ \times \mathbb{R}^n \to \mathbb{R}^+$ belongs to class Ψ if it is continuously differentiable once in t and twice in x.*

Now, we define an operator $\mathcal{L}V$ from $\mathbb{R}^+ \times \mathbb{R}^n \times \mathbb{R}^n$ to $\mathbb{R}$ by:

$$\mathcal{L}V(t,x(t-r_i),x(t)) = V_t(t,x(t)) + V_x(t,x(t))f(t,x(t-r_i),x(t)) + \frac{1}{2}trace[g^T(t)V_{xx}(t,x(t))g(t)],$$

where $g(t) = g(t,x(t-r_i),x(t))$, $V_t(t,x(t)) = \frac{\partial V(t,x(t))}{\partial t}$, $V_x(t,x(t)) = \left(\frac{\partial V(t,x(t))}{\partial x_1}, \frac{\partial V(t,x(t))}{\partial x_2},\right.$
$\left.\cdots, \frac{\partial V(t,x(t))}{\partial x_n}\right)$ and $V_{xx}(t,x(t)) = \left(\frac{\partial^2 V(t,x(t))}{\partial x_i \partial x_j}\right)_{n \times n}$.

Before stating our main result in this paper, we need the following two lemmas, which are useful in the proof of the main result. The first lemma can be found in any monograph on Markov chain theory (see e.g., [47]).

Lemma 1. *The transition probability $P_{ij}(t)$ of the Markov chain $\{\tau(t), t \geq 0\}$ can be calculated by:*

$$P_{ij}(t) = \sum_{n=0}^{\infty} \tilde{P}^{(n)}_{ij} \frac{(\lambda t)^n}{n!} e^{-\lambda t}, \tag{2}$$

where $\tilde{P}^{(n)}_{ij}$ is the i,jth component of the nth power of the matrix $\tilde{P}$, which is defined as follows:

$$\tilde{P}_{ij} = \begin{cases} 1 - \frac{q_i}{\lambda}, & if \ j = i, \\ \frac{q_i}{\lambda} R_{ij}, & if \ j \neq i. \end{cases}$$

Lemma 2. *The transition probability $P_{ij}(t)$ has the following estimate:*

$$P_{ij}(t) \leq \delta_{ij} e^{-\lambda t} + \sum_{n=1}^{\infty} \frac{q}{\lambda}\left(\frac{q}{\lambda}(m-1) + 2v\right)^{n-1} \frac{(\lambda t)^n}{n!} e^{-\lambda t}.$$

Proof. First, when $k = 1$, from Assumption 1, we know that $\tilde{P}^{(1)}_{ij} \leq \frac{q}{\lambda}$. Next, we will prove that for any r_i and r_j and $k \geq 2$, $\tilde{P}^{(k)}_{ij} \leq \frac{q}{\lambda}(\frac{q}{\lambda}(m-1) + \frac{k}{k-1}v)^{k-1}$. If $k = 2$, the conclusion is obvious. We assume that the conclusion holds for $k = n - 1$. Then, we have:

$$\tilde{P}_{ij}^{(n)} = \sum_{s} \tilde{P}_{is}^{(n-1)} \tilde{P}_{sj}^{(1)}$$

$$\leq \frac{q}{\lambda} v \left(\frac{q}{\lambda}(m-1) + \frac{n-1}{n-2}v \right)^{n-2} + (m-1)\left(\frac{q}{\lambda}\right)^2 \left(\frac{q}{\lambda}(m-1) + \frac{n-1}{n-2}v\right)^{n-2}$$

$$= \frac{q}{\lambda}\left(\frac{q}{\lambda}(m-1) + \frac{n-1}{n-2}v\right)^{n-2}\left(\frac{q}{\lambda}(m-1) + v\right)$$

$$= \frac{q}{\lambda}\left(\frac{q}{\lambda}(m-1) + \frac{n-1}{n-2}v\right)^{n-1} \frac{\frac{q}{\lambda}(m-1)+v}{\frac{q}{\lambda}(m-1)+\frac{n-1}{n-2}v}$$

$$\leq \frac{q}{\lambda}\left(\frac{q}{\lambda}(m-1) + \frac{n-1}{n-2}v\right)^{n-1}$$

$$\leq \frac{q}{\lambda}\left(\frac{q}{\lambda}(m-1) + \frac{n}{n-1}v\right)^{n-1}.$$

Substituting the above inequality into (2), we obtain:

$$P_{ij}(t) = \sum_{n=0}^{\infty} \tilde{P}_{ij}^{(n)} \frac{(\lambda t)^n}{n!} e^{-\lambda t}$$

$$= \tilde{P}_{ij}^{(0)} e^{-\lambda t} + \sum_{n=1}^{\infty} \tilde{P}_{ij}^{(n)} \frac{(\lambda t)^n}{n!} e^{-\lambda t}$$

$$\leq \delta_{ij} e^{-\lambda t} + \left[\frac{q}{\lambda} + \sum_{n=2}^{\infty} \frac{q}{\lambda}\left(\frac{q}{\lambda}(m-1) + \frac{n}{n-1}v\right)^{n-1}\right] \frac{(\lambda t)^n}{n!} e^{-\lambda t}$$

$$\leq \delta_{ij} e^{-\lambda t} + \sum_{n=1}^{\infty} \frac{q}{\lambda}\left(\frac{q}{\lambda}(m-1) + 2v\right)^{n-1} \frac{(\lambda t)^n}{n!} e^{-\lambda t}.$$

$\square$

3. Main Results

In this section, we will use the Markov chain theory, stochastic analysis theory and Lyapunov's method to obtain a criterion for pth moment exponential stability of system (1).

Theorem 1. *Let $q, m, v, \lambda, \gamma, c, c_1, c_2, p, \theta, \rho, \beta, M \geq \frac{c_2}{c_1} e^{(\gamma + \lambda - 2v\lambda - \theta - q(m-1))r} \geq 1$ be all positive numbers. If there exists a function $V \in \Psi$ such that the following conditions hold:*
(1) For all $x \in \mathbb{R}^n$:

$$c_1 |x|^p \leq V(t,x) \leq c_2 |x|^p,$$

(2) For all $k \in N$:

$$\mathbb{E}V(t_k, x(t_k)) \leq \rho \mathbb{E}V(t_k^- - \tau(t_k^-), x(t_k^- - \tau(t_k^-))),$$

(3) For all $k \in N$ and $t \in [t_{k-1}, t_k)$:

$$\max_{i \in S}\left[e^{-\lambda\delta}\mathbb{E}\mathcal{L}V(t, x(t-r_i), x(t)) + \sum_{j=1}^{m} \frac{e^{(q(m-1)+2v\lambda)\eta}-1}{m-1+\frac{2v\lambda}{q}} e^{-\lambda\delta}\mathbb{E}\mathcal{L}V(t, x(t-r_j), x(t)) \right] \leq c\mathbb{E}V(t, x(t))$$

if $\mathbb{E}V(t+u, x(t+u)) \leq \beta\mathbb{E}V(t, x(t))$ for all $u \in [-r, 0]$ and $\beta \geq Me^{(\gamma+\lambda-2v\lambda-\theta-q(m-1))r}$, where $\eta = \max_k\{t_k - t_{k-1}\} < \infty$ and $\delta = \min_k\{t_k - t_{k-1}\} > 0$.
(4) $\theta + q(m-1) + 2v\lambda - \lambda - \gamma < 0$,
(5) $\rho \sup_i e^{-(\theta+q(m-1)+2v\lambda-\lambda-\gamma)r_i} \leq \rho \frac{\sum_{i=1}^{m} e^{-(\theta+q(m-1)+2v\lambda-\lambda-\gamma)r_i}}{m-1+\frac{2v\lambda}{q}} \leq e^{(\lambda-q(m-1)-2v\lambda)\eta},$
(6) $\log\frac{c_1}{c_2} + [c + \lambda + \gamma - 2v\lambda - \theta - q(m-1)]\eta < [\lambda + \gamma - 2v\lambda - \theta - q(m-1)]r.$

Then, system (1) is p-moment exponentially stable.

Proof. First, we assume that $\mathbb{E}\|\varphi\|^p \neq 0$. We will prove that for any k and $t \in [t_k, t_{k+1})$:

$$\mathbb{E}V(t, x(t)) \leq c_2 \mathbb{E}\|\varphi\|^p e^{(\theta + q(m-1) + 2v\lambda - \lambda - \gamma)t}. \tag{3}$$

Now, we will check that (3) holds for $k = 0$. If (3) is not true, then there exists $t \in [0, t_1)$, s.t. $\mathbb{E}V(t, x(t)) > c_2 \mathbb{E}\|\varphi\|^p e^{(\theta + q(m-1) + 2v\lambda - \lambda - \gamma)t}$. Let:

$$t^* = \inf\{t \in [0, t_1) : \mathbb{E}V(t, x(t)) > c_2 \mathbb{E}\|\varphi\|^p e^{(\theta + q(m-1) + 2v\lambda - \lambda - \gamma)t}\}$$

and

$$t^{**} = \sup\{t \in [-r, t^*] : \mathbb{E}V(t, x(t)) \leq c_1 \mathbb{E}\|\varphi\|^p e^{(\theta + q(m-1) + 2v\lambda - \lambda - \gamma)(t+r)}\}.$$

Then, for any $t \in [t^{**}, t^*]$ and $u \in [-r, 0]$:

$$e^{(\gamma + \lambda - 2v\lambda - \theta - q(m-1))(t+u)} \mathbb{E}V(t + u, x(t + u))$$
$$\leq c_2 \mathbb{E}\|\varphi\|^p$$
$$\leq c_1 M e^{-(\gamma + \lambda - 2v\lambda - \theta - q(m-1))r} \mathbb{E}\|\varphi\|^p$$
$$= M e^{(\gamma + \lambda - 2v\lambda - \theta - q(m-1))t^{**}} \mathbb{E}V(t^{**}, x(t^{**}))$$
$$\leq M e^{(\gamma + \lambda - 2v\lambda - \theta - q(m-1))t} \mathbb{E}V(t, x(t)),$$

which implies that for all $u \in [-r, 0]$:

$$\mathbb{E}V(t + u, x(t + u)) \leq M e^{(\gamma + \lambda - 2v\lambda - \theta - q(m-1))r} \mathbb{E}V(t, x(t)) \leq \beta \mathbb{E}V(t, x(t)).$$

By condition (3), we obtain:

$$\mathbb{E}\mathcal{L}V(t, x(t - \tau(t)), x(t))$$
$$= \sum_{j=1}^{m} \mathbb{P}(\tau(t) = r_j) \mathbb{E}\mathcal{L}V(t, x(t - r_j), x(t))$$
$$= \sum_{i=1}^{m} \sum_{j=1}^{m} \mathbb{P}(\tau(0) = r_i) P_{ij}(t) \mathbb{E}\mathcal{L}V(t, x(t - r_j), x(t))$$
$$\leq \sum_{i=1}^{m} \sum_{j=1}^{m} \mathbb{P}(\tau(0) = r_i) \left(\delta_{ij} e^{-\lambda t} + \sum_{n=1}^{\infty} \frac{q}{\lambda} \left(\frac{q}{\lambda}(m-1) + 2v \right)^{n-1} \frac{(\lambda t)^n}{n!} e^{-\lambda t} \right) \mathbb{E}\mathcal{L}V(t, x(t - r_j), x(t))$$
$$= \sum_{i=1}^{m} \mathbb{P}(\tau(0) = r_i) \left[e^{-\lambda t} \mathbb{E}\mathcal{L}V(t, x(t - r_i), x(t)) + \sum_{j=1}^{m} \frac{e^{(q(m-1)+2v\lambda)t} - 1}{m - 1 + \frac{2v\lambda}{q}} e^{-\lambda t} \mathbb{E}\mathcal{L}V(t, x(t - r_j), x(t)) \right]$$
$$\leq \sum_{i=1}^{m} \mathbb{P}(\tau(0) = r_i) \left[e^{-\lambda \delta} \mathbb{E}\mathcal{L}V(t, x(t - r_i), x(t)) + \sum_{j=1}^{m} \frac{e^{(q(m-1)+2v\lambda)\eta} - 1}{m - 1 + \frac{2v\lambda}{q}} e^{-\lambda \delta} \mathbb{E}\mathcal{L}V(t, x(t - r_j), x(t)) \right]$$
$$\leq c \mathbb{E}V(t, x(t)).$$

Using Itô's formula and the standard stopping time technique, we derive:

$$\mathbb{E}V(t^*, x(t^*))$$
$$\leq \mathbb{E}V(t^{**}, x(t^{**}))e^{c(t^*-t^{**})}$$
$$= e^{-(\gamma+\lambda-2v\lambda-\theta-q(m-1))r}c_1\mathbb{E}\|\varphi\|^p e^{ct^*}e^{-(\gamma+\lambda-2v\lambda-\theta-q(m-1)+c)t^{**}}$$
$$\leq e^{-(\gamma+\lambda-2v\lambda-\theta-q(m-1))r}c_1\mathbb{E}\|\varphi\|^p e^{(c+\gamma+\lambda-2v\lambda-\theta-q(m-1))t^*}e^{(\theta+q(m-1)+2v\lambda-\lambda-\gamma)t^*}$$
$$\leq e^{(c+\gamma+\lambda-2v\lambda-\theta-q(m-1))\eta}e^{-(\gamma+\lambda-2v\lambda-\theta-q(m-1))r}c_1\mathbb{E}\|\varphi\|^p e^{(\theta+q(m-1)+2v\lambda-\lambda-\gamma)t^*}$$
$$= e^{(c+\gamma+\lambda-2v\lambda-\theta-q(m-1))\eta}e^{-(\gamma+\lambda-2v\lambda-\theta-q(m-1))r}\frac{c_1}{c_2}\mathbb{E}V(t^*, x(t^*)).$$

Noting that $\mathbb{E}\|\varphi\|^p \neq 0$, so $\mathbb{E}V(t^*, x(t^*)) \neq 0$. This implies

$$1 \leq \frac{c_1}{c_2}e^{(c+\gamma+\lambda-2v\lambda-\theta-q(m-1))\eta}e^{-(\gamma+\lambda-2v\lambda-\theta-q(m-1))r},$$

which contradict condition (6). So, (3) holds for $k = 0$.

Assume that (3) holds for $k \leq l - 1$, $k \in \mathbb{N}$. We will use the mathematical inductive method to prove that (3) holds for $k = l$. To this end, we divide the proof into two steps. Firstly, we will prove that it holds for $t = t_l$. From condition (2), we have:

$$\mathbb{E}V(t_l, x(t_l))$$
$$\leq \rho\mathbb{E}V(t_l^- - \tau(t_l^-), x(t_l^- - \tau(t_l^-)))$$
$$\leq \rho\sum_{j=1}^{m}\mathbb{E}V(t_l^- - r_j, x(t_l^- - r_j))\mathbb{P}(\tau(t_l^-) = r_j)$$
$$\leq \rho\sum_{j=1}^{m}c_2\mathbb{E}\|\varphi\|^p e^{(\theta+q(m-1)+2v\lambda-\lambda-\gamma)(t_l-r_j)}\mathbb{P}(\tau(t_l^-) = r_j)$$
$$= \rho\sum_{j=1}^{m}c_2\mathbb{E}\|\varphi\|^p e^{(\theta+q(m-1)+2v\lambda-\lambda-\gamma)t_{l-1}}e^{(\theta+q(m-1)+2v\lambda-\lambda-\gamma)(t_l-t_{l-1})}e^{-(\theta+q(m-1)+2v\lambda-\lambda-\gamma)r_j}\mathbb{P}(\tau(t_l^-) = r_j).$$

Using the Markov property in point t_{l-1}, we obtain:

$$\mathbb{E}V(t_l, x(t_l))$$
$$\leq \rho\sum_{j=1}^{m}\mathbb{E}V(t_l - r_j, x(t_l - r_j))\mathbb{P}(\tau(t_l^-) = r_j)$$
$$\leq \rho\sum_{j=1}^{m}c_2\mathbb{E}\|\varphi\|^p e^{(\theta+q(m-1)+2v\lambda-\lambda-\gamma)t_{l-1}}e^{(\theta+q(m-1)+2v\lambda-\lambda-\gamma)(t_l-t_{l-1})}e^{-(\theta+q(m-1)+2v\lambda-\lambda-\gamma)r_j}\mathbb{P}(\tau(t_l^-) = r_j)$$
$$= \rho\sum_{j=1}^{m}c_2\mathbb{E}\|\varphi\|^p e^{(\theta+q(m-1)+2v\lambda-\lambda-\gamma)t_{l-1}}e^{(\theta+q(m-1)+2v\lambda-\lambda-\gamma)(t_l-t_{l-1})}e^{-(\theta+q(m-1)+2v\lambda-\lambda-\gamma)r_j}$$
$$\times \mathbb{E}(\mathbb{P}(\tau(t_l^-) = r_j)|\mathcal{F}_{t_{l-1}})$$
$$= \rho\sum_{j=1}^{m}c_2\mathbb{E}\|\varphi\|^p e^{(\theta+q(m-1)+2v\lambda-\lambda-\gamma)t_{l-1}}e^{(\theta+q(m-1)+2v\lambda-\lambda-\gamma)(t_l-t_{l-1})}e^{-(\theta+q(m-1)+2v\lambda-\lambda-\gamma)r_j}$$
$$\times \mathbb{E}(P_{\tau(t_{l-1})j}(t_l - t_{l-1}))$$
$$= \rho\sum_{j=1}^{m}c_2\mathbb{E}\|\varphi\|^p e^{(\theta+q(m-1)+2v\lambda-\lambda-\gamma)t_{l-1}}e^{(\theta+q(m-1)+2v\lambda-\lambda-\gamma)(t_l-t_{l-1})}e^{-(\theta+q(m-1)+2v\lambda-\lambda-\gamma)r_j}$$
$$\times \sum_{i=1}^{m}P_{ij}(t_l - t_{l-1})\mathbb{P}(\tau(t_{l-1}) = r_i),$$

where the symbol $P_{\tau(t_{l-1})j}(t_l - t_{l-1})$ in the second equality denotes the transition probability from state $\tau(t_{l-1})$ to j during time $t_l - t_{l-1}$. Combining Lemma 2 and condition (5), we have:

$$\sum_{i=1}^{m} \mathbb{P}(\tau(t_{l-1}) = r_i) \left(\sum_{j=1}^{m} \rho c_2 \mathbb{E}\|\varphi\|^p e^{-(\theta+q(m-1)+2v\lambda-\lambda-\gamma)r_j} e^{(\theta+q(m-1)+2v\lambda-\lambda-\gamma)(t_l-t_{l-1})} \right.$$

$$\left. \times e^{(\theta+q(m-1)+2v\lambda-\lambda-\gamma)t_{l-1}} P_{ij}(t_l - t_{l-1}) \right)$$

$$\leq \sum_{i=1}^{m} \mathbb{P}(\tau(t_{l-1}) = r_i) \left[\rho c_2 \mathbb{E}\|\varphi\|^p e^{(\theta+q(m-1)+2v\lambda-\lambda-\gamma)t_{l-1}} e^{(\theta+q(m-1)+2v\lambda-\lambda-\gamma)(t_l-t_{l-1})} \right.$$

$$\times e^{-(\theta+q(m-1)+2v\lambda-\lambda-\gamma)r_i} e^{-\lambda(t_l-t_{l-1})} + \left(\sum_{j=1}^{m} e^{-(\theta+q(m-1)+2v\lambda-\lambda-\gamma)r_j} \rho c_2 \mathbb{E}\|\varphi\|^p \right.$$

$$\times e^{(\theta+q(m-1)+2v\lambda-\lambda-\gamma)t_{l-1}} e^{(\theta+q(m-1)+2v\lambda-\lambda-\gamma)(t_l-t_{l-1})}$$

$$\left. \left. \times \sum_{n=1}^{\infty} \frac{q}{\lambda} \left(\frac{q}{\lambda}(m-1) + 2v \right)^{n-1} \frac{(\lambda(t_l - t_{l-1}))^n}{n!} e^{-\lambda(t_l-t_{l-1})} \right) \right]$$

$$\leq \sum_{i=1}^{m} \mathbb{P}(\tau(t_{l-1}) = r_i) \left[\rho c_2 \mathbb{E}\|\varphi\|^p e^{(\theta+q(m-1)+2v\lambda-\lambda-\gamma)t_{l-1}} e^{(\theta+q(m-1)+2v\lambda-\lambda-\gamma)(t_l-t_{l-1})} \right.$$

$$\times e^{-\lambda(t_l-t_{l-1})} \frac{\sum_{j=1}^{m} e^{-(\theta+q(m-1)+2v\lambda-\lambda-\gamma)r_j}}{m - 1 + \frac{2v\lambda}{q}}$$

$$+ \left(\rho c_2 \mathbb{E}\|\varphi\|^p \sum_{j=1}^{m} e^{-(\theta+q(m-1)+2v\lambda-\lambda-\gamma)r_j} e^{(\theta+q(m-1)+2v\lambda-\lambda-\gamma)t_{l-1}} e^{(\theta+q(m-1)+2v\lambda-\lambda-\gamma)(t_l-t_{l-1})} \right.$$

$$\left. \left. \times \sum_{n=1}^{\infty} \frac{q}{\lambda} \left(\frac{q}{\lambda}(m-1) + 2v \right)^{n-1} \frac{(\lambda(t_l - t_{l-1}))^n}{n!} e^{-\lambda(t_l-t_{l-1})} \right) \right]$$

$$= \sum_{i=1}^{m} \mathbb{P}(\tau(t_{l-1}) = r_i) \left[\rho c_2 \mathbb{E}\|\varphi\|^p e^{(\theta+q(m-1)+2v\lambda-\lambda-\gamma)t_{l-1}} e^{(\theta+q(m-1)+2v\lambda-\lambda-\gamma)(t_l-t_{l-1})} \right.$$

$$\left. \times \sum_{j=1}^{m} e^{-(\theta+q(m-1)+2v\lambda-\lambda-\gamma)r_j} \sum_{n=0}^{\infty} \frac{q}{\lambda} \left(\frac{q}{\lambda}(m-1) + 2v \right)^{n-1} \frac{(\lambda(t_k - t_{k-1}))^n}{n!} e^{-\lambda(t_k-t_{k-1})} \right]$$

$$= e^{(\theta+q(m-1)+2v\lambda-\lambda-\gamma)t_{l-1}} e^{(\theta+q(m-1)+2v\lambda-\lambda-\gamma)(t_l-t_{l-1})} \rho c_2 \mathbb{E}\|\varphi\|^p \frac{\sum_{j=1}^{m} e^{-(\theta+q(m-1)+2v\lambda-\lambda-\gamma)r_j}}{m - 1 + \frac{2v\lambda}{q}}$$

$$\times e^{(q(m-1)+2v\lambda-\lambda)(t_l-t_{l-1})}$$

$$\leq e^{(\theta+q(m-1)+2v\lambda-\lambda-\gamma)t_{l-1}} e^{(\theta+q(m-1)+2v\lambda-\lambda-\gamma)(t_l-t_{l-1})} \rho c_2 \mathbb{E}\|\varphi\|^p \frac{\sum_{j=1}^{m} e^{-(\theta+q(m-1)+2v\lambda-\lambda-\gamma)r_j}}{m - 1 + \frac{2v\lambda}{q}}$$

$$\times e^{(q(m-1)+2v\lambda-\lambda)\eta}$$

$$\leq c_2 \mathbb{E}\|\varphi\|^p e^{(\theta+q(m-1)+2v\lambda-\lambda-\gamma)t_{l-1}} e^{(\theta+q(m-1)+2v\lambda-\lambda-\gamma)(t_l-t_{l-1})}$$

$$= c_2 \mathbb{E}\|\varphi\|^p e^{(\theta+q(m-1)+2v\lambda-\lambda-\gamma)t_l}.$$

This implies that (3) is satisfied for $t = t_l$. Next, we will prove that (3) holds for $t \in (t_l, t_{l+1})$. If (3) is not true, then there exists $t \in (t_l, t_{l+1})$, s.t. $\mathbb{E}V(t, x(t)) > c_2 \mathbb{E}\|\varphi\|^p e^{(\theta+q(m-1)+2v\lambda-\lambda-\gamma)t}$. Let:

$$t^* = \inf\{t \in [t_l, t_{l+1}) : \mathbb{E}V(t, x(t)) > c_2 \mathbb{E}\|\varphi\|^p e^{(\theta+q(m-1)+2v\lambda-\lambda-\gamma)t}\}$$

and

$$t^{**} = \sup\{t \in (t_l, t^*] : \mathbb{E}V(t, x(t)) \leq c_1 \mathbb{E}\|\varphi\|^p e^{(\theta+q(m-1)+2v\lambda-\lambda-\gamma)(t+r)}\}.$$

Then, for any $t \in [t^{**}, t^*]$ and $u \in [-r, 0]$, we obtain:

$$e^{(\gamma + \lambda - 2v\lambda - \theta - q(m-1))(t+u)} \mathbb{E}V(t + u, x(t + u))$$

$$\leq c_2 \mathbb{E}\|\varphi\|^p$$

$$< c_1 M e^{-(\gamma + \lambda - 2v\lambda - \theta - q(m-1))r} \mathbb{E}\|\varphi\|^p$$

$$= M e^{(\gamma + \lambda - 2v\lambda - \theta - q(m-1))t^{**}} \mathbb{E}V(t^{**}, x(t^{**}))$$

$$\leq M e^{(\gamma + \lambda - 2v\lambda - \theta - q(m-1))t} \mathbb{E}V(t, x(t)),$$

which implies that for all $u \in [-r, 0]$:

$$\mathbb{E}V(t + u, x(t + u)) \leq M e^{(\gamma + \lambda - 2v\lambda - \theta - q(m-1))r} \mathbb{E}V(t, x(t)) \leq \beta \mathbb{E}V(t, x(t)).$$

By condition (3), we obtain:

$$\mathbb{E}\mathcal{L}V(t, x(t - \tau(t)), x(t))$$

$$= \sum_{j=1}^{m} \mathbb{P}(\tau(t) = r_j) \mathbb{E}\mathcal{L}V(t, x(t - r_j), x(t))$$

$$= \sum_{i=1}^{m} \sum_{j=1}^{m} \mathbb{P}(\tau(t_l) = r_i) P_{ij}(t - t_l) \mathbb{E}\mathcal{L}V(t, x(t - r_j), x(t))$$

$$\leq \sum_{i=1}^{m} \sum_{j=1}^{m} \mathbb{P}(\tau(t_l) = r_i) \left(\delta_{ij} e^{-\lambda(t - t_l)} + \sum_{n=1}^{\infty} \frac{q}{\lambda} \left(\frac{q}{\lambda}(m - 1) + 2v \right)^{n-1} \frac{(\lambda(t - t_l))^n}{n!} e^{-\lambda(t - t_l)} \right)$$

$$\times \mathbb{E}\mathcal{L}V(t, x(t - r_j), x(t))$$

$$= \sum_{i=1}^{m} \mathbb{P}(\tau(t_l) = r_i) \left[e^{-\lambda(t - t_l)} \mathbb{E}\mathcal{L}V(t, x(t - r_i), x(t)) + \sum_{j=1}^{m} \frac{e^{(q(m-1)+2v\lambda)(t - t_l)} - 1}{m - 1 + \frac{2v\lambda}{q}} e^{-\lambda(t - t_l)} \right.$$

$$\left. \times \mathbb{E}\mathcal{L}V(t, x(t - r_j), x(t)) \right]$$

$$\leq \sum_{i=1}^{m} \mathbb{P}(\tau(t_l) = r_i) \left[e^{-\lambda \delta} \mathbb{E}\mathcal{L}V(t, x(t - r_i), x(t)) + \sum_{j=1}^{m} \frac{e^{(q(m-1)+2v\lambda)\eta} - 1}{m - 1 + \frac{2v\lambda}{q}} e^{-\lambda \delta} \mathbb{E}\mathcal{L}V(t, x(t - r_j), x(t)) \right]$$

$$\leq c \mathbb{E}V(t, x(t)).$$

For any $t \in [t^{**}, t^*]$, it follows that:

$$\mathbb{E}V(t^*, x(t^*))$$

$$\leq \mathbb{E}V(t^{**}, x(t^{**})) e^{c(t^* - t^{**})}$$

$$= e^{-(\gamma + \lambda - 2v\lambda - \theta - q(m-1))r} c_1 \mathbb{E}\|\varphi\|^p e^{ct^*} e^{-(\gamma + \lambda - 2v\lambda - \theta - q(m-1)+c)t^{**}}$$

$$\leq e^{-(\gamma + \lambda - 2v\lambda - \theta - q(m-1))r} c_1 \mathbb{E}\|\varphi\|^p e^{(c + \gamma + \lambda - 2v\lambda - \theta - q(m-1))t^*} e^{(\theta + q(m-1) + 2v\lambda - \lambda - \gamma)t^*}$$

$$\leq e^{(c + \gamma + \lambda - 2v\lambda - \theta - q(m-1))\eta} e^{-(\gamma + \lambda - 2v\lambda - \theta - q(m-1))r} c_1 \mathbb{E}\|\varphi\|^p e^{(\theta + q(m-1) + 2v\lambda - \lambda - \gamma)t^*}$$

$$= e^{(c + \gamma + \lambda - 2v\lambda - \theta - q(m-1))\eta} e^{-(\gamma + \lambda - 2v\lambda - \theta - q(m-1))r} \frac{c_1}{c_2} \mathbb{E}V(t^*, x(t^*)).$$

This is a contradiction to condition (6). So, (3) holds for $t \in (t_l, t_{l+1})$. In other words, we have proved that (3) holds for any k and $t \in [t_k, t_{k+1})$. According to condition (1), we have:

$$\mathbb{E}|x(t)|^p \leq \frac{c_2}{c_1} \mathbb{E}\|\varphi\|^p e^{(\theta + q(m-1) + 2v\lambda - \lambda - \gamma)t},$$

which verifies that system (1) is pth moment exponentially stable. $\square$

Remark 2. *The model we consider here combines the characteristics of the continuous-time systems and discrete-time systems. So, the stability analysis of such systems is more complex than the case of the pure continuous time-systems and discrete-time systems. In the proof of Theorem 1, the key* *step is to estimate the transition probability of Markov chain $\tau(t)$. Here, we have applied some classic results in the Markov chain theory. In order to overcome the difficulties that arise from the discrete-time systems, we have used the Markov property of $\tau(t)$. The technique applied here is new, and it is different from those methods used in the existing literature.*

Remark 3. *According to Theorem 1, we can see that if a certain subsystem is unstable, and then the switched system remains stable. In fact, from the statement of Theorem 1, we can see that if the parameters (the size of delay, the altitude of impulsive control gain, the impulsive times interval) of every subsystem are given, we can control the parameters q, v and λ in order to ensure the stability of the switched system.*

4. An Example

In this section, we will consider an example to illustrate the validity of our result.

Example 2. *Consider the following 2D stochastic neural network:*

$$\begin{cases} dx(t) = [Ax(t) + Bx(t - \tau(t))]dt + Df(x(t - \tau(t)))dW(t) \\ \Delta x(t_k) = -0.35x(t_k^- - \tau(t_k^-)), \ t = t_k, \ k = 1, 2, \cdots \\ x_0 = \varphi(u) = [0.015, -0.02]^T, \ u \in [-0.5, 0], \end{cases} \quad (4)$$

where

$$A = \begin{bmatrix} 0.02 & 0 \\ 0 & 0.02 \end{bmatrix}, \quad B = \begin{bmatrix} -0.015 & -0.001 \\ -0.002 & -0.025 \end{bmatrix},$$

$$D = \begin{bmatrix} 0.02 & 0 \\ 0 & 0.03 \end{bmatrix}$$

where $\delta = \eta = t_k - t_{k-1} = 0.2$. $f(\cdot) = \tanh(\cdot)$. The Markov process $\{\tau(t), t \geq 0\}$ takes values in $S = \{0.3, 0.5\}$ with generator:

$$Q = \begin{bmatrix} -0.4 & 0.4 \\ 0.3 & -0.3 \end{bmatrix}.$$

Now, we assert that system (4) is mean square stable. Obviously, $\lambda = 0.4$, $q = 0.4$, $v = 0.25$. Taking $V(t, x(t)) = |x(t)|^2$, then condition (1) of Theorem 1 holds for $p = 2$, $c_1 = c_2 = 1$. Let $\gamma = 0.4$ and $\theta = 0.1$. From a direct computation, it follows that $\rho = 0.422$. Using Itô's formula, we have:

$$\mathbb{E}\mathcal{L}V(t, x(t - r_i), x(t))$$
$$= 0.04\mathbb{E}|x(t)|^2 + 0.025\mathbb{E}|x(t)|^2 + 0.025\mathbb{E}|x(t - r_i)|^2 + 0.0009\mathbb{E}|x(t - r_i)|^2$$
$$\leq 0.065\mathbb{E}|x(t)|^2 + (0.026 \times 1.1)\mathbb{E}|x(t)|^2$$
$$= 0.0936\mathbb{E}|x(t)|^2,$$

$c = 0.0936\left[e^{-\lambda\delta} + \frac{m}{m-1+\frac{2v\lambda}{q}}\left(e^{(q(m-1)+2v\lambda)\delta} - 1\right)e^{-\lambda\delta}\right] = 0.1$. *Thus, conditions (1)–(3) are satisfied.*

Next, we turn to check that conditions (4)–(6) hold. In fact, $\theta + q(m - 1) + 2v\lambda - \lambda - \gamma = -0.1 < 0$. $\sup_i e^{-(\theta+q(m-1)+2v\lambda-\lambda-\gamma)r_i} = e^{0.05} = 1.05 < \dfrac{\sum_{i=1}^m e^{-(\theta+q(m-1)+2v\lambda-\lambda-\gamma)r_i}}{m-1+\frac{2v\lambda}{q}} = \dfrac{e^{0.03}+e^{0.05}}{1.5} = 1.38$, $\rho\dfrac{\sum_{i=1}^m e^{-(\theta+q(m-1)+2v\lambda-\lambda-\gamma)r_i}}{m-1+\frac{2v\lambda}{q}} = 0.422 \times \dfrac{e^{0.03}+e^{0.05}}{1.5} = 0.422 \times 1.38 = 0.59 < e^{-0.2\times0.2} = 0.96$. $(c + \lambda + \gamma - \theta - 2v\lambda - q(m - 1))\delta = 0.04 < (\lambda + \gamma - \theta - 2v\lambda - q(m - 1))r = 0.05$. That is, all the conditions of Theorem 1 hold. Therefore, from Theorem 1, we see that the neural network (4) is mean square exponentially stable.

5. Conclusions

In this paper, we have investigated the pth moment exponential stability of impulsive stochastic functional differential equations with Markovian switched delay effects. By using stochastic process theory, stochastic analysis theory, Razumikin technology and the Lyaponov method, a novel sufficient condition is obtained. Different from the previous literature, the model that we study is new and more complex. Moreover, an example is provided to show the efficiency of our result. In addition, it maybe more reasonable that if the impulse instants $\{t_k\}_{k=1}^{\infty}$ are random variables, and how to consider the stability is more challenging. In the future, we will consider this question.

Funding: This work was jointly supported by the National Natural Science Foundation of China (62103172, 12126331), and the Natural Science Foundation of Jiangsu Province(BK20191033).

Data Availability Statement: Not applicable.

Conflicts of Interest: The author declares no conflict of interest.

References

1. Mao, X. Stability of stochastic differential equations with Markovian switching. *Stoch. Process. Appl.* **1999**, *79*, 45–67. [CrossRef]
2. Mao, X. Stability of stochastic functional differential equations with Markovian switching. *Funct. Differ. Equ.* **1999**, *6*, 375–396.
3. Mao, X. Exponential stability of stochastic delay interval systems with Markovian switching. *IEEE Trans. Autom. Control* **2002**, *47*, 1604–1612.
4. Huang, L.; Mao, X. Stability of singular stochastic systems with Markovian switching. *IEEE Trans. Autom. Control* **2011**, *56*, 424–429. [CrossRef]
5. Du, N.; Dang, N.; Dieu, N. On stability in distribution of stochastic differential delay equations with Markovian switching. *Syst. Control Lett.* **2014**, *65*, 43–49. [CrossRef]
6. Wang, B.; Zhu, Q. Stability analysis of Markov switched stochastic differential equations with both stable and unstable subsystems. *Syst. Control Lett.* **2017**, *105*, 55–61. [CrossRef]
7. Zhu, F.; Han, Z.; Zheng, J. Stability analysis of stochastic differential equations with Markovian switching. *Syst. Control Lett.* **2012**, *61*, 1209–1214. [CrossRef]
8. Yan, Z.; Zhang, W.; Zhang, G. Finite-time stability and stabilization of Ito stochastic systems with Markovian switching: Mode-dependent parameter approach. *IEEE Trans. Autom. Control* **2015**, *60*, 2428–2433. [CrossRef]
9. Aberkane, S. Stochastic stabilization of a class of nonhomogeneous Markovian jump linear systems. *Syst. Control Lett.* **2011**, *60*, 156–160. [CrossRef]
10. Benjelloun, K.; Boukas, E. Mean square stochastic stability of linear time-daley system with Markovian jumping parameters. *IEEE Trans. Autom. Control* **1998**, *43*, 1456–1460. [CrossRef]
11. Bolzern, P.; Colaneri, P.; Nicolao, D. Markov jump linear systems with switching transition rate: Mean square stability with dwell time. *Automatica* **2010**, *46*, 1081–1088. [CrossRef]
12. Fragoso, M.; Costa, O. A unified approach for stochastic and mean square stability of continuous-time linear systems with Markovian jumping parameters and additive disturbances. *SIAM J. Control Optim.* **2005**, *44*, 1165–1191. [CrossRef]
13. Ugrinovskii, V.; Pota, H. Decentralized control of power systems via robust control of uncertain Markov jump parameter systems. *Int. J. Control* **2005**, *78*, 662–677. [CrossRef]
14. Wang, G.; Xu, L. Almost sure stability and stabilization of Markovian jump systems with stochastic switching. *IEEE Trans. Autom. Control* **2022**, *67*, 1529–1536. [CrossRef]
15. Antunes, D.; Hespanha, J.; Silvestre, C. Stochastic hybrid systems with renewal transitions: Moment analysis with application to networked control systems with delays. *SIAM J. Control Optim.* **2013**, *51*, 1481–1499. [CrossRef]
16. Cetinkaya, A.; Ishii, H.; Hayakawa, T. Analysis of stochastic switched systems with application to networked control under jamming attacks. *IEEE Trans. Autom. Control* **2019**, *64*, 2013–2028. [CrossRef]
17. Filipovic, V. Exponential stability of stochastic switched systems. *Trans. Inst. Meas. Control* **2009**, *31*, 205–212. [CrossRef]
18. Schioler, H.; Simonsen, M.; Leth, J. Stochastic stability of systems with semi-Markovian switching. *Automatica* **2014**, *50*, 2961–2964. [CrossRef]
19. Wang, B.; Zhu, Q. Stability analysis of semi-Markov switched stochastic systems. *Automatica* **2018**, *94*, 72–80. [CrossRef]
20. Mu, X.; Hu, Z. Stability analysis for semi-Markovian switched singular stochastic systems. *Automatica* **2020**, *118*, 109014. [CrossRef]
21. Luo, W.; Liu, X.; Yang, J. Stability of stochastic functional differential systems with semi-Markovian switching and Levy noise and its application. *Int. J. Control Autom.* **2020**, *18*, 708–718. [CrossRef]
22. Caraballo, T.; Mchiri, L.; Mohsen, B.; Rhaima, M. pth moment exponential stability of neutral stochastic pantograph differential equations with Markovian switching. *Commun. Nonlinear Sci. Numer. Simulat.* **2021**, *102*, 105916. [CrossRef]
23. Hespanha, J.; Morse, A. Stability of switched systems with average dwell time. In Proceedings of the 38th IEEE Conference on Decision and Control, Phoenix, AZ, USA, 7–10 December 1999; pp. 2655–2660

24. Xiong, J.; Lam, J.; Shu, Z.; Mao, X. Stability analysis of continuous-time switched systems with a random switching signal. *IEEE Trans. Autom. Control* **2014**, *59*, 180–186. [CrossRef]
25. Mao, X.; Yuan, C. *Stochastic Differential Equations with Markovian Switching*; Imperial College Press: London, UK, 2006.
26. Chen, Y.; Zheng, W. Stability analysis and control for switched stochastic delayed systems. *Int. J. Robust Nonlinear Control* **2016**, *26*, 303–328. [CrossRef]
27. Chen, H.; Shi, P.; Lim, C. Stability of neutral stochastic switched time delay systems: An average dwell time approach. *Int. J. Robust Nonlinear Control* **2017**, *27*, 512–532. [CrossRef]
28. Meng, X.; Lam, J.; Gao, H. Network-based H_∞ control for stochastic systems. *Int. J. Robust Nonlinear Control* **2009**, *19*, 295–312. [CrossRef]
29. Zhou, Q.; Shi, P. A new approach to network-based H_∞ control for stochastic systems. *Int. J. Robust Nonlinear Control* **2012**, *22*, 1036–1059. [CrossRef]
30. Wu, Z.; Cui, M.; Shi, P.; Karimi, H. Stability of stochastic nonlinear systems with state-dependent switching. *IEEE Trans. Autom. Control* **2013**, *58*, 1904–1918. [CrossRef]
31. Zhao, X.; Peng, S.; Yin, Y.; Nguang, S. New results on stability of slowly switched systems: A multiple discontinuous Lyapunov function approach. *IEEE Trans. Autom. Control* **2017**, *62*, 3502–3509. [CrossRef]
32. Chatterjee, D.; Liberzon, D. Stability analysis of deterministic and stochastic switched systems via a comparison principle and multiple Lyapunov functions. *SIAM J. Control Optim.* **2006**, *45*, 174–206. [CrossRef]
33. Peng, S.; Zhang, Y. Some new criteria on pth moment stability of stochastic functional differential equations with Markovian switching. *IEEE Trans. Autom. Control* **2010**, *55*, 2886–2890. [CrossRef]
34. Huang, L.; Mao, X. On input-to-state stability of stochastic retarded systems with Markovian switching. *IEEE Trans. Autom. Control* **2009**, *54*, 1898–1902. [CrossRef]
35. Yue, D.; Han, Q. Delay-dependent exponential stability of stochastic systems with time-varying delay nonlinearity and Markovian switching. *IEEE Trans. Autom. Control* **2005**, *50*, 217–222.
36. Ding, K.; Zhu, Q. Extended dissipative anti-disturbance control for delayed switched singular semi-Markovian jump systems with multi-disturbance via disturbance observer. *Automatica* **2021**, *128*, 109556. [CrossRef]
37. Teel, A.; Subbaraman, A.; Sferlazza, A. Stability analysis for stochastic hybrid systems: A survey. *Automatica* **2014**, *50*, 2435–2456. [CrossRef]
38. Wu, F.; Yin, G.; Wang, L. Moment exponential stability of random delay systems with two-time-scale Markovian switching. *Nonlinear Anal. RWA* **2012**, *13*, 2476–2490. [CrossRef]
39. Vinodkumar, A.; Senthilkumar, T.; Hariharan, S.; Alzabut, J. Exponential stabilization of fixed and random time impulsive delay differential system with applications. *Math. Biosci. Eng.* **2012**, *13*, 2476–2490. [CrossRef]
40. Rengamannar, K.; Balakrishnan, G.; Palanisamy, M.; Niezabitowski, M. Exponential stability of non-linear stochastic delay differential system with generalized delay-dependent impulsive points. *Appl. Math. Comput.* **2012**, *13*, 2476–2490. [CrossRef]
41. Rengamannar, K.; Palanisamy, M. Exponential stability of non-linear neutral stochastic delay differential system with generalized delay-dependent impulsive points. *J. Franklin Inst.* **2021**, *358*, 5014–5038.
42. Chandrasekar, A.; Rakkiyappan, R. Impulsive controller design for exponential synchronization of delayed stochastic memristor-based recurrent neural networks. *Neurocomputing* **2016**, *173*, 1348–1355. [CrossRef]
43. Hu, L.; Ren, Y.; Sakthivel, R. Stability of square-mean almost automorphic mild solutions to impulsive stochastic differential equations driven by G-Brownian motion. *Int. J. Control* **2020**, *93*, 3016–3025. [CrossRef]
44. Peng, S.; Deng, F. New criteria on pth moment input-to-state stability of impulsive stochastic delayed differential systems. *IEEE Trans. Autom. Control* **2017**, *62*, 3573–3579. [CrossRef]
45. Hu, W.; Zhu, Q.; Karimi, H. Some improved Razumikhin stability criteria for impulsive stochastic delay differential systems. *IEEE Trans. Autom. Control* **2019**, *64*, 5207–5213. [CrossRef]
46. Hu, W.; Zhu, Q. Stability criteria for impulsive stochastic functional differential systems with distributed-delay dependent impulsive effects. *IEEE Trans. Syst. Man. Cybern. Syst.* **2021**, *51*, 2027–2032. [CrossRef]
47. Ross, S. *Stochastic Processes*; John Wiley & Sons: New York, NY, USA, 1996.

 *mathematics*

Article

Neural Adaptive Fixed-Time Attitude Stabilization and Vibration Suppression of Flexible Spacecraft

Qijia Yao [1], Hadi Jahanshahi [2], Irene Moroz [3], Naif D. Alotaibi [4] and Stelios Bekiros [5,*]

1   School of Aerospace Engineering, Beijing Institute of Technology, Beijing 100081, China; qijia_yao@126.com
2   Department of Mechanical Engineering, University of Manitoba, Winnipeg, MB R3T 5V6, Canada; jahanshahi.hadi90@gmail.com
3   Mathematical Institute, University of Oxford, Oxford OX2 6GG, UK; Irene.Moroz@maths.ox.ac.uk
4   Department of Electrical and Computer Engineering, Faculty of Engineering, King Abdulaziz University, Jeddah 21589, Saudi Arabia; ndalotabi@kau.edu.sa
5   Department of Banking and Finance, FEMA, University of Malta, MSD 2080 Msida, Malta
*   Correspondence: stelios.bekiros@um.edu.mt

Abstract: This paper proposes a novel neural adaptive fixed-time control approach for the attitude stabilization and vibration suppression of flexible spacecraft. First, the neural network (NN) was introduced to identify the lumped unknown term involving uncertain inertia, external disturbance, torque saturation, and elastic vibrations. Then, the proposed controller was synthesized by embedding the NN compensation into the fixed-time backstepping control framework. Lyapunov analysis showed that the proposed controller guaranteed the stabilization of attitude and angular velocity to the adjustable small neighborhoods of zero in fixed time. The proposed controller is not only robust against uncertain inertia and external disturbance, but also insensitive to elastic vibrations of the flexible appendages. At last, the excellent stabilization performance and good vibration suppression capability of the proposed control approach were verified through simulations and detailed comparisons.

Keywords: attitude stabilization; flexible spacecraft; neural adaptive control; fixed-time control; vibration suppression; Lyapunov analysis

MSC: 37N35; 93C40; 93D15

Citation: Yao, Q.; Jahanshahi, H.; Moroz, I.; Alotaibi, N.D.; Bekiros, S. Neural Adaptive Fixed-Time Attitude Stabilization and Vibration Suppression of Flexible Spacecraft. *Mathematics* **2022**, *10*, 1667. https://doi.org/10.3390/math10101667

Academic Editor: Quanxin Zhu

Received: 17 April 2022
Accepted: 9 May 2022
Published: 12 May 2022

Publisher's Note: MDPI stays neutral with regard to jurisdictional claims in published maps and institutional affiliations.

1. Introduction

To accomplish long-duration and complicated space missions, modern spacecraft are usually installed with large and lightweight flexible appendages, such as solar panels and antennas. For instance, the Engineering Test Satellite-VIII (ETS-VIII) launched by Japan is typically a flexible spacecraft with two large deployable reflectors measuring $17 \times 19 \text{ m}^2$ and a pair of large solar array panels measuring $19 \times 2 \text{ m}^2$ [1,2]. Generally, the attitude maneuver of a spacecraft may induce elastic vibrations of the flexible appendages. This can in turn cause perturbations on the attitude dynamics of the spacecraft. Moreover, the spacecraft is inevitably influenced by uncertain inertia, external disturbance, and torque saturation due to the harsh space environment and physical limitations. Even worse, the inertia matrix of the spacecraft may be fully unknown in some extreme cases. For example, when the space manipulator captures a non-cooperative target, the inertia matrix of the combined spacecraft is difficult to be obtained accurately [3–6]. Consequently, the attitude control of a flexible spacecraft is quite challenging due to the presence of these issues.

The study on the flexible spacecraft attitude control started in the mid-1970s [7,8] and has continued ever since. To realize attitude control and vibration suppression simultaneously, an effective idea is regarding the uncertain inertia, external disturbance, torque saturation, and elastic vibrations as the lumped unknown term and then compensating it in the feedforward loop. Generally, there are three main methods to tackle the lumped unknown term. The first method is a robust control by utilizing the disturbance observer to observe the lumped unknown term. In [9,10], disturbance observer-based proportional-differential (PD) controllers were developed. In [11], a disturbance observer-based backstepping control method was proposed. In [12], a disturbance observer was integrated with the active disturbance rejection control design. The second method is adaptive control by utilizing the parametric adaptation technique to estimate the lumped unknown term. In [13,14], adaptive control and stabilization schemes were developed. The third method is intelligent control by utilizing the neural network (NN) or fuzzy logic system to identify the lumped unknown term. In [15,16], fuzzy sliding mode control approaches were proposed. In [17], an intelligent PD control scheme was proposed based on the NN identification. In [18,19], Takagi–Sugeno (T–S) fuzzy model-based optimal controllers were constructed. There have also been some related results focused on the attitude control of flexible spacecraft equipped with piezoelectric devices for active vibration suppression [20–24].

To efficiently fulfill various space missions, the flexible spacecraft is expected to realize the attitude maneuver in a specific time. However, most of the above controllers only ensure that the overall closed-loop system is asymptotically stable or uniformly ultimately bounded. Alternatively, the finite-time control guarantees the stabilization of attitude and angular velocity to zero or the small neighborhoods of zero in finite time. In [25–28], several terminal sliding mode controllers were developed for the finite-time attitude control of flexible spacecraft. Particularly, in [26], a disturbance observer was incorporated into the terminal sliding mode control design to enhance the control performance. In [27,28], adaptive terminal sliding mode control schemes were presented. Nevertheless, the finite-time control has the minor disadvantage that its settling time is dependent on the initial states of the system. To solve this weakness, the concept of fixed-time control was proposed [29–32]. The fixed-time control can be regarded as a typical class of finite-time control, whose settling time is bounded, and the upper bound of the settling time does not depend on the initial system conditions. In [33–36], several terminal sliding mode controllers were designed for the fixed-time attitude control of flexible spacecraft. Specifically, in [33], a disturbance observer-based terminal sliding mode control approach was developed. In [34–36], a parametric adaptation technique was integrated with the terminal sliding mode control design.

It should be pointed out that the above finite-time and fixed-time controllers were mainly designed based on the terminal sliding mode control technique. Unfortunately, the terminal sliding mode control exhibits the disadvantages of undesired chattering phenomenon and singularity problem. These disadvantages restrict the practical implementation of the terminal sliding mode control to some extent. Moreover, artificial intelligence has been rarely employed for the finite-time and fixed-time attitude control of flexible spacecraft. When involving the NN or fuzzy logic system into the closed-loop control design, the finite-time or fixed-time stability is difficult to be proved theoretically. Actually, the fixed-time attitude control and vibration suppression of flexible spacecraft is still an open problem which needs to be further investigated.

The above discussions motivated our research. In this paper, a novel neural adaptive fixed-time control approach is presented for the attitude stabilization of flexible spacecraft. The NN was introduced to identify the lumped unknown item involving uncertain inertia, external disturbance, torque saturation, and elastic vibrations. Then, the proposed controller was synthesized by embedding the NN compensation into the fixed-time backstepping control framework. Lyapunov analysis showed that the proposed controller guaranteed the stabilization of attitude and angular velocity to the adjustable small neighborhoods of zero in fixed time. In comparisons with the above finite-time and fixed-time controllers, the main contributions of this research lie in the following two aspects.

- Rather than the terminal sliding mode control technique, the proposed controller was developed under the fixed-time backstepping control framework. In this way, the proposed controller does not have the chattering phenomenon and singularity problem existing in the terminal sliding mode control.
- The NN was integrated with the proposed controller to compensate the lumped unknown item. Benefiting from the NN compensation, the proposed controller is not only robust against uncertain inertia and external disturbance, but also insensitive to elastic vibrations of the flexible appendages.

The rest of this paper is arranged as follows. Section 2 describes the problem and provides some preliminaries. Section 3 provides the control design and Lyapunov analysis. Section 4 presents the simulations and detailed comparisons. Lastly, Section 5 summarizes this research.

2. Problem Description and Preliminaries

2.1. Problem Description

Suppose a flexible spacecraft composed of a rigid hub and flexible appendages. By employing the modified Rodrigues parameters (MRPs), the attitude kinematics of the flexible spacecraft can be expressed as

$$\dot{\sigma} = G(\sigma)\omega, \tag{1}$$

where $G(\sigma) = \frac{1}{2}\left(\frac{1-\sigma^{\mathrm{T}}\sigma}{2}\mathbf{I}_3 + \sigma^{\times} + \sigma\sigma^{\mathrm{T}}\right) \in \mathbb{R}^{3\times3}$, $\sigma = [\sigma_1, \sigma_2, \sigma_3]^{\mathrm{T}} \in \mathbb{R}^3$ and $\omega = [\omega_1, \omega_2, \omega_3]^{\mathrm{T}} \in \mathbb{R}^3$ denote the attitude and angular velocity of the spacecraft, and $\mathbb{R}^n$ and $\mathbb{R}^{n\times m}$ stand for the sets of $n \times 1$ real vectors and $n \times n$ real matrices, respectively. The notation $\omega^{\times}$ stands for the skew-symmetric matrix of ω, denoted as

$$\omega^{\times} = \begin{bmatrix} 0 & -\omega_3 & \omega_2 \\ \omega_3 & 0 & -\omega_1 \\ -\omega_2 & \omega_1 & 0 \end{bmatrix}. \tag{2}$$

Referring to [20,21], the attitude dynamics of the flexible spacecraft can be expressed as

$$J\dot{\omega} + \omega^{\times}\left(J\omega + \delta\dot{\eta}\right) + \delta\ddot{\eta} = \mathrm{sat}(u) + d, \tag{3}$$

$$\ddot{\eta} + C\dot{\eta} + K\eta = -\delta^{\mathrm{T}}\omega, \tag{4}$$

where $J \in \mathbb{R}^{3\times3}$ denotes the inertia matrix which may be fully unknown in some extreme cases, $u \in \mathbb{R}^3$ is the control torques generated by actuators, $d \in \mathbb{R}^3$ denotes the external disturbance, $\eta \in \mathbb{R}^L$ denotes the modal variables, L is the number of elastic modes considered in the control design, $\delta \in \mathbb{R}^{3\times L}$ is the coupling matrix between the rigid hub and the flexible appendages, $C = \mathrm{diag}[2\xi_1\omega_{n1}, 2\xi_2\omega_{n2}, \ldots, 2\xi_N\omega_{nL}] \in \mathbb{R}^{L\times L}$ is the damping matrix, $K = \mathrm{diag}[\omega_{n1}^2, \omega_{n2}^2, \ldots, \omega_{nL}^2] \in \mathbb{R}^{L\times L}$ is the stiffness matrix, and ω_{ni} and ξ_i denote the natural frequencies and damping ratios of the ith mode, respectively. Moreover, the saturated control torques can be expressed as $\mathrm{sat}(u) = [\mathrm{sat}(u_1), \mathrm{sat}(u_2), \mathrm{sat}(u_3)]^{\mathrm{T}}$, whose elements are presented as

$$\mathrm{sat}(u_i) = \begin{cases} u_i, & |u_i| \leq u_{\mathrm{m}}, \\ \mathrm{sgn}(u_i)u_{\mathrm{m}}, & |u_i| > u_{\mathrm{m}}, \end{cases} \quad i = 1, 2, 3, \tag{5}$$

where u_{m} stands for the maximum acceptable input value. Then, the saturated control torques can be rewritten as

$$\mathrm{sat}(u) = u + u_{\Delta}, \tag{6}$$

where u_Δ denotes the input deviations caused by torque saturation. Subsequently, the attitude kinematics and dynamics of the flexible spacecraft can be rearranged as

$$M(\sigma)\ddot{\sigma} + C(\sigma,\dot{\sigma})\dot{\sigma} = G^{-T}(\sigma)u + \chi, \tag{7}$$

where $M(\sigma) = G^{-T}(\sigma)JG^{-1}(\sigma)$, $C(\sigma,\dot{\sigma}) = -G^{-T}(\sigma)JG^{-1}(\sigma)\dot{G}(\sigma)G^{-1}(\sigma)$ $- G^{-T}(\sigma)(J\omega)^\times G^{-1}(\sigma)$, and $\chi = G^{-T}(\sigma)(-\omega^\times\delta\dot{\eta} - \delta\ddot{\eta} + d + u_\Delta)$. According to [37], system (7) has the following fundamental properties.

Property 1. *The matrix $M(\sigma)$ is symmetric and positive definite.*

Property 2. *The matrix $\dot{M}(\sigma) - 2C(\sigma,\dot{\sigma})$ is skew symmetric.*

Property 3. *The matrices $M(\sigma)$ and $C(\sigma,\dot{\sigma})$ are bounded with $\underline{m}I_3 \leq M(\sigma) \leq \overline{m}I_3$ and $\|C(\sigma,\dot{\sigma})\| \leq \overline{c}\|\dot{\sigma}\|$, where $\underline{m}$, $\overline{m}$, and $\overline{c}$ are positive constants.*

The purpose of this research was to develop an appropriate controller to realize the fixed-time attitude stabilization of flexible spacecraft even under uncertain inertia, external disturbance, and torque saturation.

2.2. Preliminaries

The following lemmas are provided, which will be used to obtain the main results of this research.

Lemma 1 ([32]). *Consider the nonlinear system:*

$$\dot{x} = f(x), \ f(0) = 0, \ x \in \mathbb{R}^n, \tag{8}$$

where $f(x)$ is a continuous nonlinear function. If there exists a positive definite function $V(x)$ satisfying $\dot{V}(x) \leq -\kappa_1 V^p(x) - \kappa_2 V^q(x) + \zeta$, where $\kappa_1 > 0$, $\kappa_2 > 0$, $0 < p < 1$, $q > 1$, and $\zeta > 0$, then system (8) is practically fixed-time stable, and $V(x)$ will converge to the following compact set in fixed time:

$$\Omega = \left\{ V(x) \in \mathbb{R} \middle| V(x) \leq \min\left\{ \left(\frac{\zeta}{\kappa_1(1-\iota)}\right)^{\frac{1}{p}}, \left(\frac{\zeta}{\kappa_2(1-\iota)}\right)^{\frac{1}{q}} \right\} \right\}, \tag{9}$$

where $0 < \iota < 1$, and the fixed settling time is bounded as $T \leq \frac{1}{\kappa_1\iota(1-p)} + \frac{1}{\kappa_2\iota(q-1)}$.

Lemma 2 ([38]). *For a continuous nonlinear function $f(Z)$, $Z \in \mathbb{R}^n$, it can be identified by a radial basis function NN (RBFNN) as*

$$f(Z) = W^{*T}\Phi(Z) + \varepsilon(Z), \tag{10}$$

where $W^* \in \mathbb{R}^N$ is the ideal RBFNN weight, $\Phi(Z) = [\varphi_1(Z), \varphi_2(Z), \dots, \varphi_N(Z)]^T$ is the basis function vector, $\varepsilon(Z)$ is the identification error satisfying $|\varepsilon(Z)| \leq \overline{\varepsilon}$, $\overline{\varepsilon}$ is a positive constant, and N is the number of RBFNN nodes. The ideal RBFNN weight W^* is defined as

$$W^* = \arg\min_{W\in\mathbb{R}^N}\left\{ \sup_{Z\in\mathbb{R}^n}\left| f(Z) - W^{*T}\Phi(Z) \right| \right\}. \tag{11}$$

In addition, $\varphi_i(Z)$ is commonly chosen as the Gaussian function:

$$\varphi_i(Z) = \exp\left(-\|Z - c_i\|^2/w_i^2\right), \ i = 1,2,\dots,N, \tag{12}$$

where $c_i = [c_{i1}, c_{i2}, \ldots, c_{in}]^{\mathrm{T}} \in \mathbb{R}^n$ and w_i are the center and width of the Gaussian function, respectively.

Lemma 3 ([39])**.** *For $x_i \in \mathbb{R}, i = 1, 2, \ldots, n, 0 < p \leq 1,$ and $q > 1,$ the following inequalities hold:*

$$\left(\sum_{i=1}^{n} |x_i| \right)^p \leq \sum_{i=1}^{n} |x_i|^p, \quad \left(\sum_{i=1}^{n} |x_i| \right)^q \leq n^{q-1} \sum_{i=1}^{n} |x_i|^q. \tag{13}$$

Lemma 4 ([39])**.** *For $x_1 \in \mathbb{R}, x_2 \in \mathbb{R}, p > 0, q > 0,$ and $\xi > 0,$ the following inequality holds:*

$$|x_1|^p |x_2|^q \leq \frac{p}{p+q} \xi |x_1|^{p+q} + \frac{q}{p+q} \xi^{-\frac{p}{q}} |x_2|^{p+q}. \tag{14}$$

3. Control Design and Lyapunov Analysis

In this section, the main results of this research are presented. First, the proposed neural adaptive fixed-time controller is synthesized by embedding the NN compensation into the fixed-time backstepping control framework. Then, the practical fixed-time stability of the overall closed-loop system is theoretically achieved through Lyapunov analysis.

3.1. Control Design

Under the fixed-time backstepping control framework, define the following error signals:

$$x_1 = \sigma, \quad x_2 = \dot{\sigma} - \mu, \tag{15}$$

where $\mu \in \mathbb{R}^3$ is the virtual control signal designed in the sequel. The whole control design procedure involves three steps. In Step 1, the virtual control signal is designed, in Step 2, the actual control signal is designed, and in Step 3, the NN weight adaptation law is designed.

Step 1: Virtual control signal design. Construct the Lyapunov function:

$$V_1 = \frac{1}{2} x_1^{\mathrm{T}} x_1. \tag{16}$$

The time differentiation of (16) can be evaluated as

$$\begin{aligned} \dot{V}_1 &= x_1^{\mathrm{T}} \dot{x}_1 \\ &= x_1^{\mathrm{T}} (x_2 + \mu). \end{aligned} \tag{17}$$

Then, the virtual control signal is designed as

$$\mu = -k_{11} \mathrm{sig}^p(x_1) - k_{12} \mathrm{sig}^q(x_1), \tag{18}$$

where $k_{11} > 0, k_{12} > 0, 0 < p < 1, q > 1,$ and the notation $\mathrm{sig}^p(\cdot)$ is defined as $\mathrm{sig}^p(x_1) = \left[|x_{11}|^p \mathrm{sgn}(x_{11}), |x_{12}|^p \mathrm{sgn}(x_{12}), |x_{13}|^p \mathrm{sgn}(x_{13}) \right]^{\mathrm{T}}$. Substituting the virtual control signal (18) into (17) and by the aid of Lemma 3, we have

$$\begin{aligned} \dot{V}_1 &= x_1^{\mathrm{T}} (x_2 - k_{11} \mathrm{sig}^p(x_1) - k_{12} \mathrm{sig}^q(x_1)) \\ &\leq x_1^{\mathrm{T}} x_2 - \kappa_{11} V_1^{\frac{p+1}{2}} - \kappa_{12} V_1^{\frac{q+1}{2}}, \end{aligned} \tag{19}$$

where $\kappa_{11} = 2^{\frac{p+1}{2}} k_{11},$ and $\kappa_{12} = 3^{\frac{1-q}{2}} 2^{\frac{q+1}{2}} k_{12}$.

Step 2: Actual control signal design. Construct the Lyapunov function:

$$V_2 = \frac{1}{2} x_2^{\mathrm{T}} M(\sigma) x_2. \tag{20}$$

By the aid of Property 2, the time differentiation of (20) can be evaluated as

$$
\begin{aligned}
\dot{V}_2 &= x_2^{\mathrm{T}} M(\sigma)\dot{x}_2 + \tfrac{1}{2} x_2^{\mathrm{T}} \dot{M}(\sigma) x_2 \\
&= x_2^{\mathrm{T}}\left(-M(\sigma)\ddot{\sigma}_{\mathrm{d}} - C(\sigma,\dot{\sigma})\dot{\sigma} + G^{-\mathrm{T}}(\sigma)u + \chi - M(\sigma)\dot{\mu}\right) + x_2^{\mathrm{T}} C(\sigma,\dot{\sigma}) x_2 \\
&= x_2^{\mathrm{T}}\left(-M(\sigma)\ddot{\sigma}_{\mathrm{d}} - C(\sigma,\dot{\sigma})\dot{\sigma}_{\mathrm{d}} - M(\sigma)\dot{\mu} - C(q,\dot{q})\mu + G^{-\mathrm{T}}(\sigma)u + \chi\right) \\
&= x_2^{\mathrm{T}}\left(G^{-\mathrm{T}}(\sigma)u + L\right),
\end{aligned}
\tag{21}
$$

where L is the lumped unknown term involving uncertain inertia, external disturbance, torque saturation, and elastic vibrations, denoted as

$$
L = -M(\sigma)\ddot{\sigma}_{\mathrm{d}} - C(\sigma,\dot{\sigma})\dot{\sigma}_{\mathrm{d}} - M(\sigma)\dot{\mu} - C(q,\dot{q})\mu + \chi.
\tag{22}
$$

Define the input variable $Z = \left[x_1^{\mathrm{T}}, x_2^{\mathrm{T}}, u^{\mathrm{T}}\right]^{\mathrm{T}}$. The RBFNN is introduced to identify the lumped unknown term. By Lemma 2, the lumped unknown term can be expressed as

$$
L = W^{*\mathrm{T}}\Phi(Z) + \varepsilon(Z),
\tag{23}
$$

where $W^* \in \mathbb{R}^{N\times 3}$ is the ideal RBFNN weight, $\Phi(Z) \in \mathbb{R}^N$ is the basis function vector, and $\varepsilon(Z) \in \mathbb{R}^3$ is the identification error satisfying $\|\varepsilon(Z)\| \le \bar{\varepsilon}$. Subsequently, the lumped unknown term can be identified by the RBFNN as

$$
\hat{L} = \hat{W}^{\mathrm{T}}\Phi(Z),
\tag{24}
$$

where $\hat{W} \in \mathbb{R}^{N\times 3}$ is the estimation of the ideal RBFNN weight. Then, the actual control signal is designed as

$$
u = G^{\mathrm{T}}(\sigma)\left(-x_1 - \frac{1}{2}x_2 - k_{21}\mathrm{sig}^p(x_2) - k_{22}\mathrm{sig}^q(x_2) - \hat{W}^{\mathrm{T}}\Phi(Z)\right),
\tag{25}
$$

where $k_{11} > 0$, and $k_{12} > 0$. Substituting the actual control signal (25) into (21), we have

$$
\begin{aligned}
\dot{V}_2 &= x_2^{\mathrm{T}}\left(-x_1 - \tfrac{1}{2}x_2 - k_{21}\mathrm{sig}^p(x_2) - k_{22}\mathrm{sig}^q(x_2) - \hat{W}^{\mathrm{T}}\Phi(Z) + W^{*\mathrm{T}}\Phi(Z) + \varepsilon(Z)\right) \\
&= x_2^{\mathrm{T}}\left(-x_1 - \tfrac{1}{2}x_2 - k_{21}\mathrm{sig}^p(x_2) - k_{22}\mathrm{sig}^q(x_2) - \tilde{W}^{\mathrm{T}}\Phi(Z) + \varepsilon(Z)\right),
\end{aligned}
\tag{26}
$$

where $\tilde{W} = \hat{W} - W^*$ is the estimation error of the RBFNN weight. Consider the inequality $x_2^{\mathrm{T}}\varepsilon(Z) \le \tfrac{1}{2}x_2^{\mathrm{T}}x_2 + \tfrac{1}{2}\bar{\varepsilon}^2$. Substituting it into (26) and by the aid of Lemma 3, we further have

$$
\begin{aligned}
\dot{V}_2 &\le -\tfrac{1}{2}x_2^{\mathrm{T}}x_1 + x_2^{\mathrm{T}}(-k_{21}\mathrm{sig}^p(x_2) - k_{22}\mathrm{sig}^q(x_2)) - x_2^{\mathrm{T}}\tilde{W}^{\mathrm{T}}\Phi(Z) + \tfrac{1}{2}\bar{\varepsilon}^2 \\
&\le -\tfrac{1}{2}x_2^{\mathrm{T}}x_1 - \kappa_{21} V_2^{\frac{p+1}{2}} - \kappa_{22} V_2^{\frac{q+1}{2}} - x_2^{\mathrm{T}}\tilde{W}^{\mathrm{T}}\Phi(Z) + \tfrac{1}{2}\bar{\varepsilon}^2,
\end{aligned}
\tag{27}
$$

where $\kappa_{21} = \dfrac{2^{\frac{p+1}{2}} k_{21}}{\lambda_{\max}^{\frac{p+1}{2}}(M(\sigma))}$, $\kappa_{22} = \dfrac{3^{\frac{1-q}{2}} 2^{\frac{q+1}{2}} k_{22}}{\lambda_{\max}^{\frac{q+1}{2}}(M(\sigma))}$, and the notations $\lambda_{\min}(\cdot)$ and $\lambda_{\max}(\cdot)$ represent the minimum and maximum eigenvalues of a matrix, respectively.

Step 3: NN weight adaptation law design. The NN weight adaptation law is given as

$$
\dot{\hat{W}}_i = \Gamma_i \Phi_i(Z_i) x_{2i} - \gamma_i \Gamma_i \hat{W}_i, \ i = 1, 2, 3,
\tag{28}
$$

where $\Gamma_i \in \mathbb{R}^{N\times N}$ are positive definite matrices, and γ_i are small positive constants. Construct the Lyapunov function:

$$
V_3 = \frac{1}{2}\sum_{i=1}^{3} \tilde{W}_i^{\mathrm{T}} \Gamma_i^{-1} \tilde{W}_i.
\tag{29}
$$

The time differentiation of (29) can be evaluated as

$$
\begin{aligned}
\dot{V}_3 &= \sum_{i=1}^{3} \widetilde{W}_i^{\mathrm{T}} \Gamma_i^{-1} \dot{\hat{W}}_i \\
&= \sum_{i=1}^{3} \widetilde{W}_i^{\mathrm{T}} \Phi_i(Z_i) x_{2i} - \sum_{i=1}^{3} \gamma_i \widetilde{W}_i^{\mathrm{T}} \hat{W}_i.
\end{aligned}
\tag{30}
$$

Consider the inequality $-\widetilde{W}_i^{\mathrm{T}} \hat{W}_i = -\|\widetilde{W}_i\|^2 - \widetilde{W}_i^{\mathrm{T}} W_i^* \le -\frac{1}{2}\|\widetilde{W}_i\|^2 + \frac{1}{2}\|W_i^*\|^2$. Substituting it into (30), we have

$$
\begin{aligned}
\dot{V}_3 &= \sum_{i=1}^{3} \widetilde{W}_i^{\mathrm{T}} \Phi_i(Z_i) x_{2i} - \sum_{i=1}^{3} \frac{\gamma_i}{2}\|\widetilde{W}_i\|^2 + \sum_{i=1}^{3} \frac{\gamma_i}{2}\|W_i^*\|^2 \\
&= \sum_{i=1}^{3} \widetilde{W}_i^{\mathrm{T}} \Phi_i(Z_i) x_{2i} - \sum_{i=1}^{3} \left(\frac{\gamma_i}{4}\|\widetilde{W}_i\|^2\right)^{\frac{p+1}{2}} - \sum_{i=1}^{3} \left(\frac{\gamma_i}{4}\|\widetilde{W}_i\|^2\right)^{\frac{q+1}{2}} + \zeta_1,
\end{aligned}
\tag{31}
$$

where ζ_1 is defined as

$$
\zeta_1 = \sum_{i=1}^{3} \left(\frac{\gamma_i}{4}\|\widetilde{W}_i\|^2\right)^{\frac{p+1}{2}} + \sum_{i=1}^{3} \left(\frac{\gamma_i}{4}\|\widetilde{W}_i\|^2\right)^{\frac{q+1}{2}} - \sum_{i=1}^{3} \frac{\gamma_i}{2}\|\widetilde{W}_i\|^2 + \sum_{i=1}^{3} \frac{\gamma_i}{2}\|W_i^*\|^2.
\tag{32}
$$

Then, the following two cases are discussed. For the case of $\frac{\gamma_i}{4}\|\widetilde{W}_i\|^2 \ge 1$, we have

$$
\left(\frac{\gamma_i}{4}\|\widetilde{W}_i\|^2\right)^{\frac{p+1}{2}} + \left(\frac{\gamma_i}{4}\|\widetilde{W}_i\|^2\right)^{\frac{q+1}{2}} - \frac{\gamma_i}{2}\|\widetilde{W}_i\|^2 \le \left(\frac{\gamma_i}{4}\|\widetilde{W}_i\|^2\right)^{\frac{q+1}{2}} - \frac{\gamma_i}{4}\|\widetilde{W}_i\|^2.
\tag{33}
$$

For the case of $\frac{\gamma_i}{4}\|\widetilde{W}_i\|^2 < 1$, by the aid of Lemma 4, we have

$$
\begin{aligned}
\left(\frac{\gamma_i}{4}\|\widetilde{W}_i\|^2\right)^{\frac{p+1}{2}} + \left(\frac{\gamma_i}{4}\|\widetilde{W}_i\|^2\right)^{\frac{q+1}{2}} - \frac{\gamma_i}{2}\|\widetilde{W}_i\|^2 &\le \left(\frac{\gamma_i}{4}\|\widetilde{W}_i\|^2\right)^{\frac{p+1}{2}} - \frac{\gamma_i}{4}\|\widetilde{W}_i\|^2 \\
&\le (1-\bar{p})\bar{p}^{\frac{\bar{p}}{1-\bar{p}}},
\end{aligned}
\tag{34}
$$

where $\bar{p} = \frac{p+1}{2}$. Introduce a compact set Θ such that $\Theta = \left\{\widetilde{W} \in \mathbb{R}^{N \times 3} \,\middle|\, \|\widetilde{W}_i\| \le \beta_i,\ i = 1, 2, 3\right\}$, where β_i are positive constants. Combining (33) and (34), it follows that

$$
\left(\frac{\gamma_i}{4}\|\widetilde{W}_i\|^2\right)^{\frac{p+1}{2}} + \left(\frac{\gamma_i}{4}\|\widetilde{W}_i\|^2\right)^{\frac{q+1}{2}} - \frac{\gamma_i}{2}\|\widetilde{W}_i\|^2 \le \alpha_i,
\tag{35}
$$

where α_i is defined as

$$
\alpha_i =
\begin{cases}
(1-\bar{p})\bar{p}^{\frac{\bar{p}}{1-\bar{p}}}, & \beta_i < \frac{2}{\sqrt{\gamma_i}}, \\[2mm]
\left(\frac{\gamma_i}{4}\beta_i^2\right)^{\frac{q+1}{2}} - \frac{\gamma_i}{4}\beta_i^2, & \beta_i \ge \frac{2}{\sqrt{\gamma_i}}.
\end{cases}
\tag{36}
$$

Substituting (35) into (31) and by the aid of Lemma 3, we further have

$$
\dot{V}_3 \le \sum_{i=1}^{3} \widetilde{W}_i^{\mathrm{T}} \Phi_i(Z_i) x_{2i} - \kappa_{31} V_3^{\frac{p+1}{2}} - \kappa_{32} V_3^{\frac{q+1}{2}} + \zeta_2,
\tag{37}
$$

where $\kappa_{31} = \dfrac{\gamma_i^{\frac{p+1}{2}}}{2^{\frac{p+1}{2}} \lambda_{\max}^{\frac{p+1}{2}}\left(\Gamma_i^{-1}\right)}$, $\kappa_{32} = \dfrac{3^{\frac{1-q}{2}} \gamma_i^{\frac{q+1}{2}}}{2^{\frac{q+1}{2}} \lambda_{\max}^{\frac{q+1}{2}}\left(\Gamma_i^{-1}\right)}$, and $\zeta_2 = \sum_{i=1}^{3}\left(\alpha_i + \frac{\gamma_i}{2}\|W_i^*\|^2\right)$.

3.2. Lyapunov Analysis

After the above preparations, the main theorem of this research can be obtained as follows.

Theorem 1. *Suppose the flexible spacecraft modeled as (1), (3), and (4), then the overall closed-loop system is practically fixed-time stable under the virtual control signal (18), the actual control signal (25), and the NN weight adaptation law (28). Specifically, the closed-loop error signals x_1, x_2, and $\tilde{W}$ will converge to the following compact sets in fixed time:*

$$\Omega_{x_1} = \left\{ x_1 \in \mathbb{R}^3 \,\middle|\, \|x_1\| \leq \sqrt{\psi} \right\}, \tag{38}$$

$$\Omega_{x_2} = \left\{ x_2 \in \mathbb{R}^3 \,\middle|\, \|x_2\| \leq \sqrt{\frac{\psi}{\lambda_{\min}(M(\sigma))}} \right\}, \tag{39}$$

$$\Omega_{\tilde{W}} = \left\{ \tilde{W} \in \mathbb{R}^{N \times 3} \,\middle|\, \|\tilde{W}_i\| \leq \sqrt{\frac{\psi}{\lambda_{\min}\left(\Gamma_i^{-1}\right)}},\ i = 1,2,3 \right\}, \tag{40}$$

where $\psi > 0$ is defined in the sequel.

Proof. Construct the Lyapunov function:

$$V = V_1 + V_2 + V_3, \tag{41}$$

where V_1, V_2, and V_3 are defined as (15), (19), and (28), respectively. Combining (19), (27), and (37) and by the aid of Lemma 3, the time differentiation of (41) can be evaluated as

$$\begin{aligned}
\dot{V} &= \dot{V}_1 + \dot{V}_2 + \dot{V}_3 \\
&\leq -\kappa_{11} V_1^{\frac{p+1}{2}} - \kappa_{12} V_1^{\frac{q+1}{2}} - \kappa_{21} V_2^{\frac{p+1}{2}} - \kappa_{22} V_2^{\frac{q+1}{2}} + \tfrac{1}{2}\bar{\varepsilon}^2 - \kappa_{31} V_3^{\frac{p+1}{2}} - \kappa_{32} V_3^{\frac{q+1}{2}} + \zeta_2 \\
&\leq -\kappa_1 V^{\frac{p+1}{2}} - \kappa_2 V^{\frac{q+1}{2}} + \zeta_3,
\end{aligned} \tag{42}$$

where $\kappa_1 = \min\{\kappa_{11}, \kappa_{21}, \kappa_{31}\}$, $\kappa_2 = 3^{\frac{1-q}{2}} \min\{\kappa_{12}, \kappa_{22}, \kappa_{32}\}$, and $\zeta_3 = \zeta_2 + \tfrac{1}{2}\bar{\varepsilon}^2$. By Lemma 1, the overall closed-loop system is practically fixed-time stable, and V will converge to the following compact set in fixed time:

$$\Omega = \left\{ V \in \mathbb{R} \,\middle|\, V \leq \min\left\{ \left(\frac{\zeta_3}{\kappa_1(1-\iota)}\right)^{\frac{2}{p+1}}, \left(\frac{\zeta_3}{\kappa_2(1-\iota)}\right)^{\frac{2}{q+1}} \right\} \right\}, \tag{43}$$

where $0 < \iota < 1$. Moreover, the fixed settling time is bounded as $T \leq \frac{2}{\kappa_1 \iota(1-p)} + \frac{2}{\kappa_2 \varsigma(\iota-1)}$. Then, define a variable as

$$\psi = 2\min\left\{ \left(\frac{\zeta_3}{\kappa_1(1-\iota)}\right)^{\frac{2}{p+1}}, \left(\frac{\zeta_3}{\kappa_2(1-\iota)}\right)^{\frac{2}{q+1}} \right\}. \tag{44}$$

Together with the definition of V, it follows that

$$x_1^{\mathrm{T}} x_1 \leq \psi, \tag{45}$$

$$x_2^{\mathrm{T}} M(\sigma) x_2 \leq \psi, \tag{46}$$

$$\tilde{W}_i^{\mathrm{T}} \Gamma_i^{-1} \tilde{W}_i \leq \psi,\ i = 1,2,3. \tag{47}$$

Thus, the closed-loop error signals x_1, x_2, and $\tilde{W}$ will converge to the compact sets Ω_{x_1}, Ω_{x_2}, and $\Omega_{\tilde{W}}$ in fixed time, respectively. This further implies that the proposed controller guarantees the stabilization of attitude σ and angular velocity ω to the small neighborhoods of zero in fixed time. Moreover, from (43), the small neighborhoods of zero are adjustable. If we set the parameters k_{11}, k_{12}, k_{21}, and k_{22} as large as desired, the small neighborhoods can be made sufficiently small. This finishes the proof. $\square$

Remark 1. *To make the proposed controller more friendly to the users, a control parameter selection strategy was carried out. The strategy contained three steps. In Step 1, we determined the control parameters k_{11}, k_{12}, k_{21}, and k_{22}. Large k_{11}, k_{12}, k_{21}, k_{22} can realize a relatively fast convergence rate; however, they may also lead to relatively large control torques at the same time. In Step 2, we determined the control parameters Γ_i and η_i. Large Γ_i and small η_i can lead to a relatively fast convergence rate; however, they may in turn result in a relatively poor transient response of the controller. In Step 3, we determined the number of RBFNN nodes N. A large N can achieve a relatively high approximation accuracy; however, it may also cause a relatively heavy onboard computational burden. Therefore, the control parameters of the proposed controller needed to be carefully tuned by trial and error for better implementations.*

Remark 2. *The RBFNN was introduced to identify the lumped unknown term involving uncertain inertia, external disturbance, torque saturation, and elastic vibrations. Benefiting from this design, the proposed controller appeared to be not only robust against uncertain inertia and external disturbance, but also insensitive to elastic vibrations of the flexible appendages. It should be noticed that the RBFNN utilized in this paper can also be replaced by some other approximation tools, such as wavelet NN, recurrent NN, fuzzy NN, and fuzzy logic system.*

Remark 3. *The proposed controller was synthesized by embedding the NN compensation into the fixed-time backstepping control framework. To facilitate the readers' understanding of the whole control design procedure, the structure of the proposed control approach is depicted in Figure 1.*

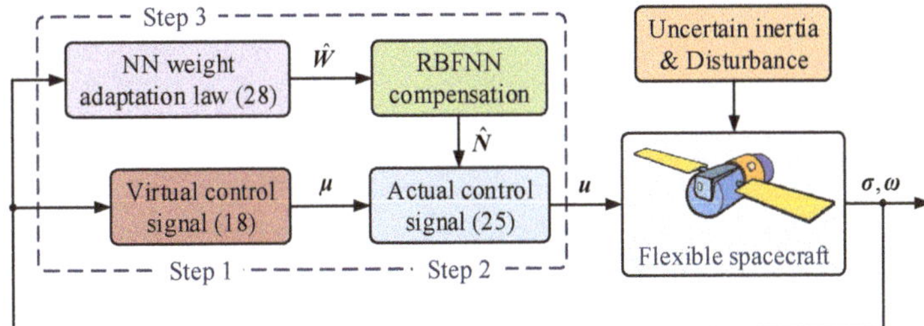

Figure 1. Diagram of the proposed neural adaptive fixed-time control approach.

4. Simulations and Comparisons

Simulations were conducted on a flexible spacecraft with two solar panels to validate the proposed control approach. Referring to [34], the inertia matrix of the flexible spacecraft was chosen as

$$J = \begin{bmatrix} 486.7 & 14.9 & -1.2 \\ 14.9 & 177.4 & -7.3 \\ -1.2 & -7.3 & 404.3 \end{bmatrix} \text{kg} \cdot \text{m}^2. \tag{48}$$

The inertia matrix was fully unknown for the control design. Moreover, the first three elastic modes were considered for the flexible spacecraft. The coupling matrix was

$$\delta = \begin{bmatrix} 1 & 0.1 & 0.1 \\ 0.5 & 0.1 & 0.01 \\ -1 & 0.3 & 0.01 \end{bmatrix} \text{kg}^{1/2} \cdot \text{m/s}^2. \tag{49}$$

The natural frequencies were chosen as $\omega_{n1} = 1.8912$, $\omega_{n2} = 2.884$, and $\omega_{n3} = 3.4181$. The damping ratios were chosen as $\xi_1 = 0.01$, $\xi_2 = 0.01$, and $\xi_3 = 0.01$. The external disturbance was

$$
d = \begin{bmatrix} 0.2\cos(0.2\pi t) - 0.1\sin(0.4\pi t) - 0.1 \\ 0.3\sin(0.2\pi t) - 0.1\cos(0.4\pi t) + 0.2 \\ 0.2\sin(0.2\pi t) - 0.2\sin(0.4\pi t) - 0.3 \end{bmatrix} \text{Nm}. \tag{50}
$$

The initial states of the flexible spacecraft were set as $\sigma(0) = [0.04, -0.06, 0.08]^{\mathrm{T}}$, $\omega(0) = [0,0,0]^{\mathrm{T}} \text{rad/s}$, $\eta(0) = [0,0,0]^{\mathrm{T}}$, and $\dot{\eta}(0) = [0,0,0]^{\mathrm{T}}$. The maximum acceptable input value was $u_m = 10 \,\text{Nm}$.

Besides the proposed neural adaptive fixed-time controller (25), the finite-time PD-like controller in [40] was also implemented for performance comparisons. Based on the homogeneous method, the compared finite-time PD-like controller was designed as

$$
u = G^{\mathrm{T}}(\sigma)\left(-k_p \mathrm{sig}^{\alpha_1}(\sigma_e) - k_d \mathrm{sig}^{\alpha_2}(\dot{\sigma}_e)\right), \tag{51}
$$

where $k_p > 0$, $k_d > 0$, $0 < \alpha_1 < 1$, and $\alpha_2 = 2\alpha_1/(1 + \alpha_1)$.

The parameters of the proposed neural adaptive fixed-time controller (25) were $k_{11} = 0.1$, $k_{12} = 0.1$, $k_{21} = 800$, $k_{22} = 800$, $p = 2/3$, $q = 4/3$, $\Gamma_i = 100\mathbf{I}_7$, and $\eta_i = 0.1$. Seven nodes were selected for the hidden layer of the RBFNN. The parameters of the RBFNN were selected as $c_i = [-3, -2, -1, 0, 1, 2, 3]^{\mathrm{T}}$ and $w_i = 6$. The initial values of the NN weight estimations were $\hat{W}_i = \mathbf{0}_7$. On the other hand, the parameters of the compared finite-time PD-like controller (51) were $k_p = 150$, $k_d = 300$, $\alpha_1 = 1/2$, and $\alpha_2 = 2/3$.

The simulation results for the proposed controller are provided in Figures 2–6. Specifically, Figures 2 and 3 show the time profiles of the attitude σ and the angular velocity ω. The time profile of the modal variables η is presented in Figure 4. Figure 5 shows the time profile of the saturated control torques u. The norms of the NN weight estimations $\|\hat{W}_i\|$ are presented in Figure 6. Moreover, the simulation results for the compared PD-like controller are provided in Figures 7–10.

As shown in Figures 2 and 3, the proposed controller guaranteed the stabilization of attitude and angular velocity to the small neighborhoods of zero rapidly and exactly. Nevertheless, Figures 7 and 8 shiw that the stabilization performance of the compared PD-like controller was relatively poor due to the presence of a lumped unknown term involving uncertain inertia, external disturbance, torque saturation, and elastic vibrations. Quantitatively, the steady-state attitude accuracy and the steady-state angular velocity accuracy under the proposed controller were $|\sigma_i| < 1 \times 10^{-4}$ and $|\omega_i| < 3 \times 10^{-4} \,\text{rad/s}$, respectively. By contrast, the steady-state attitude accuracy and the steady-state angular velocity accuracy under the compared PD-like controller were $|\sigma_i| < 3 \times 10^{-4}$ and $|\omega_i| < 8 \times 10^{-4} \,\text{rad/s}$, respectively. It was clearly seen that the proposed controller achieved a much higher control accuracy than the compared PD-like controller. In Figure 4, the elastic vibrations of the flexible appendages were damped nearly to zero within $80 \,\text{s}$ under the proposed controller. However, Figure 9 shows obvious residual vibrations of the flexible appendages under the compared PD-like controller. Figures 5 and 10 show that the control torques under both controllers always remained within the predefined saturation constraints. Moreover, in Figure 6, the NN weight estimations of the proposed controller changed with time smoothly.

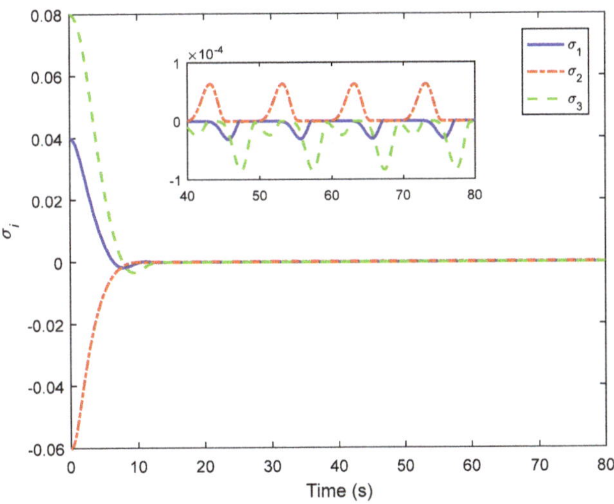

Figure 2. Time profile of the attitude tracking under the proposed controller.

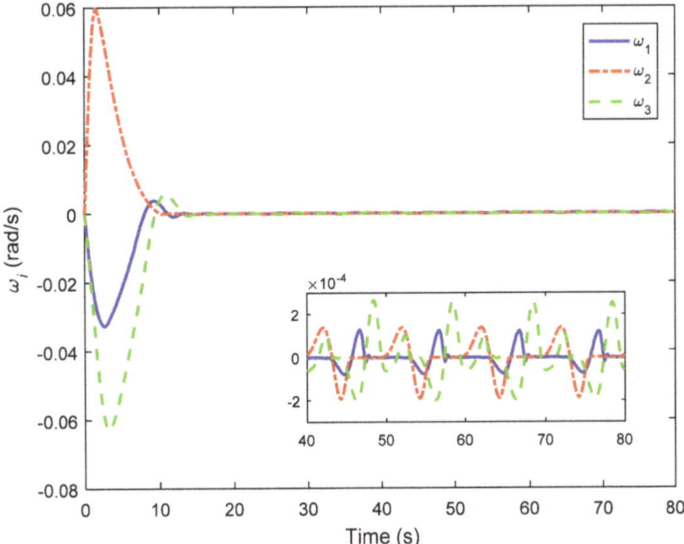

Figure 3. Time profile of the angular velocity tracking under the proposed controller.

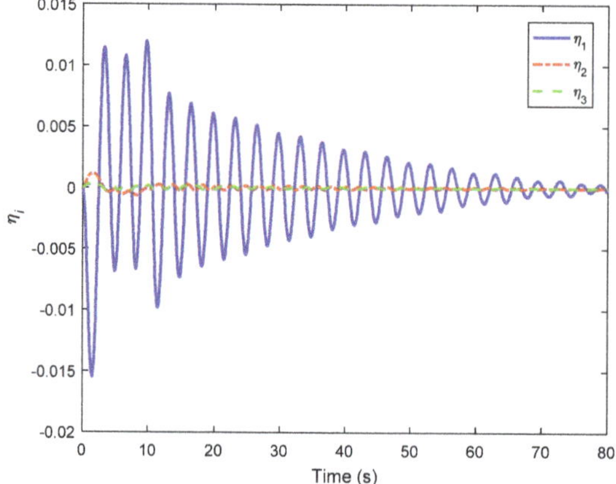

Figure 4. Time profile of the modal variables under the proposed controller.

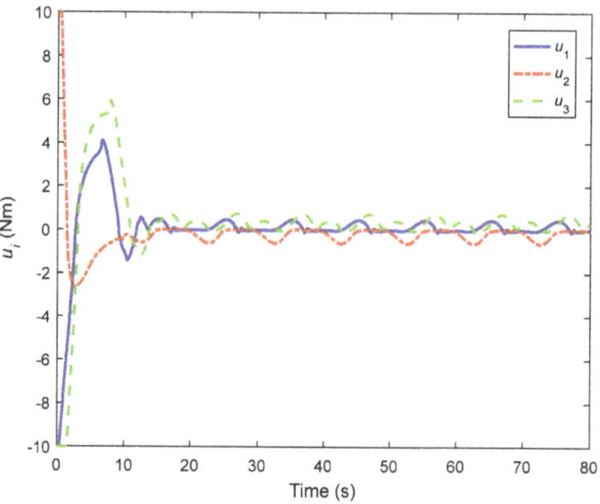

Figure 5. Time profile of the control torques under the proposed controller.

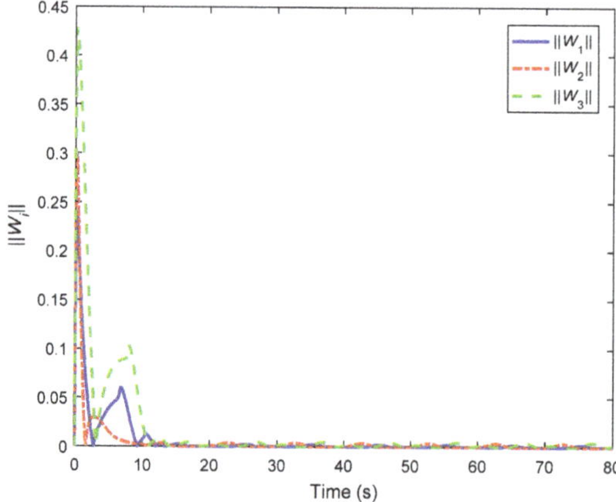

Figure 6. Norms of the NN weight estimations under the proposed controller.

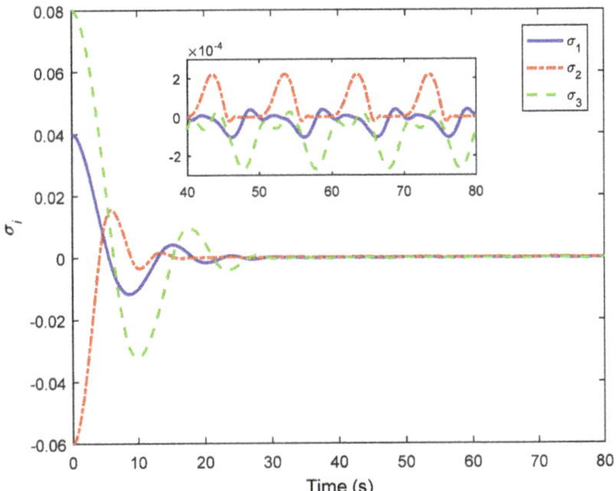

Figure 7. Time profile of the attitude tracking under the PD-like controller.

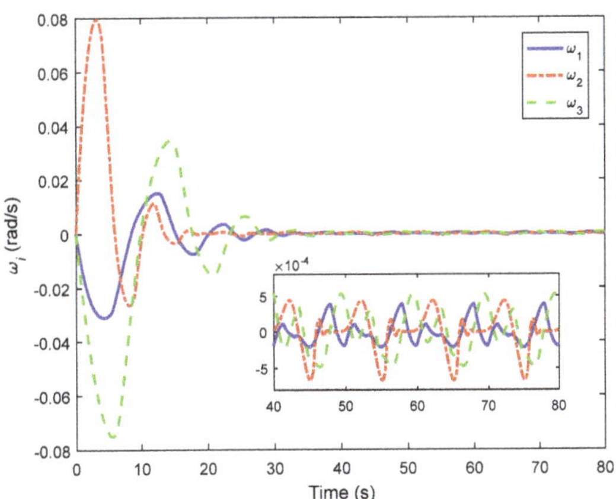

Figure 8. Time profile of the angular velocity tracking under the PD-like controller.

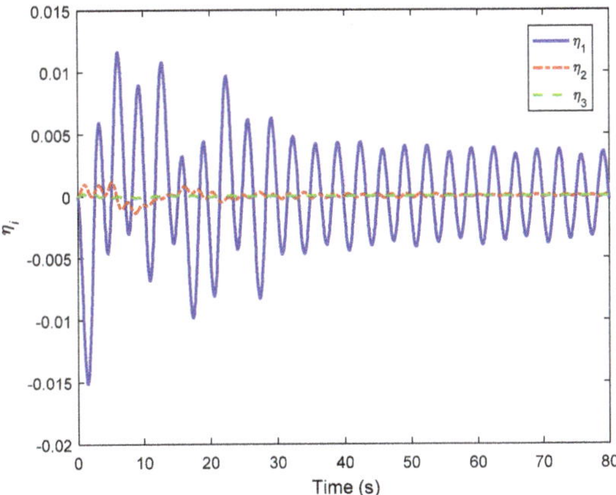

Figure 9. Time profile of the modal variables under the PD-like controller.

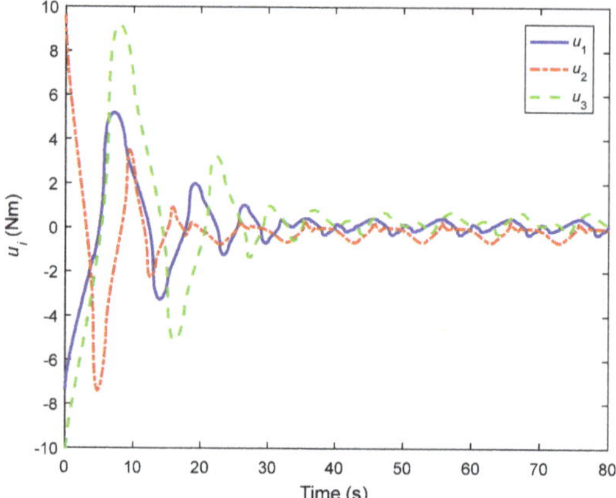

Figure 10. Time profile of the control torques under the PD-like controller.

Furthermore, some comparisons between the proposed controller and the compared PD-like controller are provided in detail in Figures 11–14. Figures 11 and 12 present the norms of the attitude σ and angular velocity ω under both controllers. Moreover, the vibration energy under both controllers are shown in Figure 13, where the vibration energy index is defined as $E_\eta = \frac{1}{2}\eta^{\mathrm{T}}\eta$. Figure 14 shows the control energy consumption under both controllers, where the control energy consumption index is defined as $E_u = \frac{1}{2}\int_0^t \|u(\tau)\|\mathrm{d}\tau$. In Figures 11–14, it is not difficult to find that the proposed controller realized attitude stabilization with higher accuracy than the compared PD-like controller, with less elastic vibration remaining and less control energy consumption. Additionally, it is obvious that the angular velocity tracking under the compared PD-like controller had a relatively large overshoot, which is unexpected in practical implementations.

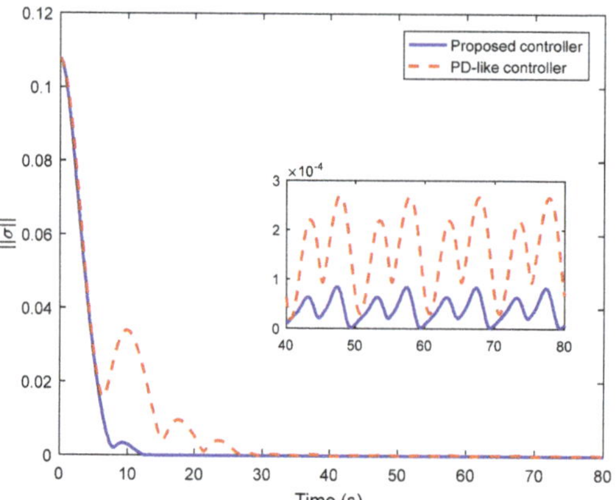

Figure 11. Norm of the attitude tracking under both controllers.

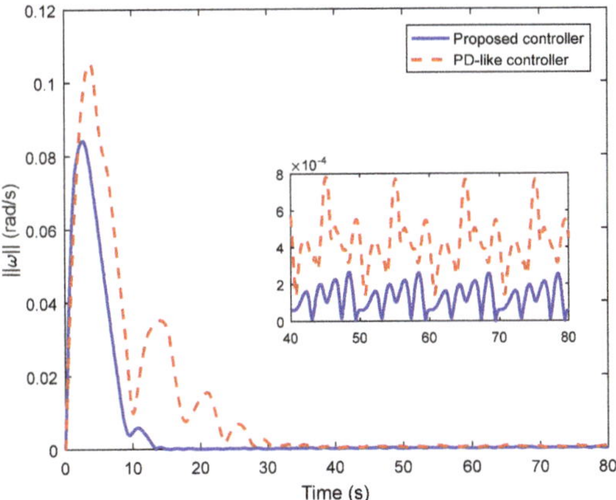

Figure 12. Norm of the angular velocity tracking under both controllers.

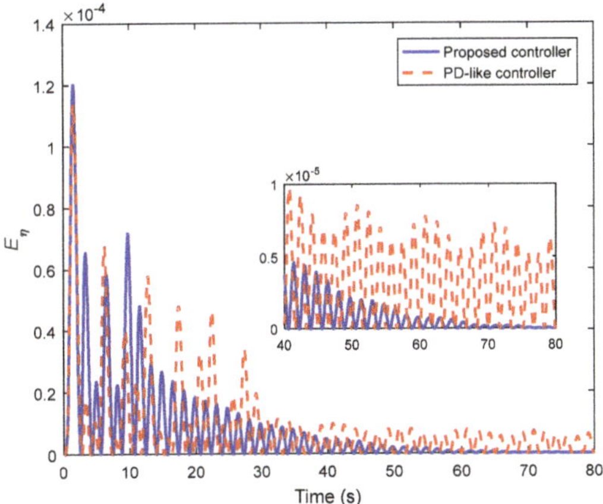

Figure 13. Vibration energy under both controllers.

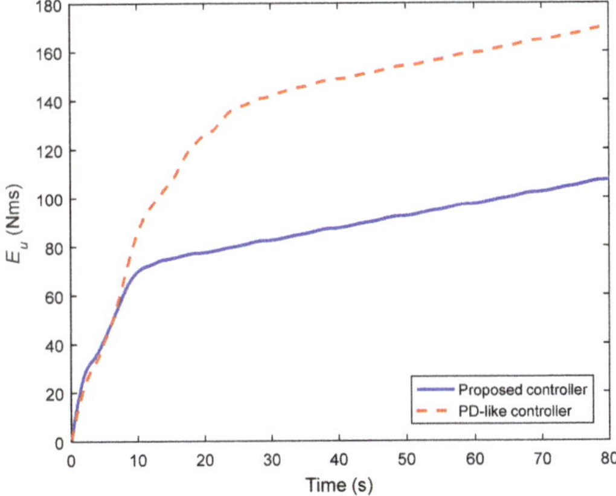

Figure 14. Control energy consumption under both controllers.

Consequently, from the simulations and detailed comparisons, the proposed controller appeared able to achieve a superior stabilization performance and better vibration suppression than the compared PD-like controller. This was mainly due to the NN compensation for the lumped unknown term. On the one hand, the robustness of the proposed controller against uncertain inertia and external disturbance was further enhanced. On the other hand, the elastic vibrations of the flexible appendages were significantly suppressed at the same time.

5. Conclusions

In this paper, a novel neural adaptive fixed-time control approach is proposed for the attitude stabilization and vibration suppression of flexible spacecraft. The NN was introduced to identify the lumped unknown term involving uncertain inertia, external disturbance, torque saturation, and elastic vibrations. After that, the proposed controller was developed by integrating with the NN compensation under the fixed-time backstepping control framework. The proposed controller guaranteed the stabilization of attitude and angular velocity to the adjustable small neighborhoods of zero in fixed time through Lyapunov analysis. It should be pointed out that the proposed controller is not only robust against uncertain inertia and external disturbance, but also insensitive to elastic vibrations of the flexible appendages. At last, the simulation results indicated that the proposed control approach was able to achieve an excellent stabilization performance and good vibration suppression.

Author Contributions: Conceptualization, Q.Y., H.J., I.M., N.D.A. and S.B.; methodology, Q.Y., H.J., I.M., N.D.A. and S.B.; formal analysis, Q.Y., H.J., I.M., N.D.A. and S.B.; writing—original draft preparation, Q.Y., H.J., I.M., N.D.A. and S.B.; writing—review and editing, Q.Y., H.J., I.M., N.D.A. and S.B.; supervision, Q.Y., H.J., I.M., N.D.A. and S.B. All authors have read and agreed to the published version of the manuscript.

Funding: The Deanship of Scientific Research (DSR) at King Abdulaziz University, Jeddah, Saudi Arabia has funded this project, under grant no. (FP-085-43).

Institutional Review Board Statement: Not applicable.

Informed Consent Statement: Not applicable.

Data Availability Statement: Not applicable.

Conflicts of Interest: The authors declare no conflict of interest.

References

1. Nagashio, T.; Kida, T.; Ohtani, T.; Hamada, Y. Design and implementation of robust symmetric attitude controller for ETS-VIII spacecraft. *Control Eng. Pract.* **2010**, *18*, 1440–1451. [CrossRef]
2. Nagashio, T.; Kida, T.; Hamada, Y.; Ohtani, T. Robust two-degrees-of-freedom attitude controller design and flight test result for engineering test satellite-VIII spacecraft. *IEEE Trans. Control Syst. Technol.* **2014**, *22*, 157–168. [CrossRef]
3. Luo, J.; Wei, C.; Dai, H.; Yin, Z.; Wei, X.; Yuan, J. Robust inertia-free attitude takeover control of postcapture combined spacecraft with guaranteed prescribed performance. *ISA Trans.* **2018**, *74*, 28–44. [CrossRef] [PubMed]
4. Gao, H.; Ma, G.; Lv, Y.; Guo, Y. Forecasting-based data-driven model-free adaptive sliding mode attitude control of combined spacecraft. *Aerosp. Sci. Technol.* **2019**, *86*, 364–374. [CrossRef]
5. Huang, X.; Duan, G. Fault-tolerant attitude tracking control of combined spacecraft with reaction wheels under prescribed performance. *ISA Trans.* **2020**, *98*, 161–172. [CrossRef] [PubMed]
6. Liu, Y.; Ma, G.; Lyu, Y.; Wang, P. Neural network-based reinforcement learning control for combined spacecraft attitude tracking maneuvers. *Neurocomputing* **2022**, *484*, 67–78. [CrossRef]
7. Modi, V.J. Attitude dynamics of satellites with flexible appendages—A brief review. *J. Spacecr. Rocket.* **1974**, *11*, 743–751. [CrossRef]
8. Likins, P. Spacecraft attitude dynamics and control—A personal perspective on early developments. *J. Guid. Control Dyn.* **1986**, *9*, 129–134. [CrossRef]
9. Li, H.; Guo, L.; Zhang, Y. An anti-disturbance PD control scheme for attitude control and stabilization of flexible spacecrafts. *Nonlinear Dyn.* **2012**, *67*, 2081–2088. [CrossRef]
10. Wang, Z.; Wu, Z.; Li, L.; Yuan, J. A composite anti-disturbance control scheme for attitude stabilization and vibration suppression of flexible spacecrafts. *J. Vib. Control* **2017**, *23*, 2470–2477. [CrossRef]

11. Sun, H.; Hou, L.; Zong, G.; Guo, L. Composite anti-disturbance attitude and vibration control for flexible spacecraft. *IET Control Theory Appl.* **2017**, *11*, 2383–2390. [CrossRef]

12. Zhu, Y.; Guo, L.; Qiao, J.; Li, W. An enhanced anti-disturbance attitude control law for flexible spacecrafts subject to multiple disturbances. *Control Eng. Pract.* **2019**, *84*, 274–283. [CrossRef]

13. Singh, S.N.; Zhang, R. Adaptive output feedback control of spacecraft with flexible appendages by modeling error compensation. *Acta Astronaut.* **2004**, *54*, 229–243. [CrossRef]

14. Maganti, G.B.; Singh, S.N. Simplified adaptive control of an orbiting flexible spacecraft. *Acta Astronaut.* **2007**, *61*, 575–589. [CrossRef]

15. Guan, P.; Liu, X.-J.; Liu, J.-Z. Adaptive fuzzy sliding mode control for flexible satellite. *Eng. Appl. Artif. Intell.* **2005**, *18*, 451–459. [CrossRef]

16. Dong, C.; Xu, L.; Chen, Y.; Wang, Q. Networked flexible spacecraft attitude maneuver based on adaptive fuzzy sliding mode control. *Acta Astronaut.* **2009**, *65*, 1561–1570. [CrossRef]

17. Hu, Q.; Xiao, B. Intelligent proportional-derivation control for flexible spacecraft attitude stabilization with unknown input saturation. *Aerosp. Sci. Technol.* **2012**, *23*, 63–74. [CrossRef]

18. Sendi, C.; Ayoubi, M.A. Robust-optimal fuzzy model-based control of flexible spacecraft with actuator constraint. *J. Dyn. Syst. Meas. Control* **2016**, *138*, 091004. [CrossRef]

19. Sendi, C. Attitude control of a flexible spacecraft via fuzzy optimal variance technique. *Mathematics* **2022**, *10*, 179. [CrossRef]

20. Di Gennaro, S. Active vibration suppression in flexible spacecraft attitude tracking. *J. Guid. Control Dyn.* **1998**, *21*, 400–408. [CrossRef]

21. Di Gennaro, S. Output stabilization of flexible spacecraft with active vibration suppression. *IEEE Trans. Aerosp. Electron. Syst.* **2003**, *39*, 747–759. [CrossRef]

22. Hu, Q.; Ma, G. Variable structure control and active vibration suppression of flexible spacecraft during attitude maneuver. *Aerosp. Sci. Technol.* **2005**, *9*, 307–317. [CrossRef]

23. Azadi, M.; Fazelzadeh, S.A.; Eghtesad, M.; Azadi, E. Vibration suppression and adaptive-robust control of a smart flexible satellite with three axes maneuvering. *Acta Astronaut.* **2011**, *69*, 307–322. [CrossRef]

24. Azadi, M.; Eghtesad, M.; Fazelzadeh, S.A.; Azadi, E. Dynamics and control of a smart flexible satellite moving in an orbit. *Multibody Syst. Dyn.* **2015**, *35*, 1–23. [CrossRef]

25. Wu, S.; Radice, G.; Sun, Z. Robust finite-time control for flexible spacecraft attitude maneuver. *J. Aerosp. Eng.* **2014**, *27*, 185–190. [CrossRef]

26. Yan, R.; Wu, Z. Finite-time attitude stabilization of flexible spacecrafts via reduced-order SMDO and NTSMC. *J. Aerosp. Eng.* **2018**, *31*, 04018023. [CrossRef]

27. Zhang, X.; Zong, Q.; Dou, L.; Tian, B.; Liu, W. Finite-time attitude maneuvering and vibration suppression of flexible spacecraft. *J. Frankl. Inst.* **2020**, *357*, 11604–11628. [CrossRef]

28. Hasan, M.N.; Haris, M.; Qin, S. Vibration suppression and fault-tolerant attitude control for flexible spacecraft with actuator faults and malalignments. *Aerosp. Sci. Technol.* **2022**, *120*, 107290. [CrossRef]

29. Polyakov, A. Nonlinear feedback design for fixed-time stabilization of linear control systems. *IEEE Trans. Autom. Control* **2012**, *57*, 2106–2110. [CrossRef]

30. Polyakov, A.; Efimov, D.; Perruquetti, W. Finite-time and fixed-time stabilization: Implicit Lyapunov function approach. *Automatica* **2015**, *51*, 332–340. [CrossRef]

31. Zuo, Z. Nonsingular fixed-time consensus tracking for second-order multi-agent networks. *Automatica* **2015**, *54*, 305–309. [CrossRef]

32. Jiang, B.; Hu, Q.; Friswell, M.I. Fixed-time attitude control for rigid spacecraft with actuator saturation and faults. *IEEE Trans. Control Syst. Technol.* **2016**, *24*, 1892–1898. [CrossRef]

33. Cao, L.; Xiao, B.; Golestani, M. Robust fixed-time attitude stabilization control of flexible spacecraft with actuator uncertainty. *Nonlinear Dyn.* **2020**, *100*, 2505–2519. [CrossRef]

34. Cao, L.; Xiao, B.; Golestani, M.; Ran, D. Faster fixed-time control of flexible spacecraft attitude stabilization. *IEEE Trans. Ind. Informat.* **2020**, *16*, 1281–1290. [CrossRef]

35. Esmaeilzadeh, S.M.; Golestani, M.; Mobayen, S. Chattering-free fault-tolerant attitude control with fast fixed-time convergence for flexible spacecraft. *Int. J. Control Autom. Syst.* **2021**, *19*, 767–776. [CrossRef]

36. Golestani, M.; Esmaeilzadeh, S.M.; Mobayen, S. Fixed-time control for high-precision attitude stabilization of flexible spacecraft. *Eur. J. Control* **2021**, *57*, 222–231. [CrossRef]

37. Spong, M.W.; Hutchinson, S.; Vidyasagar, M. *Robot Modeling and Control*; John Wiley and Sons: New York, NY, USA, 2006.

38. Sanner, R.M.; Slotine, J.-J.E. Gaussian networks for direct adaptive control. *IEEE Trans. Neural Netw.* **1992**, *3*, 837–863. [CrossRef]

39. Hardy, G.H.; Littlewood, J.E.; Pólya, G. *Inequalities*; Cambridge University Press: Cambridge, UK, 1952.

40. Du, H.; Li, S. Finite-time attitude stabilization for a spacecraft using homogeneous method. *J. Guid. Control Dyn.* **2012**, *35*, 740–748. [CrossRef]

Article

Optimal Timing Fault Tolerant Control for Switched Stochastic Systems with Switched Drift Fault

Chenglong Zhu [1,2,*], **Li He** [3], **Kanjian Zhang** [2,*], **Wei Sun** [1] **and Zengxiang He** [2]

[1] School of Mathematics Science, Liaocheng University, Liaocheng 252000, China; sunwei@lcu.edu.cn
[2] Key Laboratory of Measurement and Control of CSE Ministry of Education, School of Automation, Southeast University, Nanjing 210096, China; zxhe@seu.edu.cn
[3] New Drug Research and Development Co., Ltd., State Key Laboratory of Antibody Research & Development, NCPC, Shijiazhuang 052165, China; heli@seu.edu.cn
* Correspondence: chenglong_zhu@seu.edu.cn (C.Z.); kjzhang@seu.edu.cn (K.Z.)

Abstract: In this article, an optimal timing fault tolerant control strategy is addressed for switched stochastic systems with unknown drift fault for each switching point. The proposed controllers in existing optimal timing control schemes are not directly aimed at the switched drift fault system, which affects the optimal control performance. A cost functional with system state information and fault variable is constructed. By solving the optimal switching time criterion, the switched stochastic system can accommodate switching drift fault. The variational technique is presented for the proposed cost function in deriving the gradient formula. Then, the optimal fault tolerant switching time is calculated by combining the Armijo step-size gradient descent algorithm. Finally, the effectiveness of the proposed controller design scheme is proved by the safe trajectory planning for a four wheel drive mobile robot and numerical example.

Keywords: fault tolerant control; switching time fault; optimal timing control; switched stochastic systems; four wheel drive mobile robot

MSC: 93E20; 93C10; 93D09

Citation: Zhu, C.; He, L.; Zhang, K.; Sun, W.; He, Z. Optimal Timing Fault Tolerant Control for Switched Stochastic Systems with Switched Drift Fault. *Mathematics* **2022**, *10*, 1880. https://doi.org/10.3390/math10111880

Academic Editor: Quanxin Zhu

Received: 31 March 2022
Accepted: 27 May 2022
Published: 30 May 2022

Publisher's Note: MDPI stays neutral with regard to jurisdictional claims in published maps and institutional affiliations.

1. Introduction

The switched system is a complex kind of hybrid system, which consists of a family of subsystems and a switch rule that coordinates the sequence of the subsystems. The switch rule is triggered by switching signals [1,2]. Compared with nonswitching systems, switching systems have higher control flexibility. Switched systems with unstable subsystems can be stabilized by designing reasonable switching rules [3,4]. Switched control systems have been given considerable attention, not only to the inherent complexity, but also the wide range of practical applications. There are numerous industrial control processes that could be modeled as the switched systems, such as wind energy conversion [5], chemical reactors [6], hybrid electric vehicles [7], robot motion planning [8], etc.

For switched systems, the optimal control problems have attracted wide attention from researchers [9–11]. Different from the traditional continuous systems, the objective of switched system optimal control is to calculate the optimal switching sequence and switching rules to optimize the cost function, see [12] for a recent survey. After years of development, the optimal timing control of continuous systems has made great progress [13–16]. However, these conclusions may be infeasible when the systems are complex switched systems. For a class of autonomous systems in which the sequence of continuous dynamics is predefined, the authors of [17] proposed the optimal time switching strategy by computing the cost function and the gradient over an underlying time grid. Considering the relatively simple cost functional, the study described in [18] combined with a gradient descent algorithm gives the gradient formula for the switching time. The previous results mostly focus on the cost function containing only integral terms. Although the cost functional in [17] is relatively more general, it cannot meet the needs of some special working conditions, such as flexible satellite attitude optimization [19] or multi-agent vehicle formation planning [20]. It is necessary to study the time optimal switching problem of the generalized cost functional with the integral term and terminal term.

Since disturbance terms often exist in practical systems, it is almost impossible to construct an accurate mathematical model to describe practical switched systems [21–23]. At the same time, the stochastic disturbances lead to the stochastic characteristic of switched systems. From a practical application point of view, stochastic switched systems can model complex dynamics, uncertainty, randomness. Considering the inevitable effect of noise and stochastic disturbance, the authors of [24] investigated the time optimal switching strategy for linear stochastic switched systems. The optimal control strategy for discrete-time bilinear systems is extended to switched linear stochastic systems in [25]. For general multi-switched time-invariant stochastic systems, the authors of [26] proposed the time optimization control approach by minimizing a cost functional with different costs defined on the states. However, it is worth mentioning that the aforementioned schemes are only applicable to systems in good operating conditions (i.e., fault free). Extra efforts are needed to analyze the fault tolerant control problem for switched stochastic systems.

With the increasing demand for safety critical systems in both military and civilian applications, the performance and safety issues need to be specially considered despite the presence of faults [27]. This stimulates the research of a fault tolerant control system that can accommodate unknown system faults and maintain its prespecified performance [28–30]. In consideration of the actuator fault, the authors of [31] designed the fault tolerant controller for a class of uncertain switched nonlinear systems. The actuator saturation fault has been investigated for a class of discrete-time switched systems [32]. For the switching point perturbation, the robust optimal control of switched autonomous systems is derived in [33]. For switched parabolic systems described by partial differential equations, the boundary system fault is researched in [34]. With the above observations, the fault tolerant control for stochastic switched systems has not been well developed yet. It is a common phenomenon that the switching time fault occurs in practical switched engineering systems. The switching signals are easily subject to electromagnetic interference and unknown abrupt phenomena such as component and interconnection failures. These factors can induce the switch time to have a delay [31] and drift faults. In addition, from the optimal control point of view, the cost function may increase rapidly and serious security accidents have occurred during the control process when the switching time exceeds or lags behind the designed optimal switch time. However, as far as we know, there are few results about optimal fault tolerant control for switched stochastic systems with switched drift fault. The challenges outlined above motivate us to focus on the optimal timing fault tolerant control problem for switched stochastic systems.

The remainder of this article is arranged as follows: The problem formulation and the control objective are stated in Section 2. Section 3 presents the main results for signal switching, more switchings and optimal fault tolerant algorithm. Section 4 illustrates the obtained result applications in a four wheel drive mobile robot and numerical example. Section 5 provides some concluding remarks.

2. Problem Formulation

Consider the switched stochastic system depicted as follows:

$$\dot{x}(t) = A_i x(t) + B_i \omega(t), \quad t \in [T_{i-1}, T_i], \quad i \in [1, 2, \cdots, N+1], \tag{1}$$

where $x(t) \in \Re^n$ is the state vector, A_i and B_i are a set of given constant real matrices of appropriate dimensions. T_0 denotes the initial time, $T_1, \cdots, T_N$ ($T_0 < T_1 < \cdots < T_N < T_{N+1} = T_f$) denote the time switching signal and T_f denotes the final time. ω is the stochastic disturbance.

Initial condition $x_0 \in \Re^n$ is a stochastic vector with mean m_0 and variance matrix P_0,

$$E[x(t_0)] = m_0, \quad Var[x(t_0)] = E[(x_0 - m_0)(x_0 - m_0)^T] = P_0. \tag{2}$$

By employing the property of mathematical expectation for the stochastic initial vector, we have the mean vector $m_0 \in \Re^n$ and variance matrix $P_0 \subset \Re^n \times \Re^n$. Since the discretization of the time and dynamic input approaches can bring about computation explosions and result in inaccurate solutions, in this paper we focus on a class of switched autonomous systems. Then, the switching time signals are the system input variables.

For the stochastic disturbance, the following condition is imposed.

Assumption 1. *The stochastic disturbance ω is the zero-mean Gaussian white noise process, which is independent of $x(t_0)$. The following statistical properties are satisfied:*

$$Cov[\omega(t), \omega(\tau)] = E[\omega(t)\omega^T(\tau)] = Q_0\delta(t - \tau), \tag{3}$$
$$Cov[x(t_0), \omega(\tau)] = E[(x_0 - m_0)\omega^T(\tau)] = 0, \tag{4}$$

where $\delta(t)$ is the Dirac delta function,

$$\int_{-\infty}^{+\infty} \delta(t)dt = 1, \quad \delta(t) = \begin{cases} 0, & t \neq 0, \\ +\infty, & t = 0. \end{cases} \tag{5}$$

The normal switching time is denoted as $T = (T_1, \cdots, T_N)$. The actuator switched fault is an unpermitted deviation T_e of the designed standard switching signal input T. The unknown switched drift fault for each switching point can be described as:

$$T_\epsilon = (T_1 + \epsilon_1, \cdots, T_i + \epsilon_i \cdots, T_N + \epsilon_N), \tag{6}$$

where the ϵ_i is unknown drift parameters.

Assumption 2. *The drift fault parameters are limited to the bounded region, $-\delta_i \leq \epsilon_i \leq \delta_i$, where δ_i is a given small positive constant. For δ_i, $T_i + \delta_i \leq T_{i+1}$, the predefined triggered sequence of subsystems is continuous and there is no jump. In addition, the switched system states are continuous at the switching time which is different from the general hybrid system.*

Remark 1. *Assumption 1 is reasonable and commonly used. In fact, for the practical engineering system, the stochastic noise disturbance is generated by the equipment plant, and is independent of the initial state of the system model. An actuator switched drift fault is a common type of fault. The drift fault parameter ϵ_i is brought by electromagnetic interference, transmission delay, equipment aging and mechanical wear in modern engineering applications. It is meaningful and reasonable to limit the amplitude of the drift fault parameter ϵ_i. The subsystem triggered sequence does not jump. Assumption 2 is the foundation of fault tolerant control switch system research.*

Due to the stochastic characteristic of the system state $x(t)$, the nominal cost functional J_0 is described as:

$$J_0 = E\{\Psi(x(t_f)) + \sum_{i=0}^{N} \int_{T_i}^{T_{i+1}} L_{i+1}(x(t))dt\}, \tag{7}$$

where $L_{i+1}(x(t)) = \frac{1}{2}x^T(t)Q_i x(t)$ are the running cost functions. $\Psi(x(t_f)) = \frac{1}{2}x^T(t_f)P_T x(t_f)$ denote the terminal cost term at the final time. The coefficient matrices $P_T = P_T^T \geq 0$, $Q_i = Q_i^T \geq 0$ are the weight matrices for the present and terminal states, where $P_T \subset \Re^n \times \Re^n$ and $Q_i \subset \Re^n \times \Re^n$.

Motivated by the integral mean value theorem, a novel cost functional mechanism is investigated to achieve an appropriate compromise between drift fault compensation and the optimal process.

$$J \;=\; E\{\Psi(x(t_f)) + \frac{1}{2^N \prod_{i=1}^{N} \delta_i} \int_{-\delta_1}^{\delta_1} \cdots \int_{-\delta_N}^{\delta_N} [\int_{t_0}^{T_1+\epsilon_1} L_1(x(t))dt + \cdots$$
$$+ \int_{T_i+\epsilon_i}^{T_{i+1}+\epsilon_{i+1}} L_{i+1}(x(t))dt + \cdots + \int_{T_N+\epsilon_N}^{t_f} L_{N+1}(x(t))]d\epsilon_N \cdots d\epsilon_1\}. \tag{8}$$

Remark 2. *It is worth mentioning that the cost function (8) is the mean value of the integral over the switch fault time T_ϵ. When the drift fault parameter $\delta_i \to 0$, $\epsilon_i \to 0$, i.e., fault free, by utilizing the L'Hôpital's rule, the cost functional (8) becomes the nominal cost functional J_0. The constructed cost functional J includes system state information and a fault variable, then the optimal switching time obtained by this cost functional is a relatively accommodated switching drift fault.In addition, the proposed cost functional mixes the integral term and terminal term. Therefore, the cost functional (8) we investigate in this paper is general and powerful enough to describe many industrial process.*

Control objective: The main purpose of this paper is to deduce the gradient formula for the corresponding cost function with respect to a switched stochastic system (1). Then, under Assumptions 1 and 2, we solve the optimal switching signal criterion, such that the the proposed cost function (8) is minimized in spite of the switched drift fault (6).

3. The Main Results

In this section, we firstly take $N = 1$ as one switching time for the system. By employing the calculus of variations and some computation, the increment of the cost functional will be deduced according to the switching signal increment. Based on the gradient descent algorithm, the optimal time fault tolerant control of the switched stochastic system is proposed. Then, the multi-switchings time case can be achieved as the single switching time extension. Finally, the optimal fault tolerant algorithm is proposed with a flow chart.

3.1. Single Switching

Consider the case of a single switching for the linear switched autonomous stochastic system with switching time drift fault ϵ,

$$\dot{x}(t) = \begin{cases} A_1 x(t) + B_1 \omega, & t \in [t_0, T_1 + \epsilon], \\ A_2 x(t) + B_2 \omega, & t \in [T_1 + \epsilon, t_f]. \end{cases} \tag{9}$$

For the switching time, we take a positive variation Δt. Compared with the nominal system (9), we denote $\tilde{x}$ to represent the state trajectory of the system switching time after the increment of Δt, that is, the switching time is $T_1 + \epsilon + \Delta t$. The increment system $\tilde{x}$ is defined as:

$$\dot{\tilde{x}}(t) = \begin{cases} A_1 \tilde{x}(t) + B_1 \omega, & t \in [t_0, T_1 + \epsilon + \Delta t], \\ A_2 \tilde{x}(t) + B_2 \omega, & t \in [T_1 + \epsilon + \Delta t, t_f]. \end{cases} \tag{10}$$

A portion of the grid is presented in Figure 1 to illustrate the different switching times.

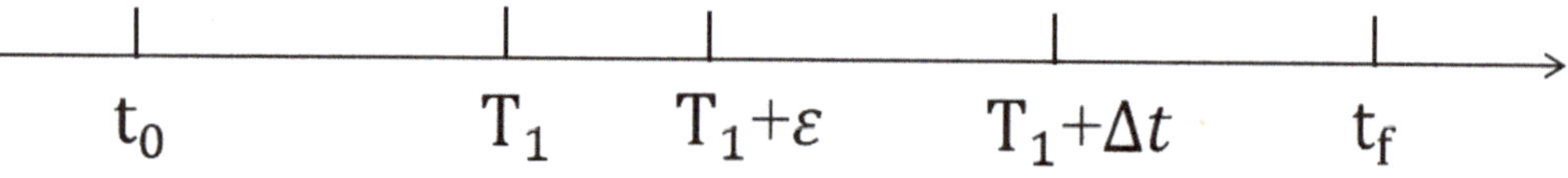

Figure 1. Switching times within the time grid.

In order to make the induced variation cost functional ΔJ clear and easy to be represented, one can consider the statistical properties of the stochastic states with the nominal system x and the increment systems $\tilde{x}$.

The second-order origin moment matrix of the system states $x(t)$ and $\tilde{x}(t)$ satisfy the following matrix differential equation:

$$\dot{m}_x(t) = \begin{cases} A_1 m_x(t) + m_x(t) A_1^T + B_1 Q_0 B_1^T, & t \in [t_0, T_1 + \epsilon], \\ A_2 m_x(t) + m_x(t) A_2^T + B_2 Q_0 B_2^T, & t \in [T_1 + \epsilon, t_f] \end{cases} \tag{11}$$

with the initial state $m_x(0) = P_0 + m_0 m_0^T$.

$$\dot{m}_{\tilde{x}}(t) = \begin{cases} A_1 m_{\tilde{x}}(t) + m_{\tilde{x}}(t) A_1^T + B_1 Q_0 B_1^T, & t \in [t_0, T_1 + \epsilon + \Delta t], \\ A_2 m_{\tilde{x}}(t) + m_{\tilde{x}}(t) A_2^T + B_2 Q_0 B_2^T, & t \in [T_1 + \epsilon + \Delta t, t_f], \end{cases} \tag{12}$$

with the same initial state $m_{\tilde{x}}(0) = P_0 + m_0 m_0^T = m_x(0)$. Then, the second-order origin moment matrix $m_x(t)$ and $m_{\tilde{x}}(t)$ have the uniform derivative equation on the interval $[t_0, T_1 + \epsilon]$.

Next, we will analyze the induced variation cost functional J. The cost functional J and $\tilde{J}$ have a main discrepancy with the nominal system state x and the increment systems state $\tilde{x}$ on the interval $[T_1 + \epsilon, T_1 + \epsilon + \Delta t]$. We subdivide the time interval according to the background grid points falling between t_0 and t_f, after the switching time $T_1 + \epsilon + \Delta t$. The nominal cost functional J can be described as

$$\begin{aligned} J &= E\{\Psi(x(t_f)) + \frac{1}{2\delta} \int_{-\delta}^{\delta} [\int_{t_0}^{T_1+\epsilon} L_1 dt + \int_{T_1+\epsilon}^{t_f} L_2 dt] d\epsilon\} \\ &= E\{\Psi(x(t_f)) + \frac{1}{2\delta} \int_{-\delta}^{\delta} [\int_{t_0}^{T_1+\epsilon} L_1 dt + \int_{T_1+\epsilon}^{T_1+\epsilon+\Delta t} L_2 dt + \int_{T_1+\epsilon+\Delta t}^{t_f} L_2 dt] d\epsilon\} \\ &\doteq J_0 + J_1 + J_2 + J_3. \end{aligned} \tag{13}$$

The increment cost functional $\tilde{J}$ can be described as

$$\begin{aligned} \tilde{J} &= E\{\Psi(\tilde{x}(t_f)) + \frac{1}{2\delta} \int_{-\delta}^{\delta} [\int_{t_0}^{T_1+\epsilon+\Delta t} L_1 dt + \int_{T_1+\epsilon+\Delta t}^{t_f} L_2 dt] d\epsilon\} \\ &= E\{\Psi(\tilde{x}(t_f)) + \frac{1}{2\delta} \int_{-\delta}^{\delta} [\int_{t_0}^{T_1+\epsilon} L_1 dt + \int_{T_1+\epsilon}^{T_1+\epsilon+\Delta t} L_1 dt + \int_{T_1+\epsilon+\Delta t}^{t_f} L_2 dt] d\epsilon\} \\ &\doteq \tilde{J}_0 + \tilde{J}_1 + \tilde{J}_2 + \tilde{J}_3. \end{aligned} \tag{14}$$

The major results in this paper are briefly summarized as the following theorem:

Theorem 1. *For the linear switched autonomous stochastic system (9) with the single switching time T_1 and the unknown switching drift fault ϵ, if the system stochastic disturbance satisfies Assumption 1 and the drift fault parameter satisfies Assumption 2, we design the general cost functional J, as presented in Equation (13). Then, the derivative dJ/dT_1 of the cost function J with respect to the switching time T_1 has the following form:*

$$\begin{aligned} \frac{dJ}{dT_1} &= \frac{1}{4\delta} \int_{-\delta}^{\delta} tr(m_x(T_1 + \epsilon)(Q_1 - Q_2)) d\epsilon \\ &\quad + \frac{1}{4\delta} \int_{-\delta}^{\delta} \int_{T_1+\epsilon}^{t_f} tr(e^{A_2(t-T_1-\epsilon)} M_1 e^{A_2^T(t-T_1-\epsilon)} Q_2) dt d\epsilon \\ &\quad + \frac{1}{2} tr((e^{A_2(t_f-T_1-\epsilon)} M_1 e^{A_2^T(t_f-T_1-\epsilon)}) P_T), \end{aligned} \tag{15}$$

where $M_1 = (A_1 - A_2)m_x(T_1 + \epsilon) + m_x(T_1 + \epsilon)(A_1^T - A_2^T) + B_1 Q_0 B_1^T - B_2 Q_0 B_2^T$, and $m_x(T_1 + \epsilon)$ takes the value of the following matrix differential equation at $t = T_1 + \epsilon$:

$$
\begin{aligned}
\dot{m}_x(t) &= A_1 m_x(t) + m_x(t)A_1^T + B_1 Q_0 B_1^T, \\
m_x(0) &= P_0 + m_0 m_0^T.
\end{aligned}
\tag{16}
$$

The cost function has the fault tolerant performance for the switching time fault.

Proof. According to the division of the time interval in Figure 1, through the following four steps, we complete the proof of the theorem.

Step 1. On the interval $t \in [t_0, T_1 + \epsilon]$, the systems (9) and (10) can be redescribed as

$$
\begin{aligned}
\dot{x}(t) &= A_1 x(t) + B_1 \omega, & t &\in [t_0, T_1 + \epsilon], \\
\dot{\tilde{x}}(t) &= A_1 \tilde{x}(t) + B_1 \omega, & t &\in [t_0, T_1 + \epsilon].
\end{aligned}
\tag{17}
\tag{18}
$$

The induced variation in the cost functional J and $\tilde{J}$,

$$
\begin{aligned}
\tilde{J}_1 - J_1 &= E\{\frac{1}{2\delta}\int_{-\delta}^{\delta}\int_{t_0}^{T_1+\epsilon} L_1 dt d\epsilon\} - E\{\frac{1}{2\delta}\int_{-\delta}^{\delta}\int_{t_0}^{T_1+\epsilon} L_1 dt d\epsilon\} \\
&= \frac{1}{2\delta}\int_{-\delta}^{\delta}\int_{t_0}^{T_1+\epsilon} E(L_1(\tilde{x}) - L_1(x)) dt d\epsilon \\
&= \frac{1}{2\delta}\int_{-\delta}^{\delta}\int_{t_0}^{T_1+\epsilon} \frac{1}{2}E(\tilde{x}^T(t)Q_1\tilde{x}(t) - x^T(t)Q_1 x(t)) dt d\epsilon.
\end{aligned}
\tag{19}
$$

Owing to the diagonal properties of weight matrices Q_1, we obtain

$$
\begin{aligned}
E(x^T(t)Q_1 x(t)) &= E(tr(x^T(t)Q_1 x(t))) = E(tr(x(t)x^T(t)Q_1)) \\
&= tr(E(x(t)x^T(t))Q_1) = tr(m_x(t)Q_1).
\end{aligned}
\tag{20}
$$

Under the same initial condition $x(0) = \tilde{x}(0)$, combining with (11), (12), (17) and (18), we can conclude that

$$
m_x(t) = m_{\tilde{x}}(t), \quad t \in [t_0, T_1 + \epsilon].
\tag{21}
$$

Then, Equation (20) is converted into

$$
E(x^T(t)Q_1 x(t)) = tr(m_x(t)Q_1) = tr(m_{\tilde{x}}(t)Q_1) = E(\tilde{x}^T(t)Q_1 x(t)).
\tag{22}
$$

Combining the above equation with (19), the following equation can be obtained:

$$
\tilde{J}_1 - J_1 = \frac{1}{2\delta}\int_{-\delta}^{\delta}\int_{t_0}^{T_1+\epsilon} \frac{1}{2}E(\tilde{x}^T(t)Q_1\tilde{x}(t) - x^T(t)Q_1 x(t)) dt d\epsilon = 0.
\tag{23}
$$

Step 2. On the interval $t \in [T_1 + \epsilon, T_1 + \epsilon + \Delta t]$, the systems in (9) and (10) are described as

$$
\begin{aligned}
\dot{x}(t) &= A_2 x(t) + B_2 \omega, & t &\in [T_1 + \epsilon, T_1 + \epsilon + \Delta t], \\
\dot{\tilde{x}}(t) &= A_1 \tilde{x}(t) + B_1 \omega, & t &\in [T_1 + \epsilon, T_1 + \epsilon + \Delta t].
\end{aligned}
\tag{24}
\tag{25}
$$

The increment of the cost function is

$$
\begin{aligned}
\tilde{J}_2 - J_2 &= \frac{1}{2\delta} \int_{-\delta}^{\delta} \int_{T_1+\epsilon}^{T_1+\epsilon+\Delta t} E(L_2(\tilde{x}) - L_1(x)) \, dt \, d\epsilon \\
&= \frac{1}{4\delta} \int_{-\delta}^{\delta} \int_{T_1+\epsilon}^{T_1+\epsilon+\Delta t} E(\tilde{x}^T(t) Q_2 \tilde{x}(t) - x^T(t) Q_1 x(t)) \, dt \, d\epsilon \\
&= \frac{1}{4\delta} \int_{-\delta}^{\delta} \int_{T_1+\epsilon}^{T_1+\epsilon+\Delta t} tr(m_{\tilde{x}} Q_1 - m_x Q_2) \, dt \, d\epsilon.
\end{aligned}
\tag{26}
$$

Consider the second-order origin moment matrix $m_x(t)$, $m_{\tilde{x}}(t)$ and Equations (11) and (12). By applying Taylor expansion, $m_x(t)$ and $m_{\tilde{x}}(t)$ at $T_1 + \epsilon$ can be calculated as:

$$
\begin{aligned}
m_x(t) &= m_x(T_1+\epsilon) + (A_2 m_x(T_1+\epsilon) + m_x(T_1+\epsilon) A_2^T \\
&\quad + B_2 Q_0 B_2^T)(t - T_1 - \epsilon) + o(t - T_1 - \epsilon) \\
&= m_x(T_1+\epsilon) + m_1(t - T_1 - \epsilon) + o(t - T_1 - \epsilon), \\
m_{\tilde{x}}(t) &= m_{\tilde{x}}(T_1+\epsilon) + (A_1 m_{\tilde{x}}(T_1+\epsilon) + m_{\tilde{x}}(T_1+\epsilon) A_1^T \\
&\quad + B_1 Q_0 B_1^T)(t - T_1 - \epsilon) + o(t - T_1 - \epsilon) \\
&= m_{\tilde{x}}(T_1+\epsilon) + \tilde{m}_1(t - T_1 - \epsilon) + o(t - T_1 - \epsilon).
\end{aligned}
\tag{27, 28}
$$

It can be seen that $m_x(T_1+\epsilon) = m_{\tilde{x}}(T_1+\epsilon)$ from (11) and (12). Note that at $t = T_1 + \epsilon + \Delta t$, the $m_x(t)$ is not equal to $m_{\tilde{x}}(t)$, then, we have

$$
\begin{aligned}
tr(m_{\tilde{x}} Q_1 - m_x Q_2) &= tr(m_x(T_1+\epsilon)(Q_1 - Q_2) + o(t - T_1 - \epsilon) \\
&\quad + (\tilde{m}_1 Q_1 - m_1 Q_2)(t - T_1 - \epsilon)).
\end{aligned}
\tag{29}
$$

Substituting the above equation into (26), one has

$$
\begin{aligned}
\tilde{J}_2 - J_2 &= \frac{1}{4\delta} \int_{-\delta}^{\delta} tr(m_x(T_1+\epsilon)(Q_1 - Q_2) \Delta t \\
&\quad + \int_{T_1+\epsilon}^{T_1+\epsilon+\Delta t} (tr(M_{11})(t - T_1 - \epsilon) + o(t - T_1 - \epsilon)) \, dt \, d\epsilon \\
&= \frac{1}{4\delta} \int_{-\delta}^{\delta} tr(m_x(T_1+\epsilon)(Q_1 - Q_2)) \Delta t + \frac{1}{2} tr(M_{11}) \Delta t^2 + o(\Delta t) \, d\epsilon \\
&= \frac{\Delta t}{4\delta} \int_{-\delta}^{\delta} tr(m_x(T_1+\epsilon)(Q_1 - Q_2)) \, d\epsilon + o(\Delta t).
\end{aligned}
\tag{30}
$$

By dividing Δt on both sides of the above equation and taking the limit operation $\Delta t \to 0$, one has

$$
\lim_{\Delta t \to 0} \frac{\tilde{J}_2 - J_2}{\Delta t} = \frac{1}{4\delta} \int_{-\delta}^{\delta} tr(m_x(T_1+\epsilon)(Q_1 - Q_2)) \, d\epsilon.
\tag{31}
$$

Step 3. On the interval $t \in [T_1 + \epsilon + \Delta t, t_f]$, the systems can be represented as

$$
\begin{aligned}
\dot{x}(t) &= A_2 x(t) + B_2 \omega, & t \in [T_1 + \epsilon + \Delta t, t_f], \\
\dot{\tilde{x}}(t) &= A_2 \tilde{x}(t) + B_2 \omega, & t \in [T_1 + \epsilon + \Delta t, t_f].
\end{aligned}
\tag{32, 33}
$$

The increment of the cost function is

$$
\begin{aligned}
\tilde{J}_3 - J_3 &= \frac{1}{2\delta} \int_{-\delta}^{\delta} \int_{T_1+\epsilon+\Delta t}^{t_f} E(L_2(\tilde{x}) - L_2(x)) dt d\epsilon \\
&= \frac{1}{4\delta} \int_{-\delta}^{\delta} \int_{T_1+\epsilon+\Delta t}^{t_f} E(\tilde{x}^T(t) Q_2 \tilde{x}(t) - x^T(t) Q_2 x(t)) dt d\epsilon \\
&= \frac{1}{4\delta} \int_{-\delta}^{\delta} \int_{T_1+\epsilon+\Delta t}^{t_f} tr((m_{\tilde{x}} - m_x) Q_2) dt d\epsilon.
\end{aligned}
\tag{34}
$$

Recalling the Taylor expansion at $T_1 + \epsilon$ for the $m_x(t)$ and $m_{\tilde{x}}(t)$,

$$
\begin{aligned}
m_{\tilde{x}}(T_1+\epsilon+\Delta t) - m_x(T_1+\epsilon+\Delta t) &= (\tilde{m}_1 - m_1)\Delta t + o(\Delta t) \\
&\doteq M_1 \Delta t + o(\Delta t).
\end{aligned}
\tag{35}
$$

By applying Taylor series expansion at $T_1 + \epsilon + \Delta t$, the $m_x(t)$ and $m_{\tilde{x}}(t)$ can be described as

$$
\begin{aligned}
m_x(t) &= m_x(T_1+\epsilon+\Delta t) + \dot{m}_x(T_1+\epsilon+\Delta t)(t - T_1 - \epsilon - \Delta t) \\
&\quad + \cdots + m_x^{(n)}(T_1+\epsilon+\Delta t) \frac{(t - T_1 - \epsilon - \Delta t)^n}{n!}, \\
m_{\tilde{x}}(t) &= m_{\tilde{x}}(T_1+\epsilon+\Delta t) + \dot{m}_{\tilde{x}}(T_1+\epsilon+\Delta t)(t - T_1 - \epsilon - \Delta t) \\
&\quad + \cdots + m_{\tilde{x}}^{(n)}(T_1+\epsilon+\Delta t) \frac{(t - T_1 - \epsilon - \Delta t)^n}{n!}.
\end{aligned}
\tag{36, 37}
$$

By employing the mathematical calculations, we have

$$
m_{\tilde{x}}(t) - m_x(t) = e^{A_2(t-T_1-\epsilon-\Delta t)} M_1 e^{A_2^T(t-T_1-\epsilon-\Delta t)} \Delta t + o(\Delta t).
\tag{38}
$$

Substituting the above equation into (34), and dividing it by Δt and taking the $\lim \Delta t \to 0$, we obtain

$$
\begin{aligned}
\lim_{\Delta t \to 0} \frac{\tilde{J}_3 - J_3}{\Delta t} &= \lim_{\Delta t \to 0} \frac{1}{4\delta \Delta t} \int_{-\delta}^{\delta} \int_{T_1+\epsilon+\Delta t}^{t_f} tr((m_{\tilde{x}} - m_x) Q_2) dt d\epsilon \\
&= \lim_{\Delta t \to 0} \frac{1}{4\delta \Delta t} \int_{-\delta}^{\delta} \int_{T_1+\epsilon+\Delta t}^{t_f} tr(e^{A_2(t-T_1-\epsilon-\Delta t)} M_{12} e^{A_2^T(t-T_1-\epsilon-\Delta t)} Q_2 \Delta t) dt d\epsilon \\
&= \frac{1}{4\delta} \int_{-\delta}^{\delta} \int_{T_1+\epsilon}^{t_f} tr(e^{A_2(t-T_1-\epsilon)} M_1 e^{A_2^T(t-T_1-\epsilon)} Q_2) dt d\epsilon.
\end{aligned}
\tag{39}
$$

Step 4. For $t = t_f$, we analyze the difference of terminal cost item of the cost functional,

$$
\begin{aligned}
\tilde{J}_0 - J_0 &= E\{\Psi(\tilde{x}(t_f))\} - E\{\Psi(x(t_f))\} \\
&= E\{\frac{1}{2}\tilde{x}^T(t_f) P_T \tilde{x}(t_f) - \frac{1}{2}x^T(t_f) P_T x(t_f)\} \\
&= \frac{1}{2} tr((m_{\tilde{x}}(t_f) - m_x(t_f)) P_T).
\end{aligned}
\tag{40}
$$

Recalling the Taylor expansion at $T_1 + \epsilon + \Delta t$ for the $m_x(t)$ and $m_{\tilde{x}}(t)$, we have

$$
m_{\tilde{x}}(t_f) - m_x(t_f) = e^{A_2(t_f-T_1-\epsilon-\Delta t)} M_1 e^{A_2^T(t_f-T_1-\epsilon-\Delta t)} \Delta t + o(\Delta t).
\tag{41}
$$

Substituting the above equation into (40), and dividing it by Δt and taking the $\lim \Delta t \to 0$, we obtain

$$
\begin{aligned}
\lim_{\Delta t \to 0} \frac{\check{J}_0 - J_0}{\Delta t} &= \lim_{\Delta t \to 0} \frac{1}{2\Delta t} tr((m_{\check{x}}(t_f) - m_x(t_f))P_T) \\
&= \lim_{\Delta t \to 0} \frac{1}{2\Delta t} tr((e^{A_2(t_f - T_1 - \epsilon - \Delta t)} M_1 e^{A_2^T(t_f - T_1 - \epsilon - \Delta t)} \Delta t + o(\Delta t))P_T) \\
&= \frac{1}{2} tr((e^{A_2(t_f - T_1 - \epsilon)} M_1 e^{A_2^T(t_f - T_1 - \epsilon)})P_T).
\end{aligned}
\tag{42}
$$

Combining the above four steps, we can complete the proof. $\square$

3.2. Multi-Switchings

In this subsection, we consider the case of more switchings ($N > 1$). Recall that the switched stochastic systems (1) have $N + 1$ linear time-invariant autonomous stochastic subsystems and the cost function (8) in Section 2. The major results in this paper with more switchings are briefly summarized as the following theorem:

Theorem 2. *For the linear switched autonomous stochastic system (1) with the multi-switching time T and the unknown switching drift fault ϵ, if the system stochastic disturbance satisfies Assumption 1 and the drift fault parameter satisfies Assumption 2, we design the general cost functional J, as presented in Equation (8). Then, the partial derivatives $\partial J(T)/\partial T_i$ ($i = 1, \ldots, N$) with respect to the ith switching time have the following form:*

$$
\begin{aligned}
\frac{\partial J}{\partial T_i} &= \frac{1}{2^{N+1} \prod_{i=1}^{N} \delta_i} \int_{-\delta_1}^{\delta_1} \cdots \int_{-\delta_N}^{\delta_N} \left(tr\left(e^{A_{N+1}(t_f - T_N - \epsilon_N)} \Gamma_{jN} e^{A_{N+1}^T(t_f - T_N - \epsilon_N)} P_T \right) \right. \\
&\quad + \sum_{i=j}^{N} \int_{T_i + \epsilon_i}^{T_{i+1} + \epsilon_{i+1}} tr\left(e^{A_{i+1}(t - T_i - \epsilon_i)} \Gamma_{ji} e^{A_{i+1}^T(t - T_i - \epsilon_i)} Q_{i+1} \right) dt \\
&\quad \left. + tr\left(m_x(T_j + \epsilon_j)(Q_j - Q_{j+1})\right) \right) d\epsilon_N \cdots d\epsilon_1,
\end{aligned}
\tag{43}
$$

where the symbol $tr(\cdot)$ is defined as the trace function

$$
\begin{aligned}
\Gamma_{jj} &= M_j, \quad j = 1, \ldots, N, \\
\Gamma_{ji} &= e^{A_i(T_i - T_{i-1})} \Gamma_{j,i-1} e^{A_i^T(T_i - T_{i-1})}, \quad i = j+1, \ldots, N, \\
M_j &= (A_j - A_{j+1}) m_x(T_j + \epsilon_j) + m_x(T_j + \epsilon_j)\left(A_j^T - A_{j+1}^T \right) + B_j Q_0 B_j^T - B_{j+1} Q_0 B_{j+1}^T.
\end{aligned}
$$

The second-order origin moment matrix $m_x(t)$ satisfies the following matrix differential equation:

$$
\dot{m}_x(t) = \begin{cases} A_i m_x(t) + m_x(t) A_i^T + B_i Q_0 B_i^T, & t \in (T_{i-1}, T_i], i = 1, \ldots, N, \\ A_{N+1} m_x(t) + m_x(t) A_{N+1}^T + B_{N+1} Q_0 B_{N+1}^T, & t \in \left(T_N, t_f \right) \end{cases}
$$

$$
m_x(t_0) = P_0 + m_0 m_0^T.
$$

The cost function has the fault tolerant performance for the switching time fault.

3.3. Optimal Fault Tolerant Algorithm

After taking into account the gradient of the cost functional in the above theorems, the next problem is to calculate the optimal switching time. In this subsection, the steepest descent algorithm with Armijo step sizes is explained in Figure 2. By denoting the initial parameters $\alpha \in (0,1)$, $\beta \in (0,1)$ and $\lambda(k) := \beta^{i(k)}$, the step size can be designed as $i(k) = \min\{i \geq 0 : J(\tau(k) - \beta^i DJ(\tau(k))) - J(\tau(k)) \leq -\alpha \beta^i \|DJ(\tau(k))\|^2\}$.

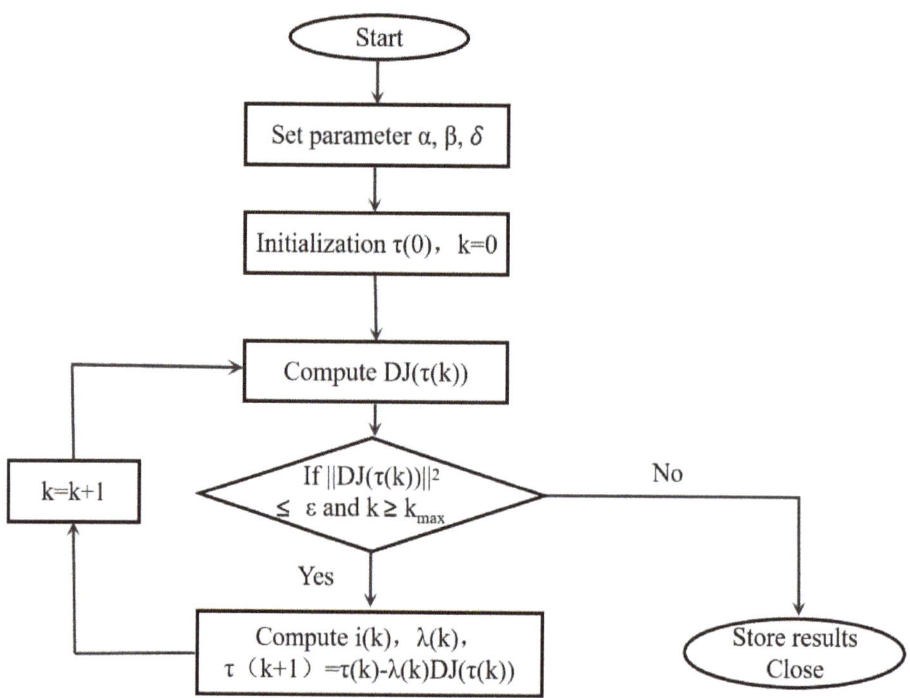

Figure 2. The steepest descent algorithm flow chart.

4. Simulation

In this section, the four wheel drive autonomous mobile robot system and numerical example are proposed to prove the feasibility of the designed optimization fault tolerant algorithm. The dynamic model of the four wheel drive mobile robot system represented in reference [35] is subject to actuator faults. It is shown that even with external stochastic disturbance and unknown switch draft fault in the actuator switched mechanism, the proposed optimization fault tolerant algorithm can explain the safety switch control of the different trajectory tasks for the autonomous mobile robot.

4.1. Practical Example

In consideration of the external stochastic disturbance, we select the lateral velocity and yaw angle of the center of gravity as the state variables. The kinematic model of the simplified four wheel drive mobile robot as shown in [35] is

$$a_x = \frac{dV_x}{dt} - V_y\frac{d\theta}{dt} = \dot{V}_x - V_y\Omega_z,$$

$$a_y = \frac{dV_y}{dt} + V_x\frac{d\theta}{dt} = \dot{V}_y + V_x\Omega_z, \tag{44}$$

where a_x is the longitudinal acceleration, a_y is the lateral acceleration, V_x and V_y are the forward velocity and lateral velocity of vehicle mass center, respectively, Ω_z is the yaw motion around the Z axis. The mobile robot vehicle dynamics equation is as follows:

$$Ma_x = M\left(\dot{V}_x - V_y\Omega_z\right) = F_{xf}\cos\delta_f + F_{xr} - F_{yf}\sin\delta_f,$$

$$Ma_y = M\left(\dot{V}_y + V_x\Omega_z\right) = F_{yf}\cos\delta_f + F_{yr} + F_{xf}\sin\delta_f,$$

$$I_z\dot{\Omega}_z = l_1 F_{yf}\cos\delta_f - l_2 F_{yr} + l_1 F_{xf}\sin\delta_f, \tag{45}$$

where δ_f is the front wheel angle, F_{xf} and F_{yf} are the longitudinal and lateral forces of the front wheel, respectively. F_{xr} and F_{yr} are the longitudinal and lateral forces of the rear wheel, respectively. l_1 is the distance from the center of mass to the front axis, and l_2 is the distance from the center of mass to the rear axis. In consideration of the lateral characteristics of the tire, we have

$$
\begin{aligned}
F_{yf} &= C_f \alpha_f, \\
F_{yr} &= C_r \alpha_r,
\end{aligned}
\tag{46}
$$

where

$$
\alpha_f = \delta_f - \frac{l_1 \Omega_z + V_y}{V_x}, \qquad \alpha_r = \frac{l_2 \Omega_z - V_y}{V_x}.
$$

By substituting the kinematic model and the tire characteristics into the vehicle dynamics equation, we can obtain

$$
\begin{aligned}
\dot{V}_y &= -\frac{1}{M}\frac{\left(C_f + C_r\right)}{V_x} V_y - \left(V_x + \frac{l_1 C_f - l_2 C_r}{M V_x}\right)\Omega_z + \frac{C_f}{M}\delta_f, \\
\dot{\Omega}_z &= -\frac{l_1 C_f - l_2 C_r}{I_z V_x} V_y - \frac{l_1^2 C_f + l_2^2 C_r}{I_z V_x}\dot{\Omega}_z + \frac{l_1 C_f}{I_z}\delta_f.
\end{aligned}
\tag{47}
$$

The forward velocity of the mobile robot along the X axis is considered constant. Then, the car has only two degrees of freedom. In order to simplify the expressions, we introduce the change in coordinates:

$$
\begin{aligned}
a_{11} &= -\frac{1}{M}\frac{\left(C_f + C_r\right)}{V_x}, \qquad a_{12} = -\left(V_x + \frac{l_1 C_f - l_2 C_r}{M V_x}\right), \qquad b_1 = \frac{C_f}{M}, \\
a_{21} &= -\frac{l_1 C_f - l_2 C_r}{I_z V_x}, \qquad a_{22} = -\frac{l_1^2 C_f + l_2^2 C_r}{I_z V_x}, \qquad b_2 = \frac{l_1 C_f}{I_z}, \\
x_1 &= V_y, \; x_2 = \Omega_z.
\end{aligned}
\tag{48}
$$

By employing the external stochastic disturbance on the front wheel angle δ_f, we select the coupling friction coefficients $b_1 = 0$, $b_2 = 1$ and $b_1 = 1$, $b_2 = 0$ to represent the the switched stochastic systems term $B_i \omega$. The forward velocity of the mobile robot along the X axis is considered constant. Thus, the four wheel drive mobile robot system has only two state variables, $x_1 = V_y$, $x_2 = \Omega_z$. In complex road conditions, the friction coefficient of tires is different. In addition, we can note that the different trajectory tasks require a different forward velocity V_x. Therefore, by different trajectory tasks, under the complex road conditions and external stochastic disturbance, the following switched stochastic systems equation is obtained for a four wheel drive mobile robot with safe trajectory planning:

$$
\dot{x}(t) = \begin{cases} A_1 x(t) + B_1 \omega, & t \in [t_0, T_1], \\ A_2 x(t) + B_2 \omega, & t \in (T_1, t_f], \end{cases}
\tag{49}
$$

where the system matrices are

$$
A_1 = \begin{bmatrix} -1 & 0 \\ 1 & 2 \end{bmatrix}, \; B_1 = \begin{bmatrix} 0 \\ 1 \end{bmatrix}, \; A_2 = \begin{bmatrix} 1 & 1 \\ 0 & -2 \end{bmatrix}, \; B_2 = \begin{bmatrix} 1 \\ 0 \end{bmatrix}.
$$

As presented in the four wheel drive mobile robot example, the robot safe trajectory planning problems can be translated into the studied switched stochastic systems. Then, the proposed optimal timing fault tolerant control strategy can solve the safe trajectory planning problem effectively.

In order to illustrate the effectiveness of the proposed algorithm with multi-switching times, the system is repeatedly switched. The system is described by three switching points, as follows:

$$
\dot{x}(t) = \begin{cases}
A_1 x(t) + B_1 \omega, & t \in [t_0, T_1], \\
A_2 x(t) + B_2 \omega, & t \in (T_1, T_2], \\
A_1 x(t) + B_1 \omega, & t \in (T_2, T_3], \\
A_2 x(t) + B_2 \omega, & t \in (T_3, t_f],
\end{cases}
\tag{50}
$$

where the initial state $x(0) = [1,0]^T$, the initial time $t_0 = 0$, the final time $t_f = 1$, the initial switching time $T_1 = 0.3$, $T_2 = 0.5$, $T_3 = 0.7$. By the switch control mechanism, the four wheel drive mobile robot system executes the desired different trajectory tasks. We need to calculate the optimal switching time T_1, T_2, T_3 to minimize the cost functional J. The weight coefficient matrices are designed as the unit matrix. The steepest descent parameters are $\alpha = \beta = 0.5$, the threshold value $\epsilon = 0.05$, $k_{max} = 200$. The experiments are implemented with Matlab2015a on a desktop PC with i7-6700 3.4 GHz CPU, 16 GB memory and Windows 1064 bit OS. The simulation results are described in Figures 3 and 4.

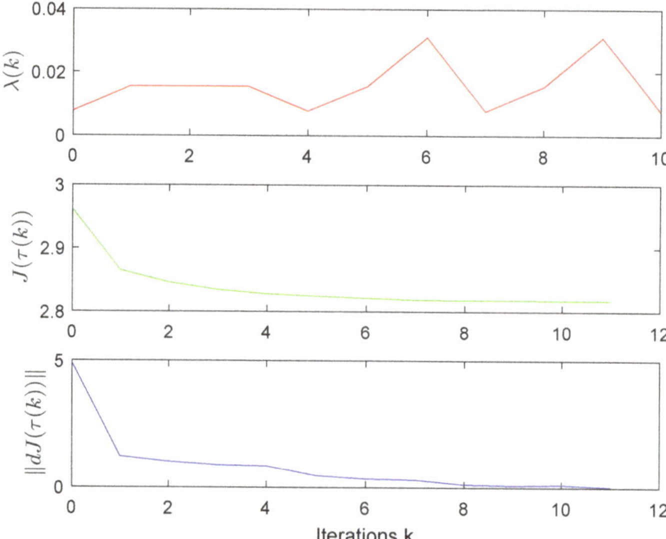

Figure 3. The designed step size $\lambda(k)$, the cost functional $J(\tau(k))$ and gradient $\|dJ(\tau(k))\|$ with k iterations.

The optimal switching time is $T = [0.2609, 0.4677, 0.7749]$ after ten iterations. Based on the proposed algorithm, we obtain the corresponding optimal cost $J = 2.8185$. From Figure 3, it is easy to see that the cost J quickly converges to a minimum value and the gradient function $\|dJ(\tau(k))\|$ reaches the termination value. In addition, the system state trajectories with respect to the switching time signal $\tau(k)$ are illustrated in Figure 4.

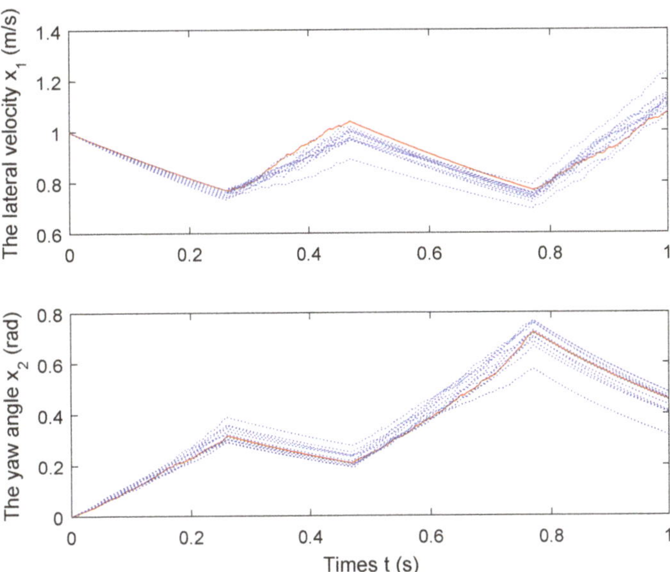

Figure 4. The trajectories of states $x_1(t)$ and $x_2(t)$.

The blue dotted lines explain the state trajectories with the iterate progress switching time vector $\tau(k)$. As a comparison, the red solid line explains the optimal trajectories with respect to the optimal switching time signal.

4.2. Numerical Example

Consider the following switched nonlinear systems:

$$\dot{x}(t) = \begin{cases} A_1 x(t) + B_1 \omega, & t \in [t_0, T_1], \\ A_2 x(t) + B_2 \omega, & t \in (T_1, T_2], \\ A_3 x(t) + B_3 \omega, & t \in (T_2, T_3], \\ A_4 x(t) + B_4 \omega, & t \in (T_3, t_f], \end{cases} \tag{51}$$

where the system matrices are

$$A_1 = \begin{bmatrix} -1 & 1 \\ 0 & 2 \end{bmatrix}, \ A_2 = \begin{bmatrix} 1 & 0 \\ 1 & -2 \end{bmatrix}, \ A_3 = \begin{bmatrix} -1 & 0 \\ 1 & 2 \end{bmatrix}, \ A_4 = \begin{bmatrix} 1 & 1 \\ 0 & -2 \end{bmatrix}$$

$$B_1 = B_3 = \begin{bmatrix} 0 \\ 1 \end{bmatrix}, \ B_2 = B_4 = \begin{bmatrix} 1 \\ 0 \end{bmatrix}.$$

We select the initial state $x(0) = [1, -1]^T$, the initial time $t_0 = 0$, the final time $t_f = 0.9$, the initial switching time $T_1 = 0.3$, $T_2 = 0.5$, $T_3 = 0.7$. The weight coefficient matrices are designed as $Q_1 = I$, $Q_2 = 2I$, $Q_3 = 3I$, $Q_4 = 4I$, $P = I$, where I denotes the unit matrix. The cost function $J(\tau(k))$ and the gradient function $\|dJ(\tau(k))\|$ with k iterations and the trajectories of the states $x_1(t)$ and $x_2(t)$ are shown in Figures 5 and 6 when employing the proposed optimal timing fault tolerant control strategy.

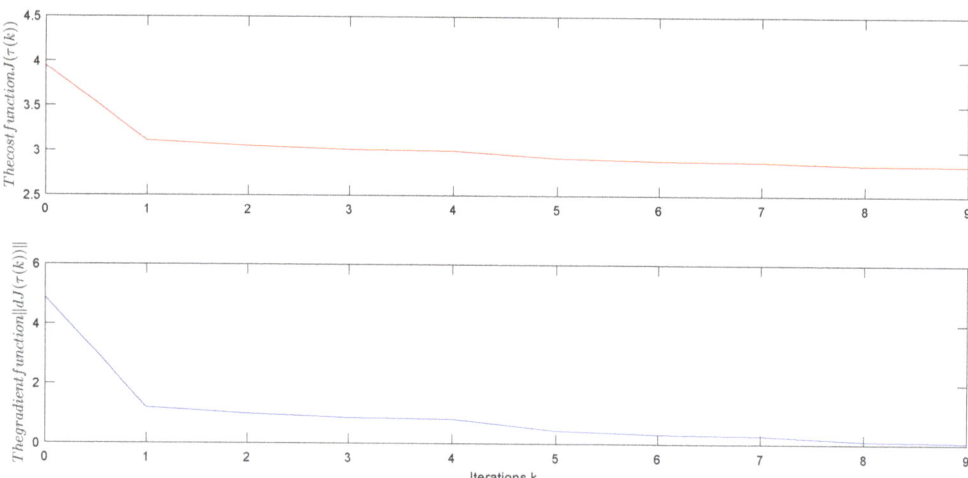

Figure 5. The cost function $J(\tau(k))$ and the gradient function $\|dJ(\tau(k))\|$ with k iterations.

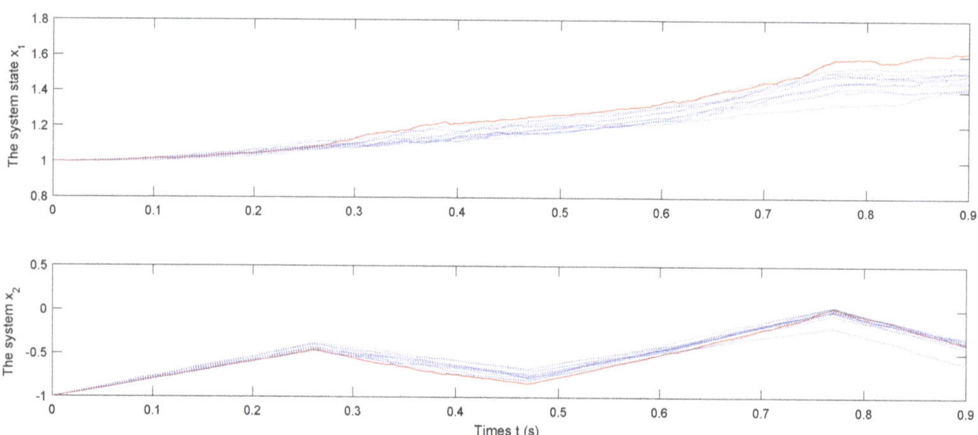

Figure 6. The trajectories of states $x_1(t)$ and $x_2(t)$.

It is worth noting that the switched subsystems are different in the numerical example which can describe the more general systems.

5. Conclusions

In this paper, an novel optimal timing fault tolerant control algorithm is proposed for switched stochastic systems with an unknown drift fault for each switching point. The designed optimal timing fault tolerant controller can not only realize the optimal performance, but also accommodate switching drift fault. Moreover, in this process, the cost functional has the general form with the integral terms and the terminal terms with the switched stochastic systems state variable. The variational technique is exploited to deduce the gradient formula. The steepest descent algorithm with Armijo step sizes is utilized to calculate the optimal switching time. The safety trajectory switching of a four wheel drive vehicle is taken as a practical application case to illustrate the effectiveness of the proposed method. Owing to the special structure of the gradient formula, how to extend the suggested methods to large-scale systems, multi-agent systems and practical systems are is a problem worthy of research.

Author Contributions: Conceptualization, C.Z.; methodology, C.Z., L.H. and K.Z.; software, C.Z. and L.H.; supervision, K.Z., W.S. and Z.H.; validation, C.Z., L.H. and Z.H.; writing, original draft, C.Z. and K.Z. All authors have read and agreed to the published version of the manuscript.

Funding: This work was funded by the Natural Science Foundation of China (No. 61973083, No. 61973077 and No. 61773118), and by the Fundamental Research Funds for the Central Universities: 2242022k30038.

Institutional Review Board Statement: Not applicable.

Informed Consent Statement: Not applicable.

Data Availability Statement: Not applicable.

Conflicts of Interest: The authors declare no conflict of interest.

References

1. Geromel, J.C.; Colaneri, P. Stability and stabilization of discrete time switched systems. *Int. J. Control* **2006**, *79*, 719–728. [CrossRef]
2. Wang, B.; Zhu, Q. Stability analysis of semi-Markov switched stochastic systems. *Automatica* **2018**, *94*, 72–80. [CrossRef]
3. Zhao, Y.; Zhu, Q. Stabilization by delay feedback control for highly nonlinear switched stochastic systems with time delays. *Int. J. Robust Nonlinear Control* **2021**, *31*, 3070–3089. [CrossRef]
4. Fan, L.; Zhu, Q. Mean square exponential stability of discrete-time Markov switched stochastic neural networks with partially unstable subsystems and mixed delays. *Inf. Sci.* **2021**, *580*, 243–259. [CrossRef]
5. Bilbao, J.; Bravo, E.; García, O.; Rebollar, C.; Varela, C. Optimising energy management in hybrid microgrids. *Mathematics* **2022**, *10*, 214. [CrossRef]
6. Ozkan, L.; Kothare, M.V. Stability analysis of a multi-model predictive control algorithm with application to control of chemical reactors. *J. Process. Control* **2006**, *16*, 81–90. [CrossRef]
7. Hamouda, M.; Menaem, A.A.; Rezk, H.; Ibrahim, M.N.; Számel, L. Comparative evaluation for an improved direct instantaneous torque control strategy of switched reluctance motor drives for electric vehicles. *Mathematics* **2021**, *9*, 302. [CrossRef]
8. Gao, F.; Wu, Y.; Zhang, Z. Global fixed-time stabilization of switched nonlinear systems: A time-varying scaling transformation approach. *IEEE Trans. Circuits Syst. II Express Briefs* **2019**, *66*, 1890–1894. [CrossRef]
9. Zhu, F.; Antsaklis, P.J. Optimal control of hybrid switched systems: A brief survey. *Discret. Event Dyn. Syst.* **2015**, *25*, 345–364. [CrossRef]
10. Li, S.; Qiu, J.; Ji, H.; Zhu, K.; Li, J. Piezoelectric vibration control for all-clamped panel using DOB-based optimal control. *Mechatronics* **2011**, *21*, 1213–1221. [CrossRef]
11. Wu, X.; Zhang, K.; Cheng, M. Optimal control of constrained switched systems and application to electrical vehicle energy management. *Nonlinear Anal. Hybrid Syst.* **2018**, *30*, 171–188. [CrossRef]
12. Fu, J.; Zhang, C. Optimal control of path-constrained switched systems with guaranteed feasibility. *IEEE Trans. Autom. Control* **2021**, *67*, 1342–1355. [CrossRef]
13. Bai, Q.; Zhu, W. Event-triggered impulsive optimal control for continuous-time dynamic systems with input time-delay. *Mathematics* **2022**, *10*, 279. [CrossRef]
14. Raisch, J.; O'Young, S.D. Discrete approximation and supervisory control of continuous systems. *IEEE Trans. Autom. Control* **1998**, *43*, 569–573. [CrossRef]
15. Dong, J.; Yang, G.H. Robust static output feedback control synthesis for linear continuous systems with polytopic uncertainties. *Automatica* **2013**, *49*, 1821–1829. [CrossRef]
16. Liu, T.; Qu, X.; Tan, W. Online optimal control for wireless cooperative transmission by ambient RF powered sensors. *IEEE Trans. Wirel. Commun.* **2020**, *19*, 6007–6019. [CrossRef]
17. Stellato, B.; Oberblobaum, S.; Goulart, P.J. Second-order switching time optimization for switched dynamical systems. *IEEE Trans. Autom. Control.* **2017**, *62*, 5407–5414. [CrossRef]
18. Seatzu, C.; Corona, D.; Giua, A.; Bemporad, A. Optimal control of continuous-time switched affine systems. *IEEE Trans. Autom. Control* **2006**, *51*, 726–741. [CrossRef]
19. Azimi, M.; Sharifi, G. A hybrid control scheme for attitude and vibration suppression of a flexible spacecraft using energy-based actuators switching mechanism. *Aerosp. Sci. Technol.* **2018**, *82*, 140–148. [CrossRef]
20. Yan, B.; Shi, P.; Lim, C.C.; Wu, C.; Shi, Z. Optimally distributed formation control with obstacle avoidance for mixed–order multi-agent systems under switching topologies. *IET Control Theory Appl.* **2018**, *12*, 1853–1863. [CrossRef]
21. Ding, K.; Zhu, Q. Extended dissipative anti-disturbance control for delayed switched singular semi-Markovian jump systems with multi-disturbance via disturbance observer. *Automatica* **2021**, *128*, 109556. [CrossRef]
22. Wu, K.; Yu, J.; Sun, C. Global robust regulation control for a class of cascade nonlinear systems subject to external disturbance. *Nonlinear Dyn.* **2017**, *90*, 1209–1222. [CrossRef]
23. Zhang, M.; Zhu, Q. Stability analysis for switched stochastic delayed systems under asynchronous switching: A relaxed switching signal. *Int. J. Robust Nonlinear Control* **2020**, *30*, 8278–8298. [CrossRef]

24. Liu, X.; Zhang, K.; Li, S.; Wei, H. Optimal timing control of discrete-time linear switched stochastic systems. *Int. J. Control Autom. Syst.* **2014**, *12*, 769–776. [CrossRef]
25. Huang, R.; Zhang, J.; Lin, Z. Optimal control of discrete-time bilinear systems with applications to switched linear stochastic systems. *Syst. Control. Lett.* **2016**, *94*, 165–171. [CrossRef]
26. Liu, X.; Li, S.; Zhang, K. Optimal control of switching time in switched stochastic systems with multi-switching times and different costs. *Int. J. Control* **2017**, *90*, 1604–1611. [CrossRef]
27. Zhu, C.; Li, C.; Chen, X.; Zhang, K.; Xin, X.; Wei, H. Event-triggered adaptive fault tolerant control for a class of uncertain nonlinear systems. *Entropy* **2020**, *22*, 598. [CrossRef]
28. Zhang, Y.; Jiang, J. Bibliographical review on reconfigurable fault-tolerant control systems. *Annu. Rev. Control* **2008**, *32*, 229–252. [CrossRef]
29. Yu, X.; Jiang, J. A survey of fault-tolerant controllers based on safety-related issues. *Annu. Rev. Control* **2015**, *39*, 46–57. [CrossRef]
30. Zhu, C.; Zhang, K.; Xin, X.; Gao, F.; Wei, H. Event-triggered adaptive fixed-time output feedback fault tolerant control for perturbed planar nonlinear systems. *Int. J. Robust Nonlinear Control* **2021**, *31*, 6934–6952. [CrossRef]
31. Zhu, Q.; Cao, J. Exponential stability of stochastic neural networks with both Markovian jump parameters and mixed time delays. *IEEE Trans. Syst. Man Cybern. Part B Cybern.* **2010**, *41*, 341–353. [CrossRef]
32. Zhang, D.; Yu, L. Fault-tolerant control for discrete-time switched linear systems with time-varying delay and actuator saturation. *J. Optim. Theory Appl.* **2012**, *153*, 157–176. [CrossRef]
33. Liu, J.; Zhang, K.; Sun, C.; Wei, H. Robust optimal control of switched autonomous systems. *IMA J. Math. Control. Inf.* **2016**, *33*, 173–189. [CrossRef]
34. Guan, Y.; Yang, H.; Jiang, B. Fault-tolerant control for a class of switched parabolic systems. *Nonlinear Anal. Hybrid Syst.* **2019**, *32*, 214–227. [CrossRef]
35. Peng, S.T. On one approach to constraining the combined wheel slip in the autonomous control of a 4ws4wd vehicle. *IEEE Trans. Control. Syst. Technol.* **2006**, *15*, 168–175.

 *mathematics*

Article

Synchronization of Epidemic Systems with Neumann Boundary Value under Delayed Impulse

Ruofeng Rao [1,*], **Zhi Lin** [1,2], **Xiaoquan Ai** [1] and **Jiarui Wu** [1]

1. Department of Mathematics, Chengdu Normal University, Chengdu 611130, China; l18728013506@163.com (Z.L.); ly6041015@163.com (X.A.); rrf2003@163.com (J.W.)
2. School of Mathematical and Computational Science, Hunan University of Science and Technology, Xiangtan 411201, China
* Correspondence: ruofengrao@cdnu.edu.cn

Abstract: This paper reports the construction of synchronization criteria for the delayed impulsive epidemic models with reaction–diffusion under the Neumann boundary value. Different from the previous literature, the reaction–diffusion epidemic model with a delayed impulse brings mathematical difficulties to this paper. In fact, due to the existence of second-order partial derivatives in the reaction–diffusion model with a delayed impulse, the methods of first-order ordinary differential equations from the previous literature cannot be effectively applied in this paper. However, with the help of the variational method and an appropriate boundedness assumption, a new synchronization criterion is derived, and its effectiveness is illustrated by numerical examples.

Keywords: Neumann boundary value; delayed impulse; synchronization; reaction–diffusion epidemic models; variational methods

MSC: 34K24; 34K45

Citation: Rao, R.; Lin, Z.; Ai, X.; Wu, J. Synchronization of Epidemic Systems with Neumann Boundary Value under Delayed Impulse. *Mathematics* **2022**, *10*, 2064. https://doi.org/10.3390/math10122064

Academic Editor: Quanxin Zhu

Received: 30 May 2022
Accepted: 13 June 2022
Published: 15 June 2022

Publisher's Note: MDPI stays neutral with regard to jurisdictional claims in published maps and institutional affiliations.

1. Introduction

The dynamics of epidemic models has always been a hot topic [1,2]. Ordinary differential equation epidemic dynamic models are the most common models, and fractional order models especially have been hot topics in recent research [3–6] whose ideas or methods have been applied to studying epidemic dynamic models. Moreover, the reaction–diffusion epidemic models have become one of the key topics because of the migration behavior of the population [7–10]. Usually, infectious diseases are controlled within a certain range, so we consider the Neumann boundary value, that is, there is no diffusion on the boundary of the infectious area because the disease area is usually isolated from the outside world by some measures, so the Neumann zero boundary value is considered in this paper. To prevent the spread of disease, the government or relevant departments often take impulse measures. This impulse management measure is not only aimed at the epidemic situation, but also considered impulse control measures for economic management, mechanical engineering and other issues [11–20]. Delayed impulse models have also been investigated by many researchers [11,12], for delayed impulse models better simulate the actual situation, that is, the impulse effect usually takes some time to appear. However, the models with a delayed impulse are usually ordinary differential systems, and reaction–diffusion systems with a delayed impulse are rarely seen in the existing literature. This inspired us to write this paper. In fact, due to the existence of second-order partial derivatives in the reaction–diffusion model with a delayed impulse, the methods of first-order ordinary differential equations in the existing literature cannot be effectively applied to partial differential equations. By means of the variational method, differential mean value theorem and convergence of sequence, the synchronization criterion of the delayed impulse reaction–diffusion epidemic model is obtained in this paper. Intuition tells us that the shorter the impulse interval, the greater the impulse frequency and the faster the impulse effect appears, and the greater the impulse intensity, the faster the synchronization between the models must be. These intuitive conclusions are affirmed in the synchronization criterion given in this paper.

The rest of this paper is organized as follows. In Section 2, we present some preliminaries about the reaction–diffusion epidemic model with a delayed impulse. In Section 3, we propose and derive the synchronization criterion for reaction–diffusion epidemic models under a delayed impulse. In Section 4, an illustrative numerical example is provided to show the effectiveness of the newly obtained criterion. Finally, some conclusions are written in Section 5.

The main contributions are as follows:

- Proposing and studying reaction–diffusion epidemic models with a delayed impulse for the first time;
- Deriving for the first time the synchronization criterion of an epidemic system with a Neumann boundary value under a delayed impulse.

2. System Description and Preliminaries

In [1], the following epidemic system was studied:

$$\begin{cases} \dfrac{dS}{dt} = -S\beta(t)I, \\ \dfrac{dI}{dt} = S\beta(t)I - I\gamma(t), \\ \dfrac{dR}{dt} = I\gamma(t), \end{cases} \tag{1}$$

where the function $S(t)$ is the fraction of the susceptible population, $I(t)$ the infected fraction, $R(t)$ the recovered fraction, and $0 < S(t) < 1, 0 < I(t) < 1, 0 < R(t) < 1$. In addition, the disease transmission rate is denoted by $\beta(t)$ and the recovery rate is $\gamma(t)$. In 2020, the authors of [7] considered the epidemic system with inevitable diffusions:

$$\begin{cases} \dfrac{\partial S(t,x)}{\partial t} = \Delta[d_1 S(t,x)] - I(t,x)\beta(t)S(t,x), \\ \dfrac{\partial I(t,x)}{\partial t} = \Delta[d_2 I(t,x)] + I(t,x)\beta(t)S(t,x) - I(t,x)\gamma(t), \\ \dfrac{\partial R(t,x)}{\partial t} = \Delta[d_3 R(t,x)] + I(t,x)\gamma(t), \end{cases} \tag{2}$$

where Δ is the Laplacian operator.

Generally speaking, $\Delta\varphi(x) = \sum\limits_{i=1}^{n} \dfrac{\partial^2 \varphi}{\partial x_i^2}$, for $x = (x_1, x_2, \cdots, x_n)^T \in \mathbb{R}^n$.

Denote $X = (X_1, X_2, X_3)^T$ with $X_1 = S, X_2 = I, X_3 = R$. The following impulsive epidemic model with a Neumann boundary value is investigated in this paper:

$$\begin{cases} \dfrac{\partial X(t,x)}{\partial t} = D\Delta X(t,x) + A(t)X(t,x) + f(t, X(t,x)), & x \in \Omega, t \geqslant t_0, \\ X(t_k^+, x) - X(t_k^-, x) = M_k X(t_k - \tau_k, x), & k \in \mathbb{N}, x \in \Omega, \\ \dfrac{\partial X(t,x)}{\partial \nu} = 0, & x \in \partial\Omega, t \geqslant 0, \\ X(0, x) = \varphi(x), & x \in \Omega, \end{cases} \tag{3}$$

where $\mathbb{N} = \{1, 2, 3, \cdots\}$, $t_0 = 0$ and t_k is the impulse moment for $k = 1, 2, \cdots$, satisfying $0 < t_1 < t_2 < \cdots < t_k < \cdots$ and $\lim\limits_{k\to\infty} t_k = +\infty$. For any impulse moment t_k, M_k is a constant parameters matrix that quantifies the impulse strength at the moment t_k. The time delay $\tau_k \in [0, \tau]$ with $(t_k - \tau_k, t_k) \subset (t_{k-1}, t_k)$ for each $k \in \mathbb{N}$, $\tau = \sup\limits_{k\in\mathbb{N}} \tau_k$, and so $t_0 \leqslant t_1 - \tau_1$. $\Omega \subset \mathbb{R}^N (N \leqslant 2)$ is a bounded smooth domain with smooth boundary $\partial\Omega$.

Denote ν the external normal direction of $\partial\Omega$.

$$D = \begin{pmatrix} d_1 & 0 & 0 \\ 0 & d_2 & 0 \\ 0 & 0 & d_3 \end{pmatrix}, \quad A(t) = \begin{pmatrix} 0 & 0 & 0 \\ 0 & -\gamma(t) & 0 \\ 0 & \gamma(t) & 0 \end{pmatrix}, \quad f(t,X) = \begin{pmatrix} -\beta(t)X_1X_2 \\ \beta(t)X_1X_2 \\ 0 \end{pmatrix}. \tag{4}$$

System (3) is the drive system, and its response system can be considered as follows,

$$\begin{cases} \dfrac{\partial Y(t,x)}{\partial t} = f(t,Y(t,x)) + D\Delta Y(t,x) + A(t)Y(t,x), & x \in \Omega, t \geqslant t_0, \\ Y(t_k^+,x) = M_k Y(t_k - \tau_k, x) + Y(t_k^-,x), & k \in \mathbb{N}, x \in \Omega, \\ \dfrac{\partial Y(t,x)}{\partial v} = 0, & x \in \partial\Omega, t \geqslant 0, \\ \phi(x) = Y(s,x), & x \in \Omega, s \in [-\tau,0], \end{cases} \tag{5}$$

and then the error system is proposed as follows,

$$\begin{cases} \dfrac{\partial e(t,x)}{\partial t} = D\Delta e(t,x) + A(t)e(t,x) + F(t,e(t,x)), & x \in \Omega, t \geqslant t_0, \\ e(t_k^+,x) - e(t_k^-,x) = M_k e(t_k - \tau_k, x), & k \in \mathbb{N}, x \in \Omega, \\ \dfrac{\partial e(t,x)}{\partial v} = 0, & t \geqslant 0, x \in \partial\Omega, \\ e(0,x) = \varphi(x) - \phi(x), & x \in \Omega, \end{cases} \tag{6}$$

where $e = X - Y$. Moreover,

$$F(e(t,x)) = f(t,X(t,x)) - f(t,Y(t,x)) = \begin{pmatrix} -\beta(t)X_1X_2 + \beta(t)Y_1Y_2 \\ \beta(t)X_1X_2 - \beta(t)Y_1Y_2 \\ 0 \end{pmatrix} \tag{7}$$

We assume in this paper that variables are left continuous at impulse moment t_k, for example, $e(t_k,x) = e(t_k^-,x)$.

Obviously, $-1 < e_i < 1$. Moreover, in consideration of the fact that population resources are limited, we can assume throughout this paper that their regional change rate is also limited, and so the change rate of the change rate is even limited:

H1 For any $i = 1,2,3$, there exists a constant $c_i > 0$ such that $|\Delta e_i(t,x)| < c_i|e_i(t,x)|$;

H2 There is a constant $\beta > 0$ such that $0 \leqslant \beta(t) \leqslant \beta$;

H3 There is a constant $\gamma > 0$ such that $|\gamma(t)| \leqslant \gamma$.

Lemma 1 (See, e.g., [21]). *$\Omega \subset \mathbb{R}^m$ is a bounded domain with its smooth boundary $\partial\Omega$ that is of class C^2. $\xi(x) \in H_0^1(\Omega)$ is a real-valued function and $\frac{\partial \xi(x)}{\partial v}|_{\partial\Omega} = 0$. Then,*

$$\int_\Omega |\nabla \xi(x)|^2 dx \geqslant \lambda_1 \int_\Omega |\xi(x)|^2 dx,$$

where λ_1 is defined by the least positive eigenvalue of the problem:

$$\begin{cases} \lambda \xi - \Delta \xi = 0, & x \in \Omega, \\ \dfrac{\partial \xi(x)}{\partial v} = 0, & x \in \partial\Omega. \end{cases}$$

3. Main Result

Theorem 1. *Assume there exists a positive definite diagonal matrix Q and a constant $q_0 > 0$ such that*

$$Q \leqslant q_0 I \tag{8}$$

and

$$\sup_{k\in\mathbb{N}}\left[\|\mathcal{I}+M_k\|+\tau_k\|M_k\|\cdot\left(\|D\|\sqrt{\lambda_{\max}C^2}+\|\tilde{A}\|+2\beta\right)\right]\sqrt{\frac{\lambda_{\max}Q}{\lambda_{\min}Q}}e^{\lambda\rho}\leqslant\rho_0<1,\quad(9)$$

then system (3) and system (5) are synchronized, where $\mathcal{I}$ is the identity matrix, $C=diag(c_1,c_2,c_3)>0$ with $c_i>0$ defined in (H1), $\rho=\sup_{k\in\mathbb{N}}(-t_{k-1}+t_k),\zeta=\inf_{k\in\mathbb{N}}(t_k-t_{k-1})>0,$

$$\tilde{A}=\begin{pmatrix}0&0&0\\0&\gamma&0\\0&\gamma&0\end{pmatrix},\qquad(10)$$

$$\lambda=\left[\frac{1}{\lambda_{\min}Q}\lambda_{\max}\left(-\lambda_1(QD+DQ)+Q\tilde{A}+\tilde{A}^TQ+Q+4q_0\beta^2\mathcal{I}\right)\right].\qquad(11)$$

Here, inequality (8) indicates that $(q_0\mathcal{I}-Q)$ is a non-negative definite matrix. For any symmetric matrix B, the real numbers $\lambda_{\min}B$ and $\lambda_{\max}B$ represent the minimum and maximum eigenvalue of B, respectively. For a matrix B, $\|B\|$ is its norm with $\|B\|=\sqrt{\lambda_{\max}(B^TB)}$.

Proof. Consider the following Lyapunov function:

$$V(t)=\int_\Omega e^T(t,x)Qe(t,x)dx,$$

where Q is a positive definite symmetric matrix. Denote $\|\eta\|^2=\sum_{i=1}^{3}\int_\Omega|\eta_i(x)|^2dx$ for any vector Lebesgue square-integrable function $\eta(x)=(\eta_1(x),\eta_2(x),\eta_3(x))^T$.

It follows from $0<X_i<1$ and $0<Y_i<1\ (i=1,2,3)$ that

$$F^T(t,e)QF(t,e)\leqslant 2q_0\beta^2\cdot 2[(X_2-Y_2)^2+(X_1-Y_1)^2]\leqslant 4q_0\beta^2e^Te$$

So,

$$D^+V(t)\leqslant-\lambda_1\int_\Omega e^T(QD+DQ)edx+\int_\Omega e^T\left(Q\tilde{A}+\tilde{A}^TQ\right)edx+\int_\Omega\left(e^TQF(t,e)+F^T(t,e)Qe\right)dx$$

$$\leqslant\frac{1}{\lambda_{\min}Q}\lambda_{\max}\left(-\lambda_1(QD+DQ)+Q\tilde{A}+\tilde{A}^TQ+Q+4q_0\beta^2\mathcal{I}\right)\int_\Omega e^T(t,x)Qe(t,x)dx,\ t\in(t_{k-1},t_k],$$

which means

$$\|e(t)\|^2\leqslant\frac{\lambda_{\max}Q}{\lambda_{\min}Q}e^{\lambda(t-t_{k-1})}\|e(t_{k-1}^+)\|^2,\quad t\in(t_{k-1},t_k].$$

Particularly,

$$\|e(t_k)\|^2=\|e(t_k^-)\|^2\leqslant\frac{\lambda_{\max}Q}{\lambda_{\min}Q}e^{\lambda(t-t_{k-1})}\|e(t_{k-1}^+)\|^2,\quad k\in\mathbb{N}.$$

On the other hand,

$$\|D\Delta e(\varsigma_k,x)\|\leqslant\|D\|\cdot\|\Delta e(\varsigma_k,x)\|\leqslant\|D\|\sqrt{\lambda_{\max}C^2}\cdot\|e(\varsigma_k)\|,$$

and

$$\|F(t,e(t))\|\leqslant\sqrt{2}\beta\cdot\sqrt{2}\sqrt{\int_\Omega[(X_2-Y_2)^2+(X_1-Y_1)^2]dx}\leqslant\|e(t)\|\cdot 2\beta.$$

Now we can see it from the differential mean value theorem and definition of $A(t)$ that there exists $\varsigma_k\in(t_k-\tau_k,t_k)\subset(t_{k-1},t_k]$ such that

$$\begin{aligned}
\|e(t_k^+)\| &= \|e(t_k^-,x) + M_k e(t_k - \tau_k, x)\| \\
&\leqslant \|\mathcal{I} + M_k\| \cdot \|e(t_k,x)\| + \|M_k\| \cdot \|e(t_k,x) - e(t_k - \tau_k, x)\| \\
&\leqslant \|\mathcal{I} + M_k\| \cdot \|e(t_k,x)\| + \tau_k \|M_k\| \cdot \left(\|D\|\sqrt{\lambda_{\max}C^2} \cdot \|e(\varsigma_k)\| + \|\tilde{A}\| \cdot \|e(\varsigma_k)\| + 2\beta\|e(\varsigma_k)\| \right) \\
&\leqslant \left[\|\mathcal{I} + M_k\| + \tau_k \|M_k\| \cdot \left(\|D\|\sqrt{\lambda_{\max}C^2} + \|\tilde{A}\| + 2\beta \right) \right] \sqrt{\frac{\lambda_{\max}Q}{\lambda_{\min}Q}} e^{\lambda(t_k - t_{k-1})} \|e(t_{k-1}^+)\| \\
&\leqslant \rho_0 \|e(t_{k-1}^+)\|, \quad \forall k = 1,2,3,\cdots
\end{aligned}$$

which means

$$\|e(t_k^+)\| \leqslant \rho_0^k \|e(0)\|, \quad \forall k = 1,2,3,\cdots. \tag{12}$$

Finally, for $t \in (t_{k-1}, t_k]$,

$$\|e(t)\|^2 \leqslant \frac{\lambda_{\max}Q}{\lambda_{\min}Q} e^{\lambda(t-t_{k-1})} \|e(t_{k-1}^+)\|^2 \leqslant \frac{\lambda_{\max}Q}{\lambda_{\min}Q} e^{\lambda\rho} \rho_0^{2(k-1)} \|e(0)\|^2,$$

which, together with $t > t_{k-1}$, implies

$$\|e(t)\| \leqslant \sqrt{\frac{\lambda_{\max}Q}{\lambda_{\min}Q} e^{\lambda\rho}} \|e(0)\| e^{-\lambda_0(t-t_0)}. \tag{13}$$

where $\lambda_0 = -\frac{1}{\zeta} \ln \rho_0 > 0$. This completes the proof. $\square$

Remark 1. *Contrary to the existing literature related to impulsive reaction–diffusion epidemic models (see, e.g., [13,14]), the delayed impulse is firstly considered in the reaction–diffusion epidemic system in this paper. Indeed, although time delays were introduced in [13], the impulse was not delayed. However, in real life, the effectiveness of many defensive measures usually takes place after a period of time. Therefore, the delayed impulse epidemic model studied in this paper clearly has practical significance.*

Remark 2. *Introducing the delayed impulse into reaction–diffusion epidemic models means bringing new mathematical difficulties to this paper. Therefore, this paper adopts a method different from [13,14] to overcome the mathematical difficulties, and a new synchronization criterion is derived.*

4. Numerical Example

Example 1. *Let* $\Omega = (0,1) \times (0,1) \subset \mathbb{R}^2$, *then* $\lambda_1 = \pi^2$. *Set* $C = \mathcal{I} = diag(1,1,1)$, $\gamma(t) = 0.1\sin^2 t$, $\beta(t) = 0.1\cos^2 t$, *and* $\beta = 0.1 = \gamma$,

$$A(t) = \begin{pmatrix} 0 & 0 & 0 \\ 0 & -0.1\sin^2 t & 0 \\ 0 & 0.1\sin^2 t & 0 \end{pmatrix}, \ f(t,X) = \begin{pmatrix} -0.1\cos^2 t X_1 X_2 \\ 0.1\cos^2 t X_1 X_2 \\ 0 \end{pmatrix}, \ \tilde{A} = \begin{pmatrix} 0 & 0 & 0 \\ 0 & 0.1 & 0 \\ 0 & 0.1 & 0 \end{pmatrix}.$$

Case 1: Let $\rho = 0.1$, $\tau_k \equiv \tau = 0.01$, $M_k \equiv -0.7\mathcal{I}$, and

$$D = \begin{pmatrix} 0.045 & 0 & 0 \\ 0 & 0.035 & 0 \\ 0 & 0 & 0.055 \end{pmatrix}.$$

Using a computer with Matlab's LMI toolbox results in the following feasibility datum:

$$Q = \begin{pmatrix} 0.0155 & 0 & 0 \\ 0 & 0.0135 & 0 \\ 0 & 0 & 0.0161 \end{pmatrix}$$

then, $q_0 = \lambda_{\max} Q = 0.0161$, $\lambda_{\min} Q = 0.0135$, and

$$\sup_{k \in \mathbb{N}} \left[\|\mathcal{I} + M_k\| + \tau_k \|M_k\| \cdot \left(\|D\| \sqrt{\lambda_{\max} C^2} + \|\tilde{A}\| + 2\beta \right) \right] \sqrt{\frac{\lambda_{\max} Q}{\lambda_{\min} Q}} e^{\lambda \rho} = 0.8196 \leqslant \rho_0 < 1,$$

where $\rho_0 = 0.8196$. According to Theorem 1, system (3) and system (5) are synchronized.

Case 2: Let $\rho = 0.05$, $\tau_k \equiv \tau = 0.01$, $M_k \equiv -0.7\mathcal{I}$, and

$$D = \begin{pmatrix} 0.045 & 0 & 0 \\ 0 & 0.035 & 0 \\ 0 & 0 & 0.055 \end{pmatrix},$$

Using Matlab's LMI toolbox results in the following feasibility datum:

$$Q = \begin{pmatrix} 0.0111 & 0 & 0 \\ 0 & 0.0131 & 0 \\ 0 & 0 & 0.0112 \end{pmatrix},$$

then, $q_0 = \lambda_{\max} Q = 0.0131$, $\lambda_{\min} Q = 0.0111$, and

$$\sup_{k \in \mathbb{N}} \left[\|\mathcal{I} + M_k\| + \tau_k \|M_k\| \cdot \left(\|D\| \sqrt{\lambda_{\max} C^2} + \|\tilde{A}\| + 2\beta \right) \right] \sqrt{\frac{\lambda_{\max} Q}{\lambda_{\min} Q}} e^{\lambda \rho} = 0.7988 \leqslant \rho_0 < 1,$$

where $\rho_0 = 0.7988$. According to Theorem 1, system (3) and system (5) are synchronized.

Remark 3. *Table 1 reveals that the bigger the impulse frequency, the faster the synchronization speed.*

Table 1. Comparison of the influence from different impulse frequencies when other data are unchanged.

	Case 1: $\rho = 0.1$	Case 2: $\rho = 0.05$
τ_k	0.01	0.01
M_k	$-0.7\mathcal{I}$	$-0.7\mathcal{I}$
ρ_0	0.8196	0.7988

Case 3: Let $\rho = 0.1$, $\tau_k \equiv \tau = 0.01$, $M_k \equiv -0.8\mathcal{I}$, and

$$D = \begin{pmatrix} 0.045 & 0 & 0 \\ 0 & 0.035 & 0 \\ 0 & 0 & 0.055 \end{pmatrix},$$

Using Matlab's LMI toolbox results in the following feasibility datum:

$$Q = \begin{pmatrix} 0.0166 & 0 & 0 \\ 0 & 0.0163 & 0 \\ 0 & 0 & 0.0165 \end{pmatrix},$$

then, $q_0 = \lambda_{\max} Q = 0.0163$, $\lambda_{\min} Q = 0.0166$, and

$$\sup_{k \in \mathbb{N}} \left[\|\mathcal{I} + M_k\| + \tau_k \|M_k\| \cdot \left(\|D\| \sqrt{\lambda_{\max} C^2} + \|\tilde{A}\| + 2\beta \right) \right] \sqrt{\frac{\lambda_{\max} Q}{\lambda_{\min} Q}} e^{\lambda \rho} = 0.5466 \leqslant \rho_0 < 1,$$

where $\rho_0 = 0.5466$. According to Theorem 1, system (3) and system (5) are synchronized.

Remark 4. *Table 2 implies that the bigger the impulse intensity, the faster the synchronization speed.*

Table 2. Comparison of the influence from different impulse intensities when other data are unchanged.

	Case 1: $M_k = -0.7\mathcal{I}$	**Case 2: $M_k = -0.8\mathcal{I}$**
ρ	0.1	0.1
τ_k	0.01	0.01
ρ_0	0.8196	0.5466

Case 4: Let $\rho = 0.1$, $\tau_k \equiv \tau = 0.001$, $M_k \equiv -0.7\mathcal{I}$, and

$$D = \begin{pmatrix} 0.045 & 0 & 0 \\ 0 & 0.035 & 0 \\ 0 & 0 & 0.055 \end{pmatrix},$$

Using Matlab's LMI toolbox results in the following feasibility datum:

$$Q = \begin{pmatrix} 0.0155 & 0 & 0 \\ 0 & 0.0135 & 0 \\ 0 & 0 & 0.0161 \end{pmatrix}$$

then, $q_0 = \lambda_{\max}Q = 0.0161$, $\lambda_{\min}Q = 0.0135$, and

$$\sup_{k\in\mathbb{N}} \left[\|\mathcal{I}+M_k\| + \tau_k\|M_k\| \cdot \left(\|D\|\sqrt{\lambda_{\max}C^2} + \|\tilde{A}\| + 2\beta \right) \right] \sqrt{\frac{\lambda_{\max}Q}{\lambda_{\min}Q}} e^{\lambda\rho} = 0.7188 \leqslant \rho_0 < 1,$$

where $\rho_0 = 0.7188$. According to Theorem 1, system (3) and system (5) are synchronized.

Remark 5. *Table 3 means that the smaller the time delays of the impulse effect, the faster the synchronization speed.*

Table 3. Comparison of the influence of different time delays of the impulse effect when other data are unchanged.

	Case 1: $\tau_k = 0.01$	**Case 2: $\tau_k = 0.001$**
ρ	0.1	0.1
M_k	$-0.7\mathcal{I}$	$-0.7\mathcal{I}$
ρ_0	0.8196	0.7188

Below, another numerical example is presented to show the validity of Theorem 1 via very simple computations.

Example 2. *Set $C = 15\mathcal{I} = diag(15, 15, 15)$, $M_k \equiv -0.9\mathcal{I}$, $D = 0.1\mathcal{I}$, $Q = \mathcal{I}$, $\tau_k \equiv \tau = 0.01$, $\beta = 0.1$, and then direct computations lead to*

$$\sqrt{\lambda_{\max}C^2} = 15, \ \|\mathcal{I}+M_k\| \equiv 0.1, \ \|M_k\| \equiv 0.9, \ \|D\| = 0.1, \ \lambda_{\max}Q = \lambda_{\min}Q = 1 = q_0. \tag{14}$$

Let $\Omega = (0,1) \times (0,1) \subset \mathbb{R}^2$; then, $\lambda_1 = \pi^2$. Set $\gamma = 0.1$; then,

$$\tilde{A} = \begin{pmatrix} 0 & 0 & 0 \\ 0 & 0.1 & 0 \\ 0 & 0.1 & 0 \end{pmatrix} \ \text{and} \ \|\tilde{A}\| = \sqrt{0.0200} = 0.1414. \tag{15}$$

Hence, it follows from (14) and (15) that

$$\lambda = \left[\frac{1}{\lambda_{\min}Q}\lambda_{\max}\left(-\lambda_1(QD+DQ)+Q\tilde{A}+\tilde{A}^TQ+Q+4q_0\beta^2\mathcal{I}\right)\right]$$

$$=\lambda_{\max}\left(-\lambda_1(QD+DQ)+Q\tilde{A}+\tilde{A}^TQ+Q+4q_0\beta^2\mathcal{I}\right)$$

$$=\lambda_{\max}\left(-\pi^2(0.1\mathcal{I}+0.1\mathcal{I})+\tilde{A}+\tilde{A}^T+\mathcal{I}+0.04\mathcal{I}\right)$$

$$=-0.6925,$$

(16)

where

$$\left(-\pi^2(0.1\mathcal{I}+0.1\mathcal{I})+\tilde{A}+\tilde{A}^T+\mathcal{I}+0.04\mathcal{I}\right)=\begin{pmatrix}-0.9339 & 0 & 0\\ 0 & -0.7339 & 0.1000\\ 0 & 0.1000 & -0.9339\end{pmatrix}.$$

Now, letting $\rho=0.3$ and $\rho_0=0.9$, we can get by (14)–(16) that

$$\sup_{k\in\mathbb{N}}\left[\|\mathcal{I}+M_k\|+\tau_k\|M_k\|\cdot\left(\|D\|\sqrt{\lambda_{\max}C^2}+\|\tilde{A}\|+2\beta\right)\right]\sqrt{\frac{\lambda_{\max}Q}{\lambda_{\min}Q}e^{\lambda\rho}}$$

$$\equiv\left[\|\mathcal{I}+M_k\|+\tau_k\|M_k\|\cdot\left(\|D\|\sqrt{\lambda_{\max}C^2}+\|\tilde{A}\|+2\beta\right)\right]\sqrt{\frac{\lambda_{\max}Q}{\lambda_{\min}Q}e^{\lambda\rho}}$$

$$=\left[0.1+0.01\times0.9\times\left(0.1\times15+0.1414+0.2\right)\right]\sqrt{e^{-0.6925\times0.3}}$$

$$<\left[0.1+0.01\times0.9\times\left(0.1\times15+0.1414+0.2\right)\right]\times1$$

$$=0.1166<\rho_0<1,$$

which implies (9) holds.

According to Theorem 1, system (3) and system (5) are synchronized (see Figures 1–3).

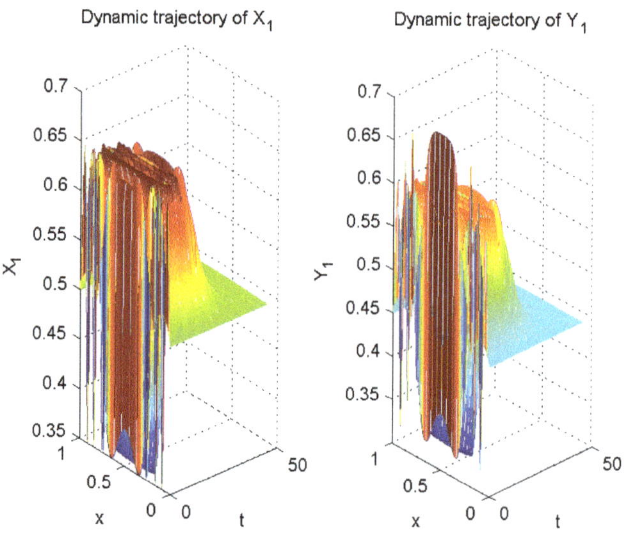

Figure 1. Computer simulation of X_1 in (3) and Y_1 in (5).

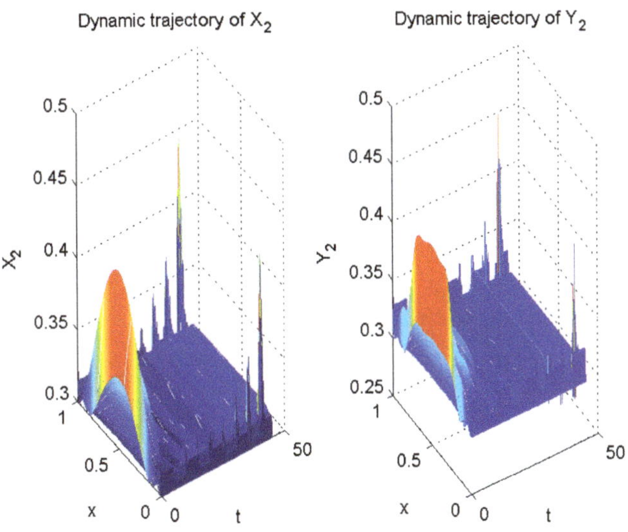

Figure 2. Computer simulation of X_2 in (3) and Y_2 in (5).

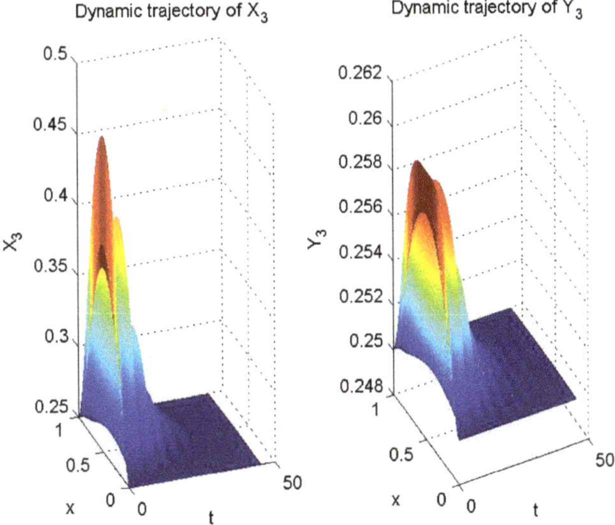

Figure 3. Computer simulation of X_3 in (3) and Y_3 in (5).

Remark 6. *Example 2 illustrates that the validity of Theorem 1 can be easily verified even without using Matlab's LMI toolbox.*

5. Conclusions

This paper reported the synchronization control of two epidemic systems with a Neumann boundary value under a delayed impulse. Different from the previous relevant literature in which the effect of the impulse control was immediate, our impulse effect was delayed, which is in line with the actual situation during an epidemic. At the same time, the newly obtained criterion and numerical examples illustrate that the shorter the time delay of the pulse effect, the faster the synchronization speed. In addition, the smaller the pulse interval, the faster the synchronization. On the other hand, Remarks 1 and 2 illustrated the novelty of this paper by comparing the related literature with this paper.

Author Contributions: Writing—original draft and revising, R.R.; participating in the discussion of the literature and polishing English, Z.L., X.A. and J.W. All authors have read and agreed to the published version of the manuscript.

Funding: This research was funded by the 2019 provincial undergraduate innovation and entrepreneurship training program of Chengdu Normal University (S201914389037).

Institutional Review Board Statement: Not applicable.

Informed Consent Statement: Not applicable.

Data Availability Statement: Not applicable.

Conflicts of Interest: The authors declare no conflict of interest.

References

1. Bacaer, N.; Gomes, M.G.M. On the final size of epidemics with seasonality. *Bull. Math. Bio.* **2009**, *71*, 1954–1966. [CrossRef] [PubMed]
2. Nabti, A.; Ghanbari, B. Global stability analysis of a fractional SVEIR epidemic model. *Math. Meth. Appl. Sci.* **2021**, *44*, 8577–8597. [CrossRef]
3. He, Z.; Abbes, A.; Jahanshahi, H.; Alotaibi, N.D.; Wang, Y. Fractional-order discrete-time sir epidemic model with vaccination: Chaos and complexity. *Mathematics* **2022**, *10*, 165. [CrossRef]
4. Rihan, F.A.; Al-Mdallal, Q.M.; AlSakaji, H.J.; Hashish, A. A fractional-order epidemic model with time-delay and nonlinear incidence rate. *Chaos Solit. Frac.* **2019**, *126*, 97–105. [CrossRef]
5. Ding, K.; Zhu, Q. Impulsive method to reliable sampled-data control for uncertain fractional-order memristive neural networks with stochastic sensor faults and its applications. *Nonlinear Dyn.* **2020**, *100*, 2595–2608. [CrossRef]
6. Xiao J.; Zhong S.; Wen S. Improved approach to the problem of the global Mittag–Leffler synchronization for fractional-order multidimension-valued BAM neural networks based on new inequalities. *Neu. Net.* **2021**, *133*, 87–100. [CrossRef]
7. Zhang, L.; Wang, Z.; Zhao, X. Time periodic traveling wave solutions for a Kermack-McKendrick epidemic model with diffusion and seasonality. *J. Evol. Equ.* **2020**, *20*, 1029–1059. [CrossRef]
8. Liu, C.; Cui, R. Qualitative analysis on an SIRS reaction-diffusion epidemic model with saturation infection mechanism. *Nonlinear Anal. RWA* **2021**, *62*, 103364. [CrossRef]
9. Wang, N.; Zhang, L.; Teng, Z. Dynamics in a reaction-diffusion epidemic model via environmental driven infection in heterogenous space. *J. Biol. Dyn.* **2021**, 1–24. [CrossRef]
10. Cui, R. Asymptotic profiles of the endemic equilibrium of a reaction-diffusion-advection SIS epidemic model with saturated incidence rate. *Discret. Contin. Dyn. Syst. B* **2021**, *26*, 2997–3022. [CrossRef]
11. Zhang, K.; Braverman, E. Time-delay systems with delayed impulses: A unified criterion on asymptotic stability. *Automatica* **2021**, *125*, 109470. [CrossRef]
12. Shi, F.; Liu, Y.; Li, Y.; Qiu, J. Input-to-state stability of nonlinear systems with hybrid inputs and delayed impulses. *Nonlinear Anal. Hybrid Sys.* **2022**, *44*, 101145. [CrossRef]
13. Hu, W.; Zhu Q. Stability criteria for impulsive stochastic functional differential systems with distributed-delay dependent impulsive effects. *IEEE Trans. Syst. Man Cyb. Syst.* **2021**, *51*, 2027–2032. [CrossRef]
14. Rao, R. Impulsive control and global stabilization of reaction-diffusion epidemic model. *Math. Meth. Appl. Sci.* **2021**. [CrossRef]
15. Hu W.; Zhu Q.; Karimi H. Some improved Razumikhin stability criteria for impulsive stochastic delay differential systems. *IEEE Trans. Auto. Control* **2019**, *64*, 5207–5213. [CrossRef]
16. Ji, Y.; Cao, J. Parameter estimation algorithms for hammerstein finite impulse response moving average systems using the data filtering theory. *Mathematics* **2022**, *10*, 438. [CrossRef]
17. Bai, Q.; Zhu, W. Event-triggered impulsive optimal control for continuous-time dynamic systems with input time-delay. *Mathematics* **2022**, *10*, 279. [CrossRef]
18. Cao, W.; Zhu, Q. Razumikhin-type theorem for pth exponential stability of impulsive stochastic functional differential equations based on vector Lyapunov function. *Nonlinear Anal. Hyb. Syst.* **2021**, *39*, 100983. [CrossRef]
19. Zhu Q.; Cao, J. Robust exponential stability of markovian jump impulsive stochastic cohen-grossberg neural networks with mixed time delays. *IEEE Trans. Neu. Net.* **2010**, *21*, 1314–1325.
20. Li, X.; Li, P. Stability of time-delay systems with impulsive control involving stabilizing delays. *Automatica* **2021**, *124*, 109336 [CrossRef]
21. Pan, J.; Liu X.; Zhong, S. Stability criteria for impulsive reaction-diffusion Cohen-Grossberg neural networks with time-varying delays. *Math. Comput. Model.* **2010**, *51*, 1037–1050. [CrossRef]

Article

Analysis of Equilibrium Points in Quantized Hill System

Abdullah A. Ansari [1], Sawsan Alhowaity [2], Elbaz I. Abouelmagd [3,*] and Shiv K. Sahdev [4]

[1] International Center for Advanced Interdisciplinary Research (ICAIR), Sangam Vihar, New Delhi 110062, India; icairndin@gmail.com
[2] Department of Mathematics, College of Science & Humanities, Shaqra University, Shaqra 15551, Saudi Arabia; salhowaity@su.edu.sa
[3] Celestial Mechanics and Space Dynamics Research Group (CMSDRG), Astronomy Department, National Research Institute of Astronomy and Geophysics (NRIAG), Helwan 11421, Cairo, Egypt
[4] Department of Mathematics, Shivaji College, University of Delhi, Delhi 110027, India; shiv_sahdev@yahoo.co.in
* Correspondence: elbaz.abouelmagd@nriag.sci.eg or eabouelmagd@gmail.com; Tel.: +20-10-2097-6040

Abstract: In this work, the quantized Hill problem is considered in order for us to study the existence and stability of equilibrium points. In particular, we have studied three different cases which give all whole possible locations in which two points are emerging from the first case and four points from the second case, while the third case does not provide a realistic locations. Hence, we have obtained four new equilibrium points related to the quantum perturbations. Furthermore, the allowed and forbidden regions of motion of the first case are investigated numerically. We demonstrate that the obtained result in the first case is a generalization to the classical one and it can be reduced to the classical result in the absence of quantum perturbation, while the four new points will disappear. The regions of allowed motions decrease as the value of the *Jacobian constant* increases, and these regions will form three separate areas. Thus, the infinitesimal body can never move from one allowed region to another, and it will be trapped inside one of the possible regions of motion with the relative large values of the *Jacobian constant*.

Keywords: Hill problem; quantum correction; equilibrium points; stability

MSC: 70F05; 70F07; 70F10; 70F15; 70H14

Citation: Ansari, A.A.; Alhowaity, S.; Abouelmagd, E.I.; Sahdev, S.K. Analysis of Equilibrium Points in Quantized Hill System. *Mathematics* **2022**, *10*, 2186. https://doi.org/10.3390/math10132186

Academic Editors: Quanxin Zhu

Received: 21 May 2022
Accepted: 21 June 2022
Published: 23 June 2022

Publisher's Note: MDPI stays neutral with regard to jurisdictional claims in published maps and institutional affiliations.

1. Introduction

Hill's problem is a particularly limiting case for the restricted three-body problem (RTBP). Researchers can obtain the Hill problem by using some scales and transformations while taking limits, as mass parameter tends to zero. Hence, it is an interesting application based problem, and many scientists have studied different versions of this problem by considering different perturbation forces in the classical Hill problem. This means the primary bodies possess point masses and move in circular orbits around their common centre of mass or in elliptical trajectory, while the third body moves in space under the effect of gravitational forces of the primary bodies without affecting their motions [1–4].

In [5], the authors have studied the Hill stability of satellites by utilising the RTBP configuration. However, in [6–8], the authors have studied the same configuration with various perturbations as radiation pressure and oblateness of the primaries. Additionally, in [9–12], the authors have explored and analysed the Hill four bodies problem with its application to the Earth–Moon–Sun system and satellite motion about binary asteroids. In this context, Hill's problem, with oblate secondary in three dimensions, has been illustrated in [13], where the equilibrium points and their stability have been determined.

Further to the precedent work, the radiation pressure effect of the bigger primary and the secondary oblateness on the new version of Hill's problem are investigated in [14], where the authors illustrated that their study is more appropriate for astronomical application. They also used iterative methods to identify the locations of equilibrium points and used the linear stability analysis method to examine their stability properties. They proved that all the equilibrium points are unstable for this model. In [15], the authors investigated Hill's problem because space missions required the knowledge of orbits with some properties, where periodic solutions are illustrated numerically due to the non-integrability of this problem.

With the continuous contributions analysing the Hill body problem, the existence of positions and stability of collinear equilibrium points in its generalized version under radiation pressure and oblateness effects are studied in [16]; the authors also performed the basins of attraction through the Newton–Raphson method for many values of used parameters. Furthermore, in [17] the author investigated the basins of convergence in the aforementioned problem; his numerical analysis revealed the extraordinary and beautiful formations on the complex plane. In [18], the authors have performed the Hill's problem by assuming the primaries as the source of radiation pressure; they have determined the asymptotic orbits at collinear points and the same to the lyapunov periodic orbits.

The spatial or planar restricted three-body problem (RTBP) under any kind of perturbation is called the perturbed model. Otherwise, it can be called the phot–gravitational, relativistic, or quantized problem in the case that the system is analysed under the effect of radiation pressure, relativistic, or quantum corrections perturbation, respectively [19–23]. The analysis of the spatial quantized RTBP (i.e., the spatial of RTBP under the effect of quantum corrections) is studied in [24], where the locations of equilibrium points and the allowed and forbidden regions of motions are examined. Furthermore, the quantized RTBP is developed to construct a new version of the Hill problem [25], where the equations of motion for the Hill problem are evaluated under the quantum corrections. Thus, the obtained system is called quantized Hill problem (QHP).

Recently, in [26], the authors investigated the Hill's problem by assuming that the infinitesimal body varies its mass according to Jeans law, they investigated numerically the location of equilibrium points, regions of motion, and basins of attraction and also examined the stability status of these points by using Meshcherskii's space–time transformation. Furthermore, in [27] the authors investigated the differences and similarities among the classical perturbation theory, Poincaré–Lindstedt technique, multiple scales method, the KB averaging method, and averaging theory, while the latter is used to find periodic orbits in the framework of the spatial QHP. They stated that this model can be utilized to develop a lunar theory and families of periodic orbits.

In the framework of RTBP, which can be reduced to the Hill model, some effective contributions are outlined in [28–30], where the effects of lack of sphericity body shape and radiation pressure on the primaries are studied. In addition, the effect of mass variation in the frame of RTBP is investigated in [31–34], where the authors have also studied the impact of these perturbations on the positions of equilibrium points, Poincaré surfaces of section, regions of possible and forbidden motion, and basins of attraction and examined the stability of these equilibrium points such that it is proven that, in most cases, these points are unstable.

In general, the Hill body problem has a great significance in both stellar and solar systems and in dynamical astronomy; it has received a considerable analysis in its own literature. Primarily, it is formulated as a model to analyse the Moon's motion around the Earth under the effect of Sun perturbation. Furthermore, its model, with simple modifications, can also serve as a model for the motion of a star in a star cluster under the created perturbations from the galaxy. The importance of this problem motivated us to study and analyse the Hill body problem under the perturbation of quantum corrections.

In this work, the QHP is considered to study the existence of equilibrium points alongside examining their stability. Under the effect of quantum corrections, the locations of equilibrium points have been analysed. In particular, we have studied three different cases which give all possible locations, where two points are emerging from the first case and they are considered a generalization for the classical two points, as well as four points from the second case, while the third case does not provide any realistic locations. Hence, we have obtained four new equilibrium points related to the quantum perturbations. Furthermore, We demonstrate that the obtained result in the first case can be reduced to the classical result, while the four new points will disappear in the absence of quantum perturbation.

The paper is organized in six sections as follows: The literature surrounding the problem is given in Section 1. The equations of motion are preformed in Section 2. In Section 3, we have determined the positions of equilibrium points. The stability of equilibrium points are studied in Section 4. Furthermore, the numerical results are estimated in Section 5. Finally, the conclusion of the work is presented in Section 6.

2. Equations of Motion

Following the same notations and procedure in [25], we can write the equations of motion of the quantized Hill problem in the synodic coordinates system as:

$$
\begin{aligned}
\ddot{x} - 2\dot{y} &= V_x, \\
\ddot{y} + 2\dot{x} &= V_y, \\
\ddot{z} &= V_z,
\end{aligned}
\tag{1}
$$

where

$$
V = \frac{1}{2}\left[3\,x^2 + 4(\alpha_1 - \alpha_{11})x - z^2\right] + \frac{1}{r}\left(1 + \frac{\alpha_{21}}{r} + \frac{\alpha_{22}}{r^2}\right),
\tag{2}
$$

and

$$
r^2 = x^2 + y^2 + z^2.
\tag{3}
$$

By integrating System (1), one can write the *Jacobian integral* as

$$
\dot{x}^2 + \dot{y}^2 + \dot{z}^2 = 2V - J_C,
\tag{4}
$$

where J_C is the *Jacobian constant*.

In System (1), the parameters α_1, α_{11}, and α_{21} represent very small amounts with order of $\mathcal{O}(1/c^2)$, but α_{22} is of order $\mathcal{O}(1/c^3)$ where c is the speed of light. Therefore, the value of $\alpha_1 - \alpha_{11}$ will tends to zero [27]. Hence $\alpha_1 - \alpha_{11} \cong 0$. In this context, System (1) can be rewritten as:

$$
\begin{aligned}
\ddot{x} - 2\dot{y} &= x[3 - q(r)], \\
\ddot{y} + 2\dot{x} &= -y\,q(r), \\
\ddot{z} &= -z[1 + q(r)],
\end{aligned}
\tag{5}
$$

where

$$
q(r) = \frac{1}{r^3}\left(1 + \frac{2\alpha_{21}}{r} + \frac{3\alpha_{22}}{r^2}\right)
$$

We would like to provide the reader with the following investigations about the aforementioned perturbation parameters. In fact, the parameters α_1, α_{11}, and α_{21} identify the size of the relativistic effect, while α_{22} estimates the quantum correction contribution. However, all of these effects tend to zero in the case of large distances [21,35].

3. Analysis of Equilibrium Points

The equilibrium points can be obtained by equating all the derivatives with respect to time by zero in system (5), hence

$$
\begin{aligned}
x[3 - q(r)] &= 0, \\
y\, q(r) &= 0, \\
z[1 + q(r)] &= 0.
\end{aligned}
\tag{6}
$$

In the classical case, we mean that the quantum effect will be neglected, the parameter $\alpha_{21} = \alpha_{22} = 0$, then the equilibrium points are given by $(x, 0, 0)$ where $r_e = 1/\sqrt[3]{3}$ and $y = z = 0$, hence $x = \pm 1/\sqrt[3]{3}$.

To find the equilibrium points under the quantized effect ($\alpha_{21} \neq 0$ and $\alpha_{22} \neq 0$), we have to find the solutions of System (6); there are some cases which can be applied to analyse the solutions of this system.

First case: $q(r) = 3$, then $y = z = 0$, and $x \neq 0$.

In this case, one obtains

$$
\frac{1}{r^3}\left(1 + \frac{2\alpha_{21}}{r} + \frac{3\alpha_{22}}{r^2}\right) = 3
\tag{7}
$$

Equation (7) gives a quintic equation in the following form

$$
3r^5 - r^2 - 2\alpha_{21}r - 3\alpha_{22} = 0
\tag{8}
$$

The solution of the fifth degree equation is generally too complicated, however the equation has at least one real root. Instead, numerical approximations can be evaluated using a root-finding algorithm for polynomials.

In fact, it is not our aim to find a solution of a quintic equation, but we aim to find the quantum corrections' impact on the locations of equilibrium points. Thus, we impose that $r = r_q = r_e + \varepsilon$, where ε is a very small quantity which embodies the effect of quantum correction on the locations of equilibrium points after substituting $r = r_q = r_e + \varepsilon$ into Equation (7) or Equation (8), keeping all terms with coefficients of ε and ε^2 only, and neglecting all terms with an order of $\mathcal{O}(\varepsilon^3)$ or more. Hence, ε will satisfy two values, ε_1 and ε_2, which are given by

$$
\varepsilon = \frac{15r_e^4 - 2r_e - 2\alpha_{21} \pm \sqrt{4\alpha_{21}^2 - 12\alpha_{22} + 180\alpha_{21}r_e^4 + 360\alpha_{22}r_e^3 - 135r_e^8 + 72r_e^5}}{2(1 - 30r_e^3)},
\tag{9}
$$

Substituting $r_e = 1/\sqrt[3]{3}$ in Equation (9), one obtains

$$
\varepsilon = \frac{1}{18}\left(2\alpha_{21} - 3^{2/3} \pm \sqrt{4\alpha_{21}^2 + 20\sqrt[3]{9}\,\alpha_{21} + 108\alpha_{22} + 3\sqrt[3]{3}}\right)
\tag{10}
$$

As α_{21} and α_{22} are very small quantities with order of $\mathcal{O}(c^2)$ and $\mathcal{O}(c^2 3)$, respectively, we keep only terms with order of $\mathcal{O}(\alpha_{21})$ and $\mathcal{O}(\alpha_{22})$ and neglect the remanning terms. Thereby, the approximated values of the perturbed parameter ε are governed by

$$
\begin{aligned}
\varepsilon_{11} &= \frac{1}{3}\left(2\alpha_{21} + 3\sqrt[3]{3}\,\alpha_{22}\right), \\
\varepsilon_{12} &= -\frac{1}{3\sqrt[3]{3}}\left(1 + \frac{4\sqrt[3]{3}}{3}\alpha_{21} + 3\sqrt[3]{9}\,\alpha_{22}\right)
\end{aligned}
\tag{11}
$$

The parameter of ε embodies the effect of the quantum corrections and it must equal zero in the absence of these corrections, i.e., when $\alpha_{21} = 0$ and $\alpha_{22} = 0$. However, the obtained solution of ε_{12} does not equal zero and gives an inconvenient solution; thus, the value of ε_{12} is rejected. Hence, the proper approximated value of the parameter ε is given by

$$\varepsilon_{11} = \frac{1}{3}\left(2\alpha_{21} + 3\sqrt[3]{3}\,\alpha_{22}\right) \tag{12}$$

Utilizing Equation (12) with relation to $r_{q_1} = re + \varepsilon_{11}$, then the distance r_{q_1} at the quantized equilibrium point is

$$r_{q_1} = \frac{1}{\sqrt[3]{3}}\left(1 + \frac{2\sqrt[3]{3}}{3}\,\alpha_{21} + \sqrt[3]{9}\,\alpha_{22}\right) \tag{13}$$

As $x = |r_{q_1}|$, we have two possible values for x and $x_1 = r_{q_1}$, $x_2 = -r_{q_1}$. Thus, the quantized equilibrium points are $(x_1, 0, 0)$ and $(x_2, 0, 0)$, which is considered a generalization of the classical case and can be reduced to the classical one when $\alpha_{21} = 0$ and $\alpha_{22} = 0$.

Second case: $q(r) = 0$, then $x = z = 0$, and $y \neq 0$.

This case could occur when the parameters of quantum corrections are negative, i.e., the values of α_{21} and α_{22} are negative [24]. Hence, $q(r) = 0$ when the solutions of the following quadratic equation are possible

$$r^2 + 2\alpha_{21}r + 3\alpha_{22} = 0. \tag{14}$$

The possible solutions of Equation (14) are

$$\begin{aligned} r_{q_2} &= -\alpha_{21} - \sqrt{\alpha_{21}^2 - 3\alpha_{22}} \\ r_{q_3} &= -\alpha_{21} + \sqrt{\alpha_{21}^2 - 3\alpha_{22}} \end{aligned} \tag{15}$$

The solutions in Equation (15) are valid if the values of r_{q_2} and r_{q_3} are positive. To investigate this property, first we remark that α_{21}, α_{22}, and α_{22}/α_{21} have values with order of $\mathcal{O}(1/c^2)$, $\mathcal{O}(1/c^3)$, and $\mathcal{O}(1/c)$. Then, the approximated series solutions of Equation (15) can be written as

$$\begin{aligned} r_{q_2} &= -2\alpha_{21}\left[1 - \frac{3}{4}\left(\frac{\alpha_{22}}{\alpha_{21}}\right) - \frac{9}{16}\left(\frac{\alpha_{22}}{\alpha_{21}}\right)^2\right] + O(\frac{1}{c^5}) \\ r_{q_3} &= -\frac{3}{2}\alpha_{22}\left[1 + \frac{3}{4}\left(\frac{\alpha_{22}}{\alpha_{21}}\right)\right] + O(\frac{1}{c^5}) \end{aligned} \tag{16}$$

It is clear that from Equation (16) the values of r_{q_2} and r_{q_3} are very small and positive when α_{21} and α_{22}, respectively, take negative values. Then, we have four new equilibrium points corresponding to the second case under the perturbation of quantum corrections, where $y_2 = |r_{q_2}|$ and $y_3 = |r_{q_3}|$. The new four points are $(0, y_2, 0)$, $(0, -y_2, 0)$, $(0, y_3, 0)$, and $(0, -y_3, 0)$

Third case: $q(r) = -1$, then $x = y = 0$, and $z \neq 0$.

In this case, one obtains

$$\frac{1}{r^3}\left(1 + \frac{2\alpha_{21}}{r} + \frac{3\alpha_{22}}{r^2}\right) = -1 \tag{17}$$

Equation (17) gives also a quintic equation in the following form

$$r^5 + r^2 + 2\alpha_{21}r + 3\alpha_{22} = 0 \tag{18}$$

To find the solution of Equation (18), we impose that $r = r_q = r_e + \delta$, where δ is very small quantity which embodies the effect of quantum correction on the locations of equilibrium points in the current case after substituting $r = r_q = r_e + \delta$ into Equation (18) and keeping all terms with coefficients of δ and δ^2 only, neglecting all terms with order of $\mathcal{O}(\delta^3)$ or more. Hence, δ will satisfy two values, ε_{31} and ε_{32}, which are given by

$$
\begin{aligned}
\varepsilon_{31} &= -\frac{3}{26}\left(\frac{11}{3\sqrt[3]{3}} + 2\alpha_{21} + \sqrt{4\alpha_{21}^2 - \frac{20\alpha_{21}}{\sqrt[3]{3}} - 52\alpha_{22} - \frac{29}{3\sqrt[3]{9}}}\right) \\
\varepsilon_{32} &= -\frac{3}{26}\left(\frac{11}{3\sqrt[3]{3}} + 2\alpha_{21} - \sqrt{4\alpha_{21}^2 - \frac{20\alpha_{21}}{\sqrt[3]{3}} - 52\alpha_{22} - \frac{29}{3\sqrt[3]{9}}}\right)
\end{aligned}
\tag{19}
$$

It is clear that from Equation (19) the obtained values of the perturbation parameter δ is complex, which mean that the assumption of the third case does not lead to realistic situations. Thus, this case does not give real equilibrium points and it is rejected.

4. Stability Status of Equilibrium Points

Next, to check the equilibrium points stability, we have to write the equations of motion in to phase space. Thus, System (5) can be rewritten in the following form

$$
\begin{aligned}
\ddot{x} - 2\dot{y} &= H_x, \\
\ddot{y} + 2\dot{x} &= H_y, \\
\ddot{z} &= H_z.
\end{aligned}
\tag{20}
$$

where

$$
H = \frac{1}{2}\left[3x^2 - z^2\right] + \frac{1}{r}\left(1 + \frac{\alpha_{21}}{r} + \frac{\alpha_{22}}{r^2}\right),
\tag{21}
$$

Here, the *Jacobian integral* can be rewritten as

$$
\dot{x}^2 + \dot{y}^2 + \dot{z}^2 = 2H - J_C,
\tag{22}
$$

The motion in the proximity of any of the equilibrium points ($a, b,$ and c) can be studied by putting $x = a + \xi$, $y = b + \eta$, and $z = c + \zeta$ in Equations (20) and (21). Then, we can rewrite the equations of motion in the phase space as

$$
\begin{aligned}
\dot{\xi} &= \xi_1, \\
\dot{\eta} &= \eta_1, \\
\dot{\zeta} &= \zeta_1, \\
\dot{\xi}_1 &= 2\eta_1 + H^0_{xx}\xi + H^0_{xy}\eta + H^0_{xz}\zeta, \\
\dot{\eta}_1 &= -2\xi_1 + H^0_{yx}\xi + H^0_{yy}\eta + H^0_{yz}\zeta, \\
\dot{\zeta}_1 &= H^0_{zx}\xi + H^0_{zy}\eta + H^0_{zz}\zeta.
\end{aligned}
\tag{23}
$$

where the superscript zero means that the second derivatives of H are evaluated at the related equilibrium point.

The characteristic polynomial of Equation (23) will be

$$
f(\lambda) = \lambda^6 + H_5\,\lambda^5 + H_4\,\lambda^4 + H_3\,\lambda^3 + H_2\,\lambda^2 + H_1\,\lambda + H_0,
\tag{24}
$$

where

$$
\begin{aligned}
H_0 ={}& H^0_{xz}\,H^0_{yy}\,H^0_{zx} - H^0_{xy}\,H^0_{yz}\,H^0_{zx} - H^0_{xz}\,H^0_{yz}\,H^0_{zy} \\
& + H^0_{xx}\,H^0_{yz}\,H^0_{zy} + H^0_{xy}\,H^0_{yx}\,H^0_{zz} - H^0_{xx}\,H^0_{yy}\,H^0_{zz}, \\
H_1 ={}& 2\,(H^0_{xz}\,H^0_{zy} - H^0_{yz}\,H^0_{zx} - H^0_{xy}\,H^0_{zz} + H^0_{yx}\,H^0_{zz}), \\
H_2 ={}& -\,4\,H^0_{zz} - H^0_{xy}\,H^0_{yx} + H^0_{xx}\,H^0_{yy} - H^0_{xz}\,H^0_{zx} \\
& - H^0_{yz}\,H^0_{zy} + H^0_{xx}\,H^0_{zz} + H^0_{yy}\,H^0_{zz}, \\
H_3 ={}& 2\,(H^0_{xy} - H^0_{yx}), \\
H_4 ={}& 4 - H^0_{xx} - H^0_{yy} - H^0_{zz}, \\
H_5 ={}& 0.
\end{aligned}
\tag{25}
$$

We will examine the stability of equilibrium points in two cases only because there are no equilibrium points in the third case.

4.1. First Case

In this case the equilibrium points $(x_1, 0, 0)$ and $(x_2, 0, 0)$ are in symmetry about the Y-axis, therefore it is enough to examine the stability of only one of these two points. In this context, we have to evaluate the values of H_{i1} corresponding to $(x_1, 0, 0)$, which are as follows:

$$
\begin{aligned}
H_{01} ={}& -\frac{3}{x_1^3} - \frac{5}{x_1^6} - \frac{2\,\alpha_{21}}{x_1^4}\left(3 + \frac{11}{x_1^3} + \frac{7}{x_1^6}\right) \\
& - \frac{3\,\alpha_{22}}{x_1^5}\left(3 + \frac{12}{x_1^3} + \frac{8}{x_1^6}\right) - \frac{2}{x_1^9}, \\
H_{11} ={}& 0, \\
H_{21} ={}& 1 - \frac{3}{x_1^3} - \frac{8\,\alpha_{21}}{x_1^4}\left(1 + \frac{2}{x_1^3}\right) \\
& - \frac{15\,\alpha_{22}}{x_1^5}\left(1 + \frac{2}{x_1^3}\right) - \frac{3}{x_1^6}, \\
H_{31} ={}& 0, \\
H_{41} ={}& 2 - \frac{2\,\alpha_{21}}{x_1^4} - \frac{6\,\alpha_{22}}{x_1^5}, \\
H_{51} ={}& 0.
\end{aligned}
\tag{26}
$$

From Equations (24) and (26), we find

$$
f(\lambda) = \lambda^6 + H_{41}\,\lambda^4 + H_{21}\,\lambda^2 + H_{01},
\tag{27}
$$

Here, $H_{41} > 0$, $H_{21} < 0$, and $H_{01} < 0$ show that the sign changes occur one at a time time, thus there exists at least one positive real root. Therefore, the equilibrium point will be unstable in this case.

4.2. Second Case

In this case, the equilibrium points $(0, y_2, 0)$ and $(0, y_3, 0)$ are symmetrical about the X-axis, hence it is sufficient to examine the stability of only two of these four points. Additionally, we have to evaluate the values of H_{i2} and H_{i3}, $i = 0, 1, 2, 3, 4$, and 5 corresponding to $(0, y_2, 0)$ and $(0, y_3, 0)$, which are as follows:

$$H_{02} = \frac{6}{y_2^3} + \frac{4}{y_2^6} + \frac{2\,\alpha_{21}}{y_2^4}\left(9 + \frac{10}{y_2^3} - \frac{7}{y_2^6}\right)$$
$$+ \frac{12\,\alpha_{22}}{y_2^5}\left(3 + \frac{3}{y_2^3} - \frac{2}{y_2^6}\right) - \frac{2}{y_2^9},$$

$$H_{12} = 0,$$

$$H_{22} = 1 + \frac{6}{y_2^3} + \frac{16\,\alpha_{21}}{y_2^4}\left(1 - \frac{1}{y_2^3}\right)$$
$$+ \frac{30\,\alpha_{22}}{y_2^5}\left(1 - \frac{1}{y_2^3}\right) - \frac{3}{y_2^6},$$

$$H_{32} = 0,$$

$$H_{42} = 2 - \frac{2\,\alpha_{21}}{y_2^4} - \frac{6\,\alpha_{22}}{y_2^5},$$

$$H_{52} = 0,$$

$$(28)$$

and

$$H_{03} = \frac{6}{y_3^3} + \frac{4}{y_3^6} + \frac{2\,\alpha_{21}}{y_3^4}\left(9 + \frac{10}{y_3^3} - \frac{7}{y_3^6}\right)$$
$$+ \frac{12\,\alpha_{22}}{y_3^5}\left(3 + \frac{3}{y_3^3} - \frac{2}{y_3^6}\right) - \frac{2}{y_3^9},$$

$$H_{13} = 0,$$

$$H_{23} = 1 + \frac{6}{y_3^3} + \frac{16\,\alpha_{21}}{y_3^4}\left(1 - \frac{1}{y_3^3}\right)$$
$$+ \frac{30\,\alpha_{22}}{y_3^5}\left(1 - \frac{1}{y_3^3}\right) - \frac{3}{y_3^6},$$

$$H_{33} = 0,$$

$$H_{43} = 2 - \frac{2\,\alpha_{21}}{y_3^4} - \frac{6\,\alpha_{22}}{y_3^5},$$

$$H_{53} = 0.$$

$$(29)$$

From Equations (24) and (28), we find

$$f(\lambda) = H_{6k}\,\lambda^6 + H_{4k}\,\lambda^4 + H_{2k}\,\lambda^2 + H_{0k}, \tag{30}$$

Here, $H_{6k} = 1 > 0$, $H_{4k} < 0$, $H_{2k} < 0$, and $H_{0k} < 0$, where $k = 2, 3$, show that the sign changes occur one at a time, thus there exists at least one positive real root. Therefore, the equilibrium point will be unstable in this case.

5. Numerical Results

In this section, we illustrate some dynamical properties numerically for the proposed system (i.e., the quantized Hill system) such as the equilibrium points and the allowed and forbidden regions of motion under the quantum corrections. In order to avoid the reparation, we will present the numerical analysis on the first case of equilibrium points; the same procedure can be carried out for the second case.

The locations of equilibrium points are shown in Figure 1, for which we have taken zero as the derivatives with respect to time in Equation (5). Then, with the help of the well-known *Mathematica Software*, the collinear equilibrium points L_1 and L_2 under the quantum corrections, as well as the unperturbed equilibrium points $\bar{L}_1$ and $\bar{L}_2$, are estimated numerically. Both points exist either side of the origin on the X-axis and are in symmetry about Y-axis. However, we mark that the distance between the perturbed points is more than the distance between the unperturbed points. Of course, this perturbation will affect the other dynamical properties.

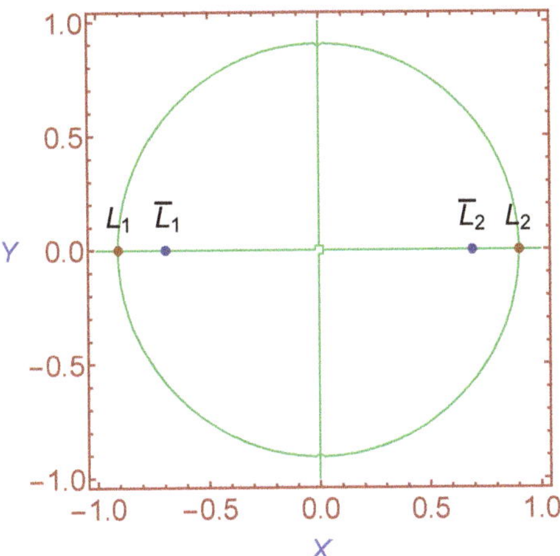

Figure 1. Locations of equilibrium points.

One of the most dynamical properties which can be identified by the *Jacobian integral* is the possible and forbidden regions of infinitesimal body motions, which are restricted to the locations of $v^2 = 2H - C_J \geq 0$ where v is the velocity of the infinitesimal body. Hence, Equation (22) can be used to determine the allowed or forbidden regions of motions, as in Figure 2, where the coloured green areas identify the regions of possible motions, while the white determine forbidden regions.

It is clear from Figure 2a that when the *Jacobian constant* is relatively small there is one large area for possible region of motion, and the body could move from any region point to another (or from L_1 ($\bar{L}_1$) to L_2 ($\bar{L}_2$)). When C_J becomes larger, the forbidden region is extended, as in Figure 2b. With further increase in the value of C_J, the forbidden region becomes larger, while the possible region of motion forms three septate areas starting from the perturbed equilibrium points L_1 and L_2, as in Figure 2c. In addition, the body cannot move from one to another, because the three areas are not connected. With further increase in the value of C_J, the inner and two outer regions decrease while the separate areas start from the unperturbed equilibrium points $\bar{L}_1$ and $\bar{L}_2$, as in Figure 2d. We remark that the infinitesimal can never move from one allowed region to another, and the body will be trapped inside one of the possible regions of motion with the relative large values of the *Jacobian constant*, as in the case of Figure 2c,d.

The condition of $v^2 \geq 0$ does not provide information about the size or shape of the orbit or the trajectory of the body; it can only identify the region where the infinitesimal body could move.

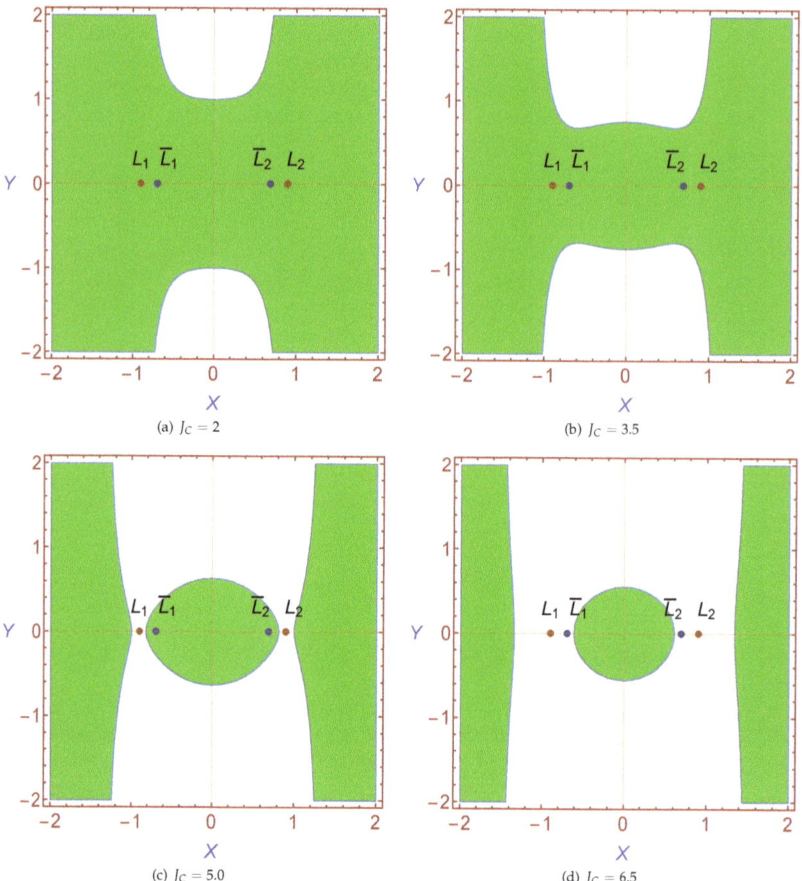

Figure 2. Regions of allowed (green area) and forbidden (white area) motion.

6. Conclusions

In this work, the quantized Hill problem is considered to study the existence of equilibrium points alongside examining their stability. Under the effect of quantum corrections, the locations of equilibrium points have been analysed, we have studied three different cases which give all possible locations, where two points emerge from the first case, taking a place on the X-axis, and four points dos so from the second case and lie on Y-axis. The third case does not provide a realistic location. Hence, we have obtained four new equilibrium points related to the quantum perturbations.

In this context, we have tested the stability status of all of the equilibrium points and we have found that all points are unstable. Further, we have illustrated the locations of equilibrium points for the first case and the related allowed regions of motion numerically. Similarly, we can perform these illustrations for the second case. Here, we found two equilibrium points which are either side of the origin on the X-axis and in symmetry about the Y-axis, as in Figure 1. The regions of possible and forbidden motion are investigated for different values of *Jacobian constant*, as in Figure 2.

Finally, we demonstrate that the obtained result in the first case is a generalization of the classical one, and it can be reduced to the classical result, while the four new points will disappear in the absence of quantum perturbation. The regions of possible motions decrease with the increasing value of *Jacobian constant* and these regions will form three separate areas. Thus, the infinitesimal body can never move from one allowed region to another, and it will be trapped inside one of the possible regions of motion with the relative large values for the *Jacobian constant*.

Author Contributions: Formal analysis, A.A.A., S.A., E.I.A. and S.K.S.; investigation, A.A.A., S.A., E.I.A. and S.K.S.; methodology, A.A.A., S.A., E.I.A. and S.K.S.; project administration, E.I.A.; software, E.I.A.; Validation, A.A.A., S.A., E.I.A. and S.K.S.; visualization, E.I.A.; Writing—original draft, E.I.A.; and writing—review and editing, A.A.A., S.A., E.I.A. and S.K.S. All authors have read and agreed to the published version of the manuscript.

Funding: This work was funded by the National Research Institute of Astronomy and Geophysics (NRIAG), Helwan 11421, Cairo, Egypt. The third author, therefore, acknowledges his gratitude for NRIAG's technical and financial support. Moreover, this paper was supported by the National Natural Science Foundation of China (NSFC), grant no. 12172322.

Institutional Review Board Statement: Not applicable.

Informed Consent Statement: Not applicable.

Data Availability Statement: The study does not report any data.

Conflicts of Interest: The authors declare no conflict of interest.

References

1. Kummer, M. On the stability of Hill's solutions of the plane restricted three body problem. *Am. J. Math.* **1979**, *101*, 1333–1354. [CrossRef]
2. Perdiou, A.; Perdios, E.; Kalantonis, V. Periodic orbits of the Hill problem with radiation and oblateness. *Astrophys. Space Sci.* **2012**, *342*, 19–30. [CrossRef]
3. Makó, Z. Connection between Hill stability and weak stability in the elliptic restricted three-body problem. *Celest. Mech. Dyn. Astron.* **2014**, *120*, 233–248. [CrossRef]
4. Gong, S.; Li, J. Analytical criteria of Hill stability in the elliptic restricted three body problem. *Astrophys. Space Sci.* **2015**, *358*, 1–10. [CrossRef]
5. Szebehely, V. Stability of artificial and natural satellites. *Celest. Mech.* **1978**, *18*, 383–389. [CrossRef]
6. Markellos, V.V.; Roy, A.E. Hill stability of satellite orbits. *Celest. Mech.* **1981**, *23*, 269–275. [CrossRef]
7. Markellos, V.V.; Roy, A.E.; Velgakis, M.J.; Kanavos, S.S. A photogravitational Hill problem and radiation effects on Hill stability of orbits. *Astrophys. Space Sci.* **2000**, *271*, 293–301. [CrossRef]
8. Markellos, V.V.; Roy, A.E.; Perdios, E.A.; Douskos, C.N. A Hill problem with oblate primaries and effect of oblateness on Hill stability of orbits. *Astrophys. Space Sci.* **2001**, *278*, 295–304. [CrossRef]
9. Scheeres, D. The Restricted Hill Four-Body Problem with Applications to the Earth-Moon-Sun System. *Celest. Mech. Dyn. Astron.* **1998**, *70*, 75–98. [CrossRef]
10. Scheeres, D.J.; Bellerose, J. The Restricted Hill Full 4-Body Problem: Application to spacecraft motion about binary asteroids. *Dyn. Syst. Int. J.* **2005**, *20*, 23–44. [CrossRef]
11. Burgos-García, J.; Gidea, M. Hill's approximation in a restricted four-body problem. *Celest. Mech. Dyn. Astron.* **2015**, *122*, 117–141. [CrossRef]
12. Liu, C.; Gong, S. Hill stability of the satellite in the elliptic restricted four–body problem. *Astrophys. Space Sci.* **2018**, *363*, 1–9. [CrossRef]
13. Perdiou, A.E.; Markellos, V.V.; Douskos, C.N. The Hill problem with oblate secondary: Numerical Exploration. *Earth Moon Planets* **2006**, *97*, 127–145. [CrossRef]
14. Markakis, M.P.; Perdiou, A.E.; Douskos, C.N. The photogravitational Hill problem with oblateness: Equilibrium points and Lyapunov families. *Astrophys. Space Sci.* **2008**, *315*, 297–306. [CrossRef]
15. Batkhin, A.B.; Batkhina, N.V. Hierarchy of periodic solutions families of spatial Hills problem. *Sol. Syst. Res.* **2009**, *43*, 178. [CrossRef]
16. Douskos, C.N. Collinear equilibrium points of Hill's problem with radiation pressure and oblateness and their fractal basins of attraction. *Astrophys. Space Sci.* **2010**, *326*, 263–271. [CrossRef]
17. Zotos, E.E. Comparing the fractal basins of attraction in the Hill problem with oblateness and radiation. *Astrophys. Space Sci.* **2017**, *362*, 190. [CrossRef]
18. Kalantonis, V.S.; Perdiou, A.E.; Douskos, C.N. Asymptotic orbits in Hills problem when the larger primary is a source of radiation. *Appl. Nonlinear Anal.* **2018**, *134*, 523–535.
19. Nartikoev, P. *Photogravitational Restricted Three-Body Problem*; Sev.-Oset. Univ.: Ordzhonikidze, Russia, 1987.
20. Ragos, O.; Perdios, E.; Kalantonis, V.; Vrahatis, M. On the equilibrium points of the relativistic restricted three-body problem. *Nonlinear Anal. Theory Methods Appl.* **2001**, *47*, 3413–3418. [CrossRef]

21. Mohr, R.; Furnstahl, R.; Hammer, H.W.; Perry, R.; Wilson, K. Precise numerical results for limit cycles in the quantum three-body problem. *Ann. Phys.* **2006**, *321*, 225–259. [CrossRef]
22. Donoghue, J.F. Leading quantum correction to the Newtonian potential. *Phys. Rev. Lett.* **1994**, *72*, 2996. [CrossRef] [PubMed]
23. Battista, E.; Esposito, G. Restricted three-body problem in effective-field-theory models of gravity. *Phys. Rev. D* **2014**, *89*, 084030. [CrossRef]
24. Alshaery, A.; Abouelmagd, E.I. Analysis of the spatial quantized three-body problem. *Results Phys.* **2020**, *17*, 103067. [CrossRef]
25. Abouelmagd, E.I.; Kalantonis, V.S.; Perdiou, A.E. A Quantized Hill's Dynamical System. *Adv. Astron.* **2021**, *2021*, 9963761. [CrossRef]
26. Bouaziz, F.; Ansari, A.A. Perturbed Hill's problem with variable mass. *Astron. Nachr.* **2021**, *342*, 666–674. [CrossRef]
27. Abouelmagd, E.I.; Alhowaity, S.; Diab, Z.; Guirao, J.L.; Shehata, M.H. On the Periodic Solutions for the Perturbed Spatial Quantized Hill Problem. *Mathematics* **2022**, *10*, 614. [CrossRef]
28. Abouelmagd, E.I.; El-Shaboury, S.M. Periodic orbits under combined effects of oblateness and radiation in the restricted problem of three bodies. *Astrophys. Space Sci.* **2012**, *341*, 331–341. [CrossRef]
29. Abouelmagd, E.I.; Mostafa, A. Out of plane equilibrium points locations and the forbidden movement regions in the restricted three-body problem with variable mass. *Astrophys. Space Sci.* **2015**, *357*, 58. [CrossRef]
30. Ansari, A.A. Triaxial primaries in circular Hill problem. *Astron. Rep.* **2021**, *65*, 1178–1183. [CrossRef]
31. Singh, J.; Leke, O. Equilibrium points and stability in the restricted three-body problem with oblateness and variable masses. *Astrophys. Space Sci.* **2012**, *340*, 27–41. [CrossRef]
32. Ansari, A.A. Effect of Albedo on the motion of the infinitesimal body in circular restricted three-body problem with variable masses. *Ital. J. Pure Appl. Math.* **2017**, *38*, 581–600.
33. Abouelmagd, E.I.; Ansari, A.A.; Ullah, M.S.; Guirao, J.L.G. A Planar Five-body Problem in a Framework of Heterogeneous and Mass Variation Effects. *Astron. J.* **2020**, *160*, 216. [CrossRef]
34. Suraj, M.S.; Aggarwal, R.; Asique, M.C.; Mittal, A. On the modified circular restricted three-body problem with variable mass. *New Astron.* **2021**, *84*, 101510. [CrossRef]
35. Bjerrum-Bohr, N.E.J.; Donoghue, J.F.; Holstein, B.R. Quantum gravitational corrections to the nonrelativistic scattering potential of two masses. *Phys. Rev. D* **2003**, *67*, 084033. [CrossRef]

MDPI
St. Alban-Anlage 66
4052 Basel
Switzerland
Tel. +41 61 683 77 34
Fax +41 61 302 89 18
www.mdpi.com

Mathematics Editorial Office
E-mail: mathematics@mdpi.com
www.mdpi.com/journal/mathematics